AF389363

Growth and Allocation of Woody Biomass in Forest Trees Based on Environmental Conditions

Growth and Allocation of Woody Biomass in Forest Trees Based on Environmental Conditions

Editors

Luigi Todaro
Angelo Rita
Alessio Collalti

MDPI • Basel • Beijing • Wuhan • Barcelona • Belgrade • Manchester • Tokyo • Cluj • Tianjin

Editors
Luigi Todaro
University of Basilicata
Italy

Angelo Rita
University of Basilicata
Italy

Alessio Collalti
National Research Council of Italy
Italy

Editorial Office
MDPI
St. Alban-Anlage 66
4052 Basel, Switzerland

This is a reprint of articles from the Special Issue published online in the open access journal *Forests* (ISSN 1999-4907) (available at: https://www.mdpi.com/journal/forests/special_issues/Growth_Allocation_Woody_Biomass).

For citation purposes, cite each article independently as indicated on the article page online and as indicated below:

LastName, A.A.; LastName, B.B.; LastName, C.C. Article Title. *Journal Name* **Year**, *Volume Number*, Page Range.

ISBN 978-3-0365-0436-0 (Hbk)
ISBN 978-3-0365-0437-7 (PDF)

Contents

About the Editors

Luigi Todaro (Ph.D. in Wood Arboriculture) is a researcher at the University of Basilicata, head of the research strategy in the field of forest-wood at the School of Agricultural, Forestry, Food, and Environmental Sciences. He has a background in wood technology particularly with regard to developing technologies aimed at maximizing the use of underused wood species for both industrial and other high value-added purposes.

Angelo Rita holds a Ph.D. in bio-ecosystems and bioecology from the University of Basilicata and a research fellowship in forest ecology from the same institute (2015 to 2020); he is a researcher at the Department of Agricultural Sciences at the University of Naples "Federico II", Italy (2020 – present). His research activities are settled on different spatial scales ranging from individual to the community level focusing on drivers of within/between tree species variability in behavior, life-history, growth pattern, and ecophysiology. The main work is explicitly linking plant physiology with population and community ecology to address pending questions about the past, present, and future of Mediterranean forests, with particular emphasis on the role of climate change and anthropogenic disturbances.

Alessio Collalti (Ph.D. in Forest Ecology) has a background concerning Forest Ecology, Carbon and Nitrogen Cycle, Forest and Vegetation Modelling, particularly with regard to vegetation numerical modelling and response under natural and anthropogenic stress, including climate change impacts and forest management scenarios. He is the Forest Modelling Lab. head and a researcher at the National Research Council of Italy.

Preface to "Growth and Allocation of Woody Biomass in Forest Trees Based on Environmental Conditions"

Terrestrial ecosystems and forests in particular, are important components because of their key role in reducing atmospheric greenhouse gas concentrations by storing a large amount of carbon in biomass and soils. Increasing attention is being paid to forestland area, which accounts for 30% of the total land surface and acts as the main C store in the land system. In their life cycle, plants uptake, process, allocate, and remobilize resources from the environment, including basic materials, such as CO_2, water, and nutrients, and other materials, such as sugars, proteins, and defensive chemicals. The relative amount of above- and belowground biomass allocated among leaves, branches, stems, roots, and reproductive tissues is a functional indicator of the forest stand and reflects the material flow, the wood quality, a plant's survival strategy, and the primary production processes. The way in which plants share their labile products across their compartments is influenced by plant size and is not fixed but likely varies over time, across growth environments, and among species. It follows that the whole allocation process would be modulated under strong natural selection. Obtaining a qualitative/quantitative understanding of the influence that these factors have on growth and biomass allocation is of fundamental importance for both understanding plant ecology and evolution and developing environmental policies and forest management practices, such as:

- sequestration to increase stocks in more recalcitrant woody carbon pools, characterized by a slow build-up of carbon with a potentially slower release of carbon to the atmosphere;

- conservation to prevent emissions from existing forest carbon pools in regions with high C stocks and where natural disturbances are less frequent to cause large immediate reductions in C stocks;

- substitution of energy-intensive products with products derived from renewable resources; and

- the improvement of practices that aim to increase wood quality for social purposes.

Luigi Todaro, Angelo Rita , Alessio Collalti
Editors

Editorial

Growth and Allocation of Woody Biomass in Forest Trees Based on Environmental Conditions

Alessio Collalti [1,*,†], **Luigi Todaro** [2,*,†] and **Angelo Rita** [2,3,*,†]

1   Forest Modelling Lab., Institute for Agriculture and Forestry Systems in the Mediterranean, National Research Council of Italy (CNR–ISAFOM), Via Madonna Alta 128, 06128 Perugia, PG, Italy

2   School of Agricultural, Forestry, Food and Environmental Science (SAFE), University of Basilicata, V.le dell'Ateneo Lucano 10, 85100 Potenza, PZ, Italy

3   Department of Agricultural Sciences, University of Naples "Federico II", Via Università 100, 80055 Portici, NA, Italy

*   Correspondence: alessio.collalti@cnr.it (A.C.); luigi.todaro@unibas.it (L.T.); angelo.rita@unibas.it (A.R.)

†   All authors contributed equally.

Academic Editor: Timothy A. Martin

Received: 27 January 2021; Accepted: 28 January 2021; Published: 28 January 2021

Terrestrial ecosystems, and forests in particular, are important components of land processes because of their key role in reducing atmospheric greenhouse gas concentrations by storing a large amount of carbon in tree biomass and soils. Increasing attention is being paid to forestland area, which accounts for 30% of the total land surface and acts as the main C store in the Earth system. In their life cycle, plants uptake, process, allocate (i.e., the distribution of net primary production among the different plant organs), and remobilize the product of photosynthesis. The relative amount of above- and below-ground biomass partitioned among leaves, branches, stems, roots, non-structural pools, and reproductive tissues is a good indicator for forest productivity and reflects the material flow, the health, the wood quality, and the plant's survival strategies. How plants share their labile products across their compartments is not fixed, rather is influenced by plant size and likely varies over time, among species and growth environments, and is affected by natural and anthropogenic disturbances (e.g., forest management). Accordingly, the whole allocation process would be constrained under strong natural selection. Our understanding of the mechanisms governing these processes is, however, still patchy, with some processes and their responses to the environmental conditions much more well understood than others. Getting a qualitative/quantitative insight into the impact that the above-mentioned factors have on tree growth and above- and below-ground biomass allocation is essential both for understanding plant ecology and evolution and for developing environmental policies and forest management practices to cope with climate change. In this regard, new insights for Amazonian forests come from the modeling work of De Faria et al. [1], who quantified the loss in above-ground biomass and the changes in recovery time (i.e., the time required for a forest to return to its former or usual condition following a disturbance) of forests affected by droughts, wildfire, and their combination. Their findings provide a valuable gaze and alarming prevision on the impacts that climate change will likely have on the Amazonian regions housing more than half of the world's remaining rainforests in the world. However, even European forests are likely to face an increasing number of extreme events in the future. The work of Schäfer et al. [2] analyzed the effect of drought and mixture of forest composition in a mature temperate forest of Norway spruce (*Picea abies* (L.) Karst.) and European beech (*Fagus sylvatica* L.) located in southern Germany. They reported differential allocation of tree biomass related to the drought condition period, where trees prioritized the stem growth at the beginning of the growing season, and root growth during the remaining growing season. Interestingly, spruces exhibited less tree water deficit than beech trees, while mixture seems to enhance the water supply of spruce trees, which should increase the stability of this species in a time of climatic warming. Similarly, but in a Mediterranean context, Ogaya and Peñuelas [3] studied the growth and the allocation in a Mediterranean holm oak (*Quercus ilex* L.) forest experimentally exposed to partial rainfall exclusion during 21 consecutive years in the Prades holm oak forest, in Catalonia, NE Spain. The authors aimed to study the effects of the expected decrease in water availability in the Mediterranean in the following decades and found that allocation in woody structures and total above-ground biomass were correlated with annual rainfall, whereas canopy allocation and the ratio of wood/canopy allocation were not dependent on rainfall. Their results highlight that water deficits characterized by lower soil moisture and higher atmospheric aridity are leading to several changes in the ecosystem functioning of the Mediterranean forests, causing a strong decrease in the capacity of these forests to mitigate climate change because of the high decrease in wood growth and suggesting that progressive substitution of the species most sensitive to water scarcity by other species more adapted to drought is expected, transforming the current Mediterranean forest. In the context of climate change, the effect of increasing temperature on tree phenology and growth, even including biomass partitioning, is also the focus of the work of Mura and colleagues [4]. Mura et al. analyzed, under field conditions in managed Norway spruce plots regenerating after clearcuts in central Norway, whether the trees growing at different elevations invest similarly in their various organs. They found that different local environmental conditions affect both the growth rate and phenology but the biomass partitioning among different parts of the tree remains essentially unchanged, giving support to the hypothesis that maintaining specific allometric trajectories is fundamental for tree functioning. High temperatures in warm months were also found to be a key environmental factor for the net primary productivity (NPP) in *Pinus massoniana* (Lamb.), a major planted tree species in southern China due to its important role in the development of forestry both for economic and ecological benefit. Huang et al. [5] established a large biomass database for *P. massoniana* including stems, branches, leaves roots, above-ground organs, and entire tree, thanks to published literature, to find out potential geographical trends in NPP for each tree compartment and their influencing factors in carbon allocation. Huang et al. found that the NPP of tree components showed no clear relationships with longitude and elevation, but a statistically significant inverse relationship with latitude, but with different sensitivities to environmental conditions, mostly temperature, and stand variables. Temperature and precipitation are also the focus for the transcontinental work of Usoltsev et al. [6] based on a database of 413 sample plots for stand biomass, ranging from France to Japan to southern China, for the genus *Populus* spp, which is overall the most widely cultivated fast-growing tree species in the middle latitude plain. They found significant changes in the structure of the forest stand biomass (stems, above- and below-ground biomass). However, while a positive and statistically significant relationship with winter temperature emerged for all components of the biomass, a less clear relationship with the precipitation was found. Poplar plantations are also the focus of the work led by Zhang and colleagues [7]. The authors, based on a field trial established in 2007 in Sihong forest farm, Jiangsu Province (China), in order to understand the response of growth, biomass production, carbon storage, planting spacing, and their interaction, they destructively harvested 24 sample trees for biomass measurements and stem analyses. Not surprisingly, they found that biomass production and carbon storage for the single tree of three clones was enhanced as planting spacing increased, with carbon concentration decreasing from stem to leaves. With these data, Zhang et al. established a Chapman–Richards empirical model for predicting tree volume growth for Chinese poplar clones. With the aim to quantify total tree biomass and its allocation to components in common aspen (*Populus tremula* L.), at both the tree and stand levels, in the forested mountainous area in central Slovakia, Konôpka et al. [8] measured, through destructive sampling, leaves, branches, stem, and roots. By these measurements, the authors derived allometric biomass models with stem base diameter as an independent variable for individual tree components. Moreover, biomass stock of the woody parts and foliage, as well as the leaf area index, were modeled using mean stand diameter as an independent variable, showing that foliage contribution to total tree biomass decreased with tree size.

Ritchie et al.'s [9] work evaluate the tree growth and total above-ground productivity (even including shrubs) of a twelve-year-old ponderosa pine (*Pinus ponderosa* (Lawson and C. Lawson)) plantation under three separate treatments representing a range of management intensities in the southern Cascade Range in northeastern California subjected to wildfire events. The authors found a significant effect of the manual grubbing release from shrub competition on tree growth when compared with the no release control and that the total above-ground biomass or carbon was only marginally influenced because shrub biomass dominated both sets of plots in this young plantation. Their results show that a broader tradeoff for controlling competing shrubs between using herbicides and grubbing or other means should be evaluated if biomass production or carbon sequestration is one of the goals of prevention or for a post-fire reforestation program.

A novel approach, aiming at comparing structural carbon allocation to tree growth and to the climate in a dendrochronological analysis, is presented in Ivanusic et al. [10] for hybrid white spruce (*Picea glauca* (Moench) × *engelmannii* (Parry)) grown in British Columbia, Canada. With this new approach, the authors found significant differences between the percent structural carbon of wood in individual natural and planted stands. Some significant relationships were found between percent carbon, ring widths, early wood, late wood, and the cell wall thickness and density values. Carbon accumulation in planted stands and natural stands was found in some cases to correlate with increasing temperatures where warmer late-season conditions appear to enhance growth and carbon accumulation in these sites.

The below-ground biomass, especially fine roots, and the relationship with forest structure in a mature European beech forest in central Italy is the core of the analyses presented by D'Andrea et al. [11]. The authors investigated the spatial variability of fine root production, soil CO_2 efflux, forest structural traits, and their reciprocal interactions and found, unexpectedly, that, in the year of study (2007–2008), fine root production resulted in the main component of NPP explaining about 70% of the spatial variability of soil respiration. The authors also found that fine root production was strictly driven by leaf area index and soil water content, suggesting close interactions between forest structure and functional forest characteristics to optimize carbon source–sink relationships.

Migolet and colleagues [12] implemented local and regional methods for estimating palm biomass in a mature plantation, using destructive sampling in the Congo Basin (West Equatorial Africa). Using data from eighteen 35-year-old oil palms in a plantation located in Makouké, central Gabon, they derived allometric equations for estimating stem, leaf, and total above-ground biomass. With a comparison with existing allometric models for oil palms generated elsewhere, the authors showed that their site-level model was a better predictor.

The current Special Issue groups a selection of works representing the most recent advances and insights linking growth and carbon allocation with, among others, environmental forcing, forest structure, and potential wood supply, including soil characteristics. We hope that new further research and scientific questions may come about in the near future by reading this collection of papers. We would like to thank the authors for their invaluable efforts and also the reviewers and the editorial board who helped us in significantly improving the quality of each of the published papers.

Funding: This research received no external funding.

Conflicts of Interest: The authors declare no conflict of interest.

Forests **2021**, *12*, 154

References

1. De Faria, B.; Marano, G.; Piponiot, C.; Silva, C.A.; Dantas, V.D.L.; Rattis, L.; Rech, A.; Collalti, A. Model-Based Estimation of Amazonian Forests Recovery Time after Drought and Fire Events. *Forests* **2020**, *12*, 8. [CrossRef]
2. Schäfer, C.; Rötzer, T.; Thurm, E.A.; Biber, P.; Kallenbach, C.; Pretzsch, H. Growth and Tree Water Deficit of Mixed Norway Spruce and European Beech at Different Heights in a Tree and under Heavy Drought. *Forests* **2019**, *10*, 577. [CrossRef]
3. Ogaya, R.; Peñuelas, J. Wood vs. Canopy Allocation of Aboveground Net Primary Productivity in a Mediterranean Forest during 21 Years of Experimental Rainfall Exclusion. *Forests* **2020**, *11*, 1094. [CrossRef]
4. Mura, C.; Strømme, C.B.; Anfodillo, T. Stable Allometric Trajectories in *Picea abies* (L.) Karst. Trees along an Elevational Gradient. *Forests* **2020**, *11*, 1231. [CrossRef]
5. Huang, X.; Huang, C.; Teng, M.; Zhou, Z.; Wang, P. Net Primary Productivity of Pinus massoniana Dependence on Climate, Soil and Forest Characteristics. *Forests* **2020**, *11*, 404. [CrossRef]
6. Usol'Tsev, V.A.; Chen, B.; Shobairi, S.O.R.; Tsepordey, I.S.; Chasovskikh, V.P.; Anees, S.A. Patterns for *Populus* spp. Stand Biomass in Gradients of Winter Temperature and Precipitation of Eurasia. *Forests* **2020**, *11*, 906. [CrossRef]
7. Zhang, Y.; Tian, Y.; Ding, S.; Lv, Y.; Samjhana, W.; Fang, S. Growth, Carbon Storage, and Optimal Rotation in Poplar Plantations: A Case Study on Clone and Planting Spacing Effects. *Forests* **2020**, *11*, 842. [CrossRef]
8. Konôpka, B.; Pajtík, J.; Šebeň, V.; Surový, P.; Merganičová, K. Biomass Allocation into Woody Parts and Foliage in Young Common Aspen (*Populus tremula* L.)—Trees and a Stand-Level Study in the Western Carpathians. *Forests* **2020**, *11*, 464. [CrossRef]
9. Ritchie, M.W.; Zhang, J.; Hammett, E. Aboveground Biomass Response to Release Treatments in a Young Ponderosa Pine Plantation. *Forests* **2019**, *10*, 795. [CrossRef]
10. Ivanusic, A.; Wood, L.; Lewis, K. Structural Carbon Allocation and Wood Growth Reflect Climate Variation in Stands of Hybrid White Spruce in Central Interior British Columbia, Canada. *Forests* **2020**, *11*, 879. [CrossRef]
11. D'Andrea, E.; Guidolotti, G.; Scartazza, A.; De Angelis, P.; Matteucci, G. Small-Scale Forest Structure Influences Spatial Variability of Belowground Carbon Fluxes in a Mature Mediterranean Beech Forest. *Forests* **2020**, *11*, 255. [CrossRef]
12. Migolet, P.; Goïta, K.; Ngomanda, A.; Biyogo, A.P.M. Estimation of Aboveground Oil Palm Biomass in a Mature Plantation in the Congo Basin. *Forests* **2020**, *11*, 544. [CrossRef]

Publisher's Note: MDPI stays neutral with regard to jurisdictional claims in published maps and institutional affiliations.

 *forests*

Article

Model-Based Estimation of Amazonian Forests Recovery Time after Drought and Fire Events

Bruno L. De Faria [1,2,*], **Gina Marano** [3], **Camille Piponiot** [4], **Carlos A. Silva** [5], **Vinícius de L. Dantas** [6], **Ludmila Rattis** [7,8], **Andre R. Rech** [1] and **Alessio Collalti** [9,10]

[1] Programa de Pós-Graduação em Ciência Florestal, Universidade Federal Vales do Jequitinhonha e Mucuri Campus JK, Diamantina 39100-000, MG, Brazil; andre.rech@ufvjm.edu.br
[2] Federal Institute of Technology North of Minas Gerais (IFNMG), Diamantina 39100-000, MG, Brazil
[3] Department of Agriculture, University of Napoli Federico II, 80055 Portici (Naples), Italy; gina.marano@unina.it
[4] Smithsonian Tropical Research Institute, 03092 Panamá, Panama; PiponiotC@si.edu
[5] School of Forest Resources and Conservation, University of Florida, Gainesville, FL 32611, USA; c.silva@ufl.edu
[6] Institute of Geography, Federal University of Uberlandia (UFU), Av. João Naves de Ávila 2121, Uberlandia 38400-902, Minas Gerais, Brazil; viniciusdantas@ufu.br
[7] Woodwell Climate Research Center, Falmouth, MA 02540, USA; lrattis@woodwellclimate.org
[8] Instituto de Pesquisa Ambiental da Amazônia, Canarana 78640-000, MT, Brazil
[9] Institute for Agriculture and Forestry Systems in the Mediterranean, National Research Council of Italy, 06128 Perugia, Italy; alessio.collalti@cnr.it
[10] Department of Innovation in Biological, Agro-food and Forest Systems, University of Tuscia, 01100 Viterbo, Italy
* Correspondence: blfaria@gmail.com

Received: 24 September 2020; Accepted: 21 December 2020; Published: 23 December 2020

Abstract: In recent decades, droughts, deforestation and wildfires have become recurring phenomena that have heavily affected both human activities and natural ecosystems in Amazonia. The time needed for an ecosystem to recover from carbon losses is a crucial metric to evaluate disturbance impacts on forests. However, little is known about the impacts of these disturbances, alone and synergistically, on forest recovery time and the resulting spatiotemporal patterns at the regional scale. In this study, we combined the 3-PG forest growth model, remote sensing and field derived equations, to map the Amazonia-wide (3 km of spatial resolution) impact and recovery time of aboveground biomass (AGB) after drought, fire and a combination of logging and fire. Our results indicate that AGB decreases by 4%, 19% and 46% in forests affected by drought, fire and logging + fire, respectively, with an average AGB recovery time of 27 years for drought, 44 years for burned and 63 years for logged + burned areas and with maximum values reaching 184 years in areas of high fire intensity. Our findings provide two major insights in the spatial and temporal patterns of drought and wildfire in the Amazon: (1) the recovery time of the forests takes longer in the southeastern part of the basin, and, (2) as droughts and wildfires become more frequent—since the intervals between the disturbances are getting shorter than the rate of forest regeneration—the long lasting damage they cause potentially results in a permanent and increasing carbon losses from these fragile ecosystems.

Keywords: Amazon; recovery time; aboveground biomass; climate change; 3-PG; fire; logging

---

1. Introduction

Natural disturbances have a key role in forest ecosystem dynamics [1], yet global changes in climate and land-uses have intensified disturbances rates in several biomes with important consequences on the ecosystems resilience [2]. Events like droughts and wildfires are becoming widespread phenomena in vast areas of the globe, potentially affecting the ecosystem services they provide [3,4] even in humid biomes with high rainfall rates, such as Amazonia [5–7]. Housing more than half of the world's remaining rainforest areas, Amazonian forests account for considerable carbon storage in living biomass and soils, estimated at around 150–200 Pg [8,9]. In addition, the region represents one of the most important biodiversity hotspots of the planet [10,11]. Amazonian forests are under considerable pressure due to the increased frequency and intensity of disturbances in moist tropical regions [12]. Forest fires and large-scale drought events are both directly dependent on climate [13] and their effects are expected to become more severe with climate change effects (i.e., mostly warming and reduction in precipitation). In combination with human activities, such as selective logging and other land-use changes, increasing fire and drought severity are expected to cause significant forest losses [14].

The Amazon Basin's historical baseline of disturbances has been heavily altered in the last 20 years as a result of anthropogenic activities, increasing the rates of deforestation, drought and wildfire and their impacts [15]. In the early 2000s, logging activities affected ca. 10,000–20,000 km^2 $year^{-1}$ of tropical forests in the Brazilian Amazon and it is estimated that understory fires destroyed ca. 85,000 km^2 of standing forests in the period 1999–2010 [16,17]. Moreover, recent studies have shown that Amazonian forests are becoming more exposed to droughts [18,19], including extreme drought events that would not be expected to take place more than once in a century (e.g., the three devastating droughts of 2005, 2010 and 2016; [20,21]). Altogether, droughts, wildfires and logging activities increase the susceptibility of forests to successive burning by increasing ignition rates, wind speed, creating drier microclimatic conditions near the soil surface and promoting exotic grass invasion. The effect of fire in forest ecosystems contrasts with that observed at larger spatial scales (i.e., global scale) and in fire-prone regions in which anthropogenic influences often reduce fire spread [22]. Therefore, the increasing risk of wildfires is an additional driver of change in the Amazon region [23].

Forest degradation due to more frequent and intense disturbances in the Amazon [24,25] results in long-term reduction in carbon stocks [26] with potential release of the C stored in Amazonian forests. The degree of degradation of the forest C stocks depends on four major factors: (1) the type of disturbance (e.g., logging, droughts and wildfires); (2) intensity (i.e., percentage of C loss); (3) the time return interval (i.e., years from one event to the next one) [25,27,28]; and (4) disturbance synergisms (i.e., the interacting effects between disturbances).

Several studies have analyzed forest recovery after disturbances at either broad or at multiple scales disturbances [29,30], but few of them have been conducted in tropical forests and specifically in the Amazon Basin. When conducted, these studies are usually limited in temporal scale (usually <20 years) [25,31,32] and focus on the effects of a single disturbance and in relatively small areas [33–35]. There is a lack of studies looking at recovery beyond 30–40 years. As a result, we still have a limited understanding on forest aboveground biomass (AGB) resilience to disturbance in Amazonian forests (i.e., how much time does it take for the forest to return to its pre-disturbance status), especially at the regional scale and taking interacting effects of multiple disturbance into consideration.

One straightforward way of addressing the consequences of disturbance in forest AGB is by integrating geospatial techniques with remote sensing and process-based forest growth models [36,37]. Specifically, remote sensing and GIS technologies allow the assessment of forest AGB at broad scales [38] whereas process-based forest growth models can provide insights on the mechanisms and processes involved in forest recovery and their relationship with spatiotemporal climate (including human)-induced scenarios. Models can help in assessing the recovery time of vegetation using climatic variables to predict vegetation productivity and its spatial variability [38]. At a regional scale, net primary productivity (NPP) is often used as an indicator of inherent plant growth potential [39]. Several studies have indeed assumed a strong relationship between productivity and biomass [40] with the first one being a function of the second. Indeed, the targeted parameter AGB is also influenced by climate, water availability and soil fertility [39–41]. In this study, we assessed the recovery time (i.e., the time necessary for a forest to recover its pre-disturbance AGB levels) of Brazilian Amazon forests AGB from drought, fire and a combination of logging and fire disturbances, using a dynamic forest carbon model that simulates vegetation recovery time as a function of climate scenarios and geospatial data. With the present study, we aim to investigate the recovery time of AGB in the Amazon forests when subject to a disturbance caused by: (1) an extreme drought, (2) a catastrophic fire and (3) a combination of logging and fire disturbances by integrating the existing knowledge [24,42–44] within our modeling framework.

2. Materials and Methods

We used a spatially implicit forest productivity model based on the net primary productivity of the 3-PG model (Figure 1) (see Section 2.1 *The Model*) to estimate forest recovery time (here defined as the time necessary for a forest to recover at its pre-disturbance AGB levels). Analysis of AGB recovery was carried out for the Brazilian Amazon biome, which encompasses about 3.5 million km^2 located between 15° S–5° N and 40° W–80° W. The region consists of one of the largest preserved forests in the world that has been experiencing strong human disturbances in recent times, especially in "the arch of deforestation" (Figure 2).

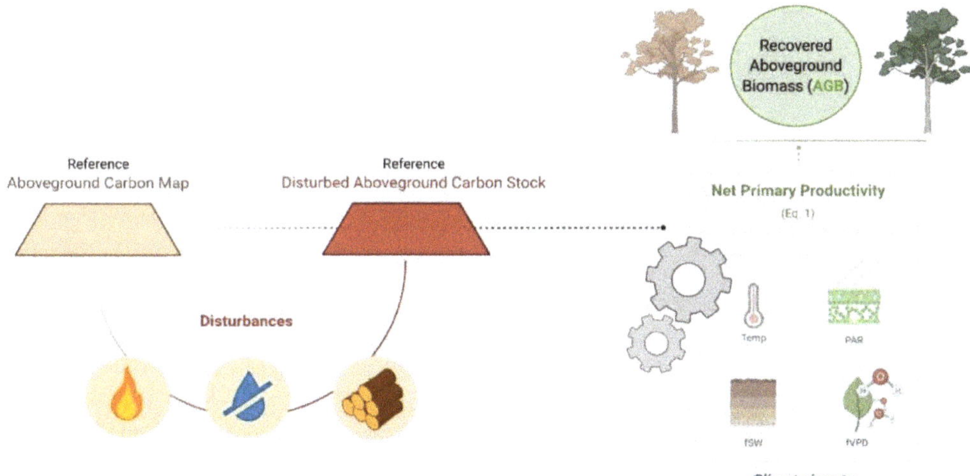

Figure 1. Proof–of–concept vegetation recovery time simulations as a function of climate variables (i.e., soil-plant available water (*fSW*), photosynthetically active radiation (PAR), vapor pressure deficit (*fVPD*), and air temperature (*fTemp*), see *The Model* for description). Aboveground biomass (AGB) losses resulting from drought stress and fire are a function of the maximum climatological water deficit (MCWD, see *The Model* for description).

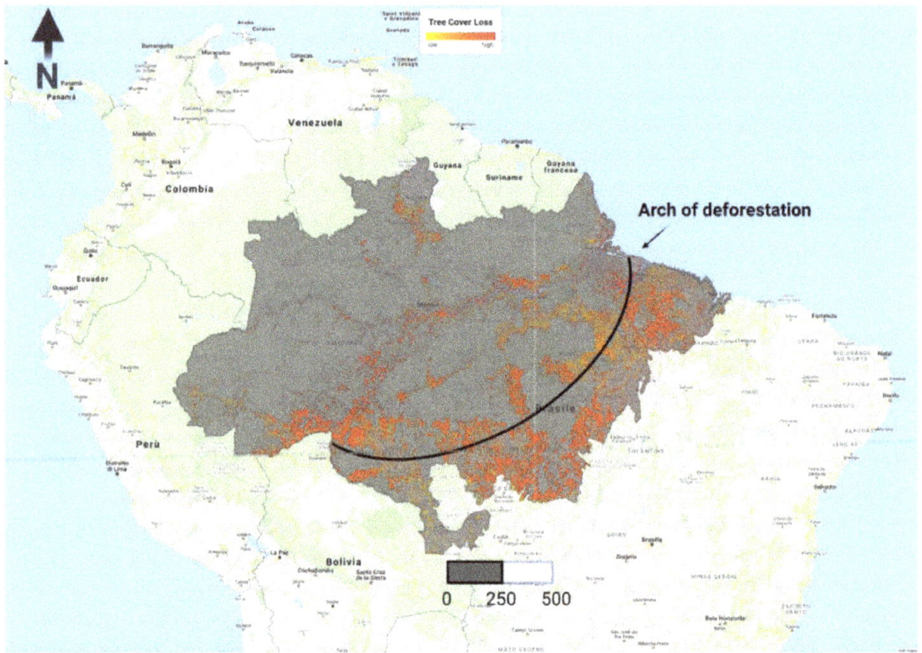

Figure 2. Study area: Amazonian forest in Brazil. Amazon biome extent (gray area). Forest loss map (yellow-red) has been displayed according to [45,46] (Global Forest Change dataset in Google Earth Engine). Red pixels identify areas of where tree cover loss has been detected.

2.1. The Model

In this study, recovery time dynamics are simulated using the 3-PG model (Physiological Principles in Predicting Growth; [47]), as embedded and parameterized into the CARLUG model by [48], driven by four monthly climatic variables: photosynthetically active radiation (PAR, mol PAR m^{-2} month^{-1}), vapor pressure deficit (VPD, KPa), precipitation (mm month^{-1}) and air temperature (°C), respectively. The 3-PG model was used to estimate gross and net primary productivity (GPP and NPP, both in g C m^{-2} month^{-1}) as follows:

$$NPP = GPP \times Y \tag{1}$$

where Y is the carbon use efficiency (i.e., the fraction of GPP not used to support autotrophic respiration, known as CUE [49–51]). GPP is computed as:

$$GPP = \alpha_x \times modifiers \times PAR \times \left(1 - e^{-k \times LAI}\right) \tag{2}$$

where α_x is the maximum quantum canopy efficiency (i.e., the maximum capacity in converting light into photosynthates without environmental or other functional limitations, in mol C mol PAR^{-1} m^{-2} month^{-1}), *modifiers* comprise environmental limitations to maximum photosynthetic rate (temperature, *fTEMP*; soil water, *fSW*; and vapor pressure deficit, *fVPD*), with values ranging from zero (complete limitation) to one (no limitation). For an in-depth description of modifiers algorithms see also [48,52]. The last two terms in

Equation (2) reflect the incident PAR effectively absorbed by the canopies (i.e., APAR) based on their leaf area index (LAI, $m^2\ m^{-2}$) and the leaf light extinction coefficient (k, unitless) as in Beer's Law [53].

Each month, the model assumes that leaf, wood, and root carbon pools increase by an overall amount equal to the NPP, which are, respectively, allocated proportionally in their three pools as in the standard 3-PG carbon partitioning-allocation scheme [54]. The partitioning of NPP is the outcome of the climate and soil conditions interacting with vegetation through a series of differential equations that describe the flow of C within the tree compartments [48]. Therefore, the model predicts the distribution of forest biomass from carbon stocks, but in order to obtain biomass we converted C to biomass assuming that one ton of biomass contains 0.5 tons of C [55]. We assume that the re-equilibration of forest carbon after disturbances (i.e., steady state undisturbed conditions) is when the AGB growth and the decay rates stabilize. We also estimated the average time to recover 90% of old-growth forests' carbon levels. The 90% threshold has often been used in similar studies (e.g., [56]) and can thus more easily be compared to previous results; the 100% threshold corresponds to a full recovery of carbon stocks, but it may take significantly longer.

The study conducted by [48] uses the recalibrated 3-PG model parameters for the Amazonian forests (the overall parameters description and their values are shown in Supporting Information, see Table S1). The 3-PG calculates NPP as a constant fraction of GPP, using an NPP/GPP ratio (Y = 0.47) based on empirical evidence [47]. For Brazilian Amazon forests other studies suggest Y to be closer to 0.3 [57] while others report much higher values at some tropical sites, even including Amazonian ones (i.e., Y > 0.5; [51]). However, the issue of whether Y is a constant value, its actual value, even including its top-down limits, is a much-debated issue as described in [51,58].

An overall 3-PG model parameter sensitivity analysis has been performed already by a number of authors (e.g., [59]) showing how the 3-PG model is mostly sensitive to stem allometric parameters (i.e., those used to obtain from trees structure the tree biomass), ratios for biomass partitioning and allocation, maximum canopy conductance, turnover time of wood, and maximum canopy quantum efficiency. For an in-depth 3-PG model parameter sensitivity analysis we refer to the works of [48,59] and this will be not considered and discussed further here. In addition, we used the pan-tropical biomass map generated by Avitabile et al. [60] as reference (pre-impact) levels to initialize the model and combining it with two comprehensive recent estimates of carbon density (i.e., estimations of [55,61] and covering a wide 250–500 $Mg\ ha^{-1}$ range (Figure S1).

2.2. Estimating Drought, Fire and Logging Impacts on AGB Stocks

The loss of AGB due to drought events was modeled as a function of the MCWD (Maximum Climatological Water Deficit index, representing the maximum climatological water deficit reached in the year), a common index used to measure the cumulative water stress in Amazonia (e.g., [42,62,63]). The MCWD reflects the intensity and length of the dry season, when evapotranspiration exceeds precipitation (i.e., negative balance). A measure of water deficit related to tree mortality in Amazonian forests that is denoted as in Lewis et al. [42], that is:

$$\Delta AGB = 0.378 - 0.052 \times \Delta MCWD \tag{3}$$

We estimated the MCWD anomalies (namely, ΔMCWD) for the year 2010 by first estimating the mean MCWD for the baseline period from 1998 to 2015, without considering both the years 2005 and 2010. The ΔMCWD have been shown to be strong predictors of drought-associated tree mortality in the Amazon [62]. Specifically, a monthly water deficit was calculated as the difference between precipitation and evapotranspiration (with ground measurements estimated at 100 mm per month [63,64], i.e., evapotranspiration is fixed at 100 mm $month^{-1}$). As a result, we assume that the forest is in water deficit when monthly precipitation falls below 100 mm. MCWD was calculated as the sum of sequential monthly water deficits, where more negative MCWD values indicate higher drought stress. We quantified the MCWD for the year of 2010 using the product 3B43 of TRMM (Tropical Rainfall Measuring Mission at 0.25° grid-resolution), and then, the average of carbon losses for each pixel using Equation (3). The 2010 drought is one of the most intense and spatially extensive drought events ever recorded in the Brazilian Amazon [42].

Effects of wildfire were estimated by using the CARLUC-Fire model [44]. This model specifically accounts for the effects of fire by estimating forest carbon losses after a fire event as a function of its intensity (FI). FI is defined as the energy released per unit length of fire-line (kWm^{-2}), which is a key factor in estimating how vegetation responds to fire events. The relationship between fire intensity and fire-induced biomass losses was derived from a large-scale fire experiment in southeast Amazonia [24,44] (Equation (4)). Based on this experiment, AGB losses were calculated as a function of FI as follows:

$$AGB_{losses} = \frac{1}{\left(1 + e^{(2.45 - 0.002373 \times FI)}\right)} \tag{4}$$

We limit our fire analysis to areas that burned between 2003 and 2016 [65] using information at 500 m resolution from the Moderate Resolution Imaging Spectroradiometer (MODIS) Collection 6 MCD64A1 burned area product over the period 2003–2016.

As a substantial proportion of fires occurred in areas likely to have been previously logged, we accounted for this effect in the estimation of the initial AGB by incorporating an additional loss in fire effects of 40% in burned areas that were also cleared. We assumed this based on findings of Berenguer et al. [43] that an average forest under selective logging stores about 40% less carbon. Logged areas were defined using data from the annual Landsat-based Project for Monitoring Amazonian Deforestation (PRODES, http://www.obt.inpe.br/prodes). Because edge effects from logging have been shown to affect forests up to 2–3 km from the border [66], we include forests located within 3 km from a deforested pixel, as a selective logging influence zone and they were defined using data from PRODES with cumulative deforestation up to 2017.

2.3. Experimental Runs

We ran the 3-PG model at 3 km × 3 km spatial resolution under mean monthly climate conditions for the 1980–2009 period, to estimate the forest recovery time for both drought, fire and logging + fire impacts (includes loss from logging and losses from fire). Climate input variables used to calculate the climatic means consisted of monthly series of temperature and mean vapor pressure deficit from the Climate Research Unit (CRU TS; [67]), while PAR was obtained from the GOES–9 satellite product [68]. In each pixel, AGB recovery was assessed by simulating AGB dynamics with the model after an AGB loss corresponding to disturbance impact.

2.4. Assessing Model Results

Light detection and ranging (Lidar) remote sensing is widely used for monitoring forest structure and biomass dynamics [69,70] in many forest ecosystems [71]. For instance, airborne lidar (ALS) technologies help quantify changes in canopy structure, carbon stocks and recovery time at the local-to-regional scale under different types of forest degradation (e.g., [25,72,73]).

In the present study, we compare our modeled recovery time from fire in logged areas with airborne lidar-derived aboveground carbon density (ACD) recovery estimates in forest stands (2891.45 Ha) located in Feliz Natal (Mato Grosso, Brasil) that were logged and burned once. For computing the recovery time of ACD from lidar, we applied a model developed by Rappaport et al. [25] that used multiple linear regression to model the recovery time of ACD (Kg C m^{-2}) in degraded forest stands based on degradation type. In their study, the model was calibrated using a chronosequence of ACD maps derived from lidar and degradation history data (from 2013 to 2018) across degraded forests stands [25]. The model is presented in Equation (5) and shows adjusted R^2 of 0.89. Herein, we chose to compare our results with those provided in Rappaport et al. [25] due to lack of available field data on the time scale addressed here to assess recovery time.

$$RT = 62.259 + 11.395 \times \log(t) - 10.268 \times CF1 \tag{5}$$

where RT refers to recovery time, t refers to time (years) and CF1 refers to degradation history, once-burned stands.

2.5. Disturbance Return Interval

In order to inquire whether global changes could determine an increase in future drought and fire frequency we projected the areal extent and spatial patterns of future drought and fire impacts up to the year 2100 in order to understand whether global changes could determine an increase in future drought and fire frequency in the study area. We analyzed both future precipitations (based on Representative Concentration Pathways, i.e., RCP 8.5—representing unmitigated climate change scenario) and a land use changes scenario (based on Aguiar et al. [74]) with a decrease in the extension and level of protection of the areas and increases in deforestation rates from 2014 to 2020 and continuing until 2100.

We built drought scenarios (2040–2070 and 2071–2100) using precipitation (related with water stress, MCWD) from the ensemble of 35 climate models participating in the Coupled Model Intercomparison Project phase 5 (CMIP-5, [75]). In detail, we derived the forcing from the mean monthly simulated precipitation anomalies first averaged for all 35 models and then bias corrections with Tropical Rainfall Measuring Mission (TRMM data product 3B43 [63]). To investigate frequency of future Amazonian droughts we assumed severe drought condition when MCWD anomalies (subtraction between future projections and the historical average) is <-40 mm (threshold derived by Phillips et al. [62]), below this threshold water stress is assumed to induce losses in AGB. We also used maps of predicted change in fire recurrence in response to global changes obtained from Fonseca et al. [76] based on future land-use change data by Aguiar et al. [74]. The fire scenarios (2040–2070 and 2071–2100) developed by Fonseca et al. [76] combine the effects of future land-use and climate change on fire relative probability in the Brazilian Amazon in the best-case and worst-case scenarios. We assume fire relative probability to equal fire relative frequency and then determine the mean fire return interval as the inverse of fire relative frequency.

3. Results

Results show that disturbances have substantially affected biomass in Brazilian Amazonia. In the locations affected by drought, fire and logging + fire, AGB decreased by 4%, 19% and 46%, respectively (Figure 3). Our results suggest that during the 2010 drought, about 1.5 million km^2 of the Brazilian Amazon lost a considerable amount of AGB (we considered losses $\geq$10% of the initial AGB). Fire could also produce substantial losses in above-ground carbon affecting 550,000 km^2 especially in southern Brazilian Amazon. Approximately 150,000 km^2 of the burned forest patches were located within 3 km from a logged forest.

Average AGB recovery time was 27 years for drought-impacted, 44 years for burned, and 63 years for logged + burned areas (includes loss from logging and loss from fire). Recovery time from drought revealed a northwest-to-southeast gradient in the study area (Figure 4a). Roughly 20% of these drought-affected areas, corresponding to ca. 364,000 km^2, were estimated to recover in the first 10 years, with maximum values reaching 90 years in parts of southeastern Brazilian Amazonia (Figure 4). Forest fires were widespread across the "arch of deforestation" (the region in southern and eastern Amazonia where the rates of deforestation are higher) during the period 2003–2016 (Figure 4b). The longest recovery times during this period were concentrated along the eastern and southwestern extent of Amazon forests in

Brazil, where the maximum was about 150 years after fire disturbance. Subsequent wildfires events (i.e., multiple fires in the same location) accounted for 10% of all forest fires during the period 2003–2016, delaying forest recovery times within these areas (Figure 4b). The longest recovery times were found in logged-and-burned forests with maximum values reaching 184 years (Figure 4c,c1). These results consider a recovery of the carbon stock corresponding to 100% (i.e., recovery time ~184 years) (see *The Model*) resulting in a difference of about 122 years in logged and burned forest which would be much faster if we would consider a recovery threshold of 90% (i.e., recovery time ~62 years) (Figure S2).

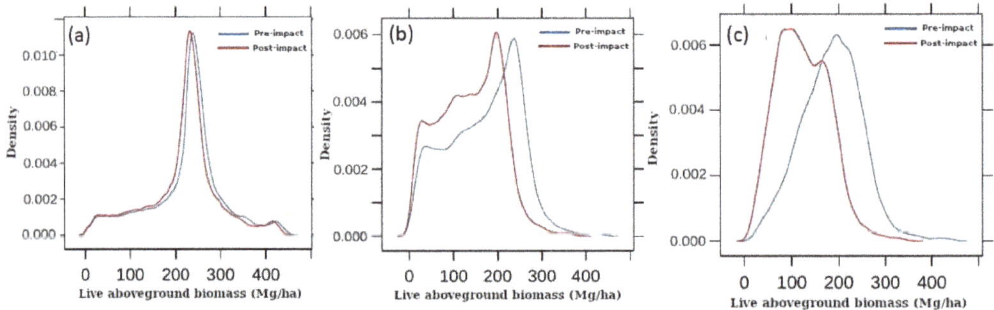

Figure 3. Biomass density plots describing patterns before and after drought (**a**), fire (**b**) and logging + fire (**c**) impacts. Only areas that burned between 2003 and 2016 are considered and, for (**c**), only burned areas up to 3 km from logging areas. Recovery is defined as 100% of pre-disturbance AGB.

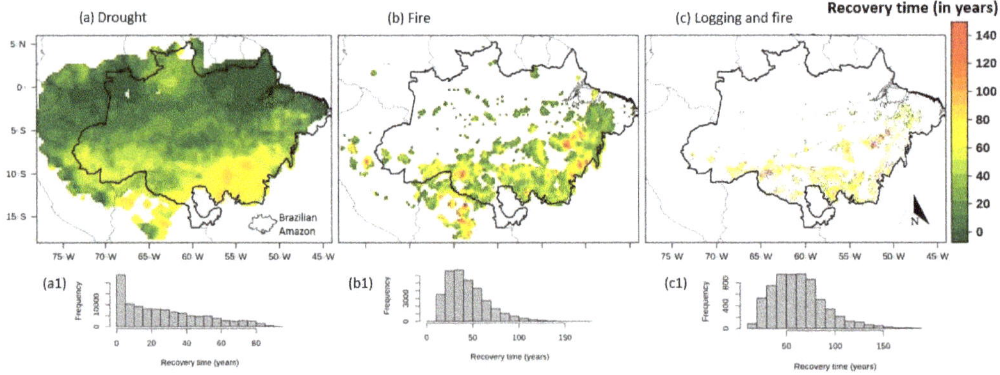

Figure 4. Aboveground recovery time (in years) for 2010 drought (**a**), fire areas that burned between 2003 and 2016 (**b**) and in areas that were both burned and logged (**c**). Histogram plots summarize AGB recovery pixels distributions (in years), for drought (**a1**), fire (**b1**) and logging + fire (**c1**).

We compared our results with a lidar-derived model of recovery time in stands that were logged and burned once (Figure 5a). Our estimations show smaller AGB decreases in comparison with lidar-based estimates of carbon losses from fire (loss of AGB of 46% vs. 55%). However, recovery rates were shown to be strongly correlated (Figure 5b).

Increases in the extent and frequency of drought and fire (Figure 6) suggest that these future disturbances could undermine the full forest recovery. Our results suggest that by 2070 the area affected by drought will increase approximately three-fold (Figure 6—top panel). Moreover, from the middle to the end of the century, the mean fire return intervals (FRI) was projected to decrease from 10 to 8 years and the median FRI to decrease from 8 to 6 years from the 2040–2070 period to the 2070–2100 periods, respectively, in a worst case land use change scenario (Figure 6 bottom panel). However, in a more optimistic scenario the area subject to high fire frequency would be smaller (Figure 6 middle panel).

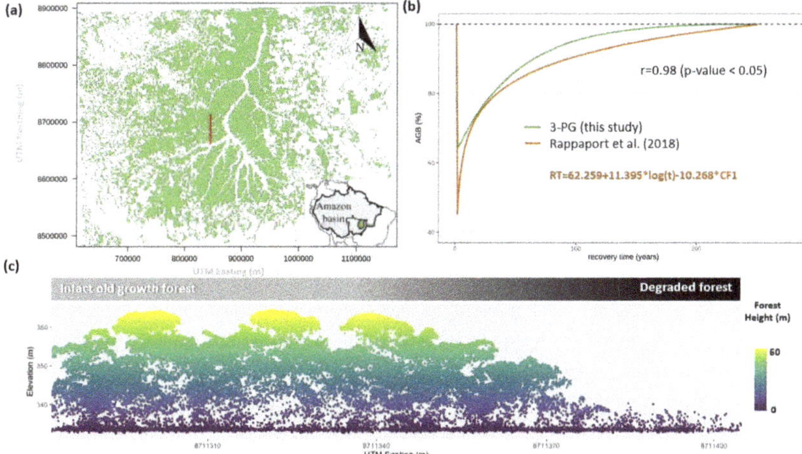

Figure 5. Airborne light detection and ranging (lidar) data were sampled (red line) in Feliz Natal, within the Xingu basin (light green), Brazilian state of Mato Grosso (**a**). The forest growth model (3-PG green line) shows the relationship between aboveground biomass (%) and recovery time in years. We compared it with a lidar-derived model of recovery time in stands that were logged and burned once (CF1 refers to once-burned) [25] (orange line) (**b**). A sample of vertical profile of a recovering forest which was degraded by fire and selective logging (**c**). The discrete return lidar data used for creating the transect figure were acquired in 2018 with a point density of 22.98 points m^{-2} covering an area of 2891.25 ha in Feliz Natal, Mato Grosso, Brazil [25], as part of the Sustainable Landscapes Brazil project program (data available from: https://www.paisagenslidar.cnptia.embrapa.br/webgis/; details of airborne lidar (ALS) data acquisitions are presented in the supplementary material, Table S2).

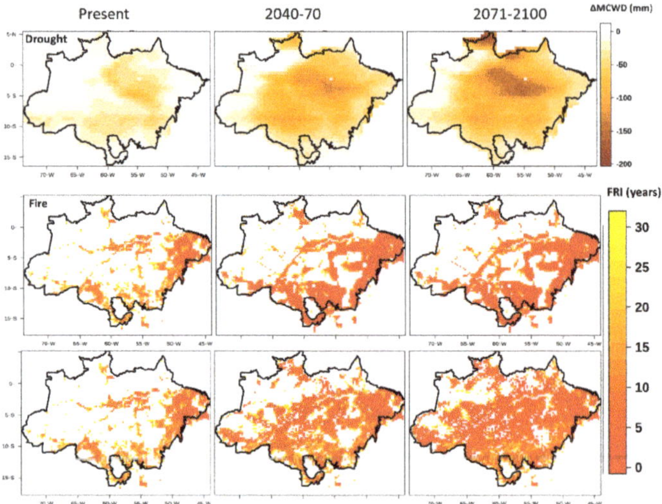

Figure 6. Projected changes in droughts (as maximum climatological water deficit anomalies, ΔMCWD) (**upper panel**) and fire return interval based on an optimistic land use scenario (**mid panel**) and in the unmitigated scenarios with the worst-case land-use scenario (**bottom panel**).

4. Discussion

In the present study, we explored the AGB changes after drought, fire and a combination of logging and fire disturbances and the time needed for complete recovery as a function of both climatic conditions and AGB in the Brazilian Amazon forest, using a modeling-based approach. Our results suggest that fire is a much greater threat than drought for the forest resilience, especially if logging occurs. These results highlight the key threat imposed by fire to Amazon forests. The intensity of the disturbance event is strongly related to both the amount of AGB lost and the recovery time of the forest. The biomass recovery rates estimates reported here are consistent with those from Poorter et al. [56] that showed AGB of Neotropical second growth forest took a median time of 66 years to recover to 90% of previous growth values after multiple disturbances events, including land use changes. On the other hand, recent evidence [77] suggests that recovery time might take at least 150 years until secondary forests (re)gain carbon levels similar to primary forests, after drought disturbances thus indicating that these biomes have recovery rates that are much lower than previously suggested.

Our results also suggest that by the end of the century, especially after 2070, the Brazilian Amazon will be affected by more frequent droughts with the southern area being more vulnerable since it will need a longer time to recover after these events. Thus, climate change will greatly increase the threat imposed to the forest, potentially jeopardizing forest resilience. The interplay between longer forest recovery times and more frequent droughts has been previously evidenced in the Amazonia, where longer recovery times have been documented [78]. Moreover, if on the one hand the extreme droughts of 2005, 2010 and 2016 have prevented the full recovery of the forests, on the other, drought effects on forest canopy carbon fixation capacity could potentially persist for several years during recovery processes [78], leading to forest degradation and changes in forest species composition, and evidence suggests that taller tree species have significantly higher mortality than small tree species, when subject to drought [79,80].

Our findings also confirm that the land carbon sink in the Brazilian Amazon will be strongly impacted by a regime of a chronic state of incomplete recovery [78], with adverse consequences also on the GPP due to shifts in precipitation patterns caused by anthropogenic emissions [81–83]. Indeed, across Amazon forests, GPP is modeled to decrease linearly with increasing seasonal water deficit [82]. Longer and more intense dry seasons have been forecasted, together with an increased frequency and severity of drought events [84–86] and future Amazon droughts are expected to become even more frequent [87,88]. Our projections suggest about one extreme drought per decade (drought return interval ranging from 4 to 16 years depending on the scenario of climate change). If drought frequency increases, Amazon forest, both as species composition and regional carbon sink, will be affected, which will thereby have an impact on global carbon cycling and contribute further to climate change [62,80,89,90]. Previous studies have shown increased fire occurrence and tree mortality during and after Amazon droughts [6,89,91–93]. If these events continue to increase in frequency, large parts of the Amazon could potentially shift from rainforest vegetation to a fire-maintained degraded forest and may promote the persistence of degraded forests with a savanna-like structure [94,95]. This change in forest type, structure and ecology would most likely reduce both the forest sink capacity and even its biodiversity and ecosystem services [94]. The net increase in areas that are more susceptible to wildfires, induced by either drought events increase, or potentially intensified by climate change, could lead to significant biomass losses [9,96].

Human pressures play a crucial role in fire ignitions, wildfires could break out also in non-dry years as in 2019, when more than 69,000 km^2 burnt despite the absence of anomalous drought [97]. As droughts and wildfires are expected to become more frequent, the time of occurrence between these disturbances may even get shorter than forest recovery time, determining permanently damaged ecosystems and widespread degradation [95]. Although forest growth models are powerful tools that can be applied in simulating the C dynamics in forests [98,99], our results are subject to some uncertainty and a number of caveats [100,101]. In this study, we modeled vegetation recovery time as a function of climate only. This approach does not account for regional variation in growth rates depending on soils types (due to their inner physico-chemical properties such as water retention or local-scale variation based on prior land use [92,93]) growth rates are also known to vary significantly by species [43]. In addition to the mechanisms mentioned above, CO_2 fertilization of Amazonian vegetation and nitrogen deposition could play an important, but yet often neglected, role in forest regeneration [102]. It has also been suggested that atmospheric CO_2 generally stimulates plant growth with increased rates in photosynthetic activity and indirectly through increased water-use efficiency [103], but not in all cases [104]. As CO_2 accumulates in the atmosphere, Amazonian trees may also accumulate more biomass resulting in denser canopies and faster growth [105]. But an increased atmospheric CO_2 concentration necessarily implies an increase in mean air temperature which is in turn speculated to increase plants' respiration and should result in a levelled-off forest carbon use efficiency [83]. Recent studies indicate that the ability of intact tropical forests to remove carbon from the atmosphere may be already saturating [9,106] while others indicate for tropical species higher thermal acclimation capacities to buffer C–losses by respiration [51], thus, calling for more studies on the possible consequences of warming and increased atmospheric CO_2 concentration on forest dynamics. However, in the Amazon phosphorus is an important limiting nutrient over large parts and its low availability may limit positive CO_2 fertilization effects.

Future Possibilities for Model Improvement

Lidar-derived 3D-point cloud and biomass products can be used to enhance models' representation of complex and heterogeneous forest ecosystems, such as those found in Amazonia [107], and therefore can be used as input or to initialize vegetation models [108]. For instance, Longo et al. [109] have used lidar to obtain initial conditions for an ecosystem model that requires an initial state for forest structure.

Their method to derive the vertical structure of the canopy from high-resolution airborne lidar successfully characterized the diversity of forest structure variability caused by human-induced forest degradation (such as logging and fire).

This new approach has strong implications on modeling recovery time and the successional trajectories of the Amazonian disturbed forest because it does not require any assumption on the successional stage of the forest, but only the vertical distribution of returns. Moreover, it could be adapted to space-borne lidar data, including NASA's Global Ecosystem Dynamics Investigation (GEDI, [110]). Fusion of GEDI and optical data [111] will further expand the spatial extent of available lidar data and potentially provide tools capable of mapping drought, fire and logging impacts helping models to assess recovery time. Moreover, integration of GEDI with either optical or radar [112] wall-to-wall data could allow large-scale characterization of forest ecosystems structure providing accurate measurements of biomass stock that could be used for assessing recovery time via repeated measurements.

5. Conclusions

This study shows how forest growth models can be used as tools for complementing field-based studies on recovery time by investigating the spatial and temporal dynamics and processes of forest recovery. Indeed, our biomass recovery map illustrates both spatial and climatic variability in carbon sequestration potential due to forest re-growth. By mapping potential for biomass recovery across Amazonia, policy makers could focus their efforts on specific areas that require special protection and need to be preserved. Moreover, such recovery maps could also help by identifying areas with higher carbon sequestration potential thus supporting policies and concrete actions to mitigate forest degradation in areas where biomass resilience is under increasing stress (such as southeastern Amazonia). The capability and timing of forest recovery after drought, fire and logging are urgent and hot topics for applied research calling upon conservation and policy actions in Amazonia. Future changes in fire regimes could push some Amazonian regions into a permanently drier climate regime and weaken the resilience of the region to possible large-scale drought–fire interactions driven by climate change. We are far from an integrated view of forest recovery processes, yet the results presented in this study may provide some new insights about forest recovery time after disturbances. The consequences that an extreme climatic event, such as a drought, may cause in the forest can result in a net loss of ecosystem services compromising these ecosystems dynamics in the long term. As a major result of projected increases in fire and drought frequency and intensity in the region, Amazonian forest resilience appears, in the medium and long term, to be severely jeopardized.

Supplementary Materials: The following are available online at http://www.mdpi.com/1999-4907/12/1/8/s1. Figure S1. Pre-disturbance reference biomass map [44]. Figure S2. The ABG dynamic as reproduced by the forest growth model (3-PG green line) showing the relationship between aboveground biomass (%) and recovery time in years to reach recovery threshold. Red dotted line 90% threshold and black dotted line 100% threshold. Table S1: Parameters description and their values used in 3-PG model (modified from Hirsch et al., [48]). Table S2: Details of ALS data acquisitions.

Author Contributions: Conceptualization, B.L.D.F.; methodology, B.L.D.F., A.C. and C.P.; software, B.L.D.F. and C.A.S.; draft preparation, G.M., A.C., B.L.D.F., C.P., L.R., V.d.L.D. and A.R.R.; writing—review and editing, G.M., A.C., C.P., L.R., C.A.S., V.d.L.D., B.L.D.F. and A.R.R.; visualization, B.L.D.F., C.A.S. and G.M.; supervision, G.M., A.C. and A.R.R. All authors have read and agreed to the published version of the manuscript.

Funding: This research received no external funding.

Acknowledgments: B.L.D.F. would like to thank the IFNMG and CNPq for financial support. He would like to especially thank the IFNMG campus Pirapora for giving him the opportunity to pursue a PhD. A.R.R. and B.L.D.F. thanks the support received from PPGCF/UFVJM.

Conflicts of Interest: The authors declare no conflict of interest.

References

1. Seidl, R.; Fernandes, P.M.; Fonseca, T.F.; Gillet, F.; Jönsson, A.M.; Merganičová, K.; Netherer, S.; Arpaci, A.; Bontemps, J.D.; Bugmann, H.; et al. Modelling natural disturbances in forest ecosystems: A review. *Ecol. Model.* **2011**, *222*, 903–924. [CrossRef]
2. D'Andrea, E.; Rezaie, N.; Prislan, P.; Gričar, J.; Collalti, A.; Muhr, J.; Matteucci, G. Frost and drought: Effects of extreme weather events on stem carbon dynamics in a Mediterranean beech forest. *Plant Cell Environ.* **2020**, *43*, 2365–2379. [CrossRef] [PubMed]
3. Pyne, S. The Ecology of Fire. Available online: https://www.nature.com/scitable/knowledge/library/the-ecology-of-fire-13259892/ (accessed on 10 June 2020).
4. Noce, S.; Collalti, A.; Valentini, R.; Santini, M. Hot spot maps of forest presence in the Mediterranean basin. *IForest* **2016**, *9*, 766–774. [CrossRef]
5. Mishra, A.K.; Singh, V.P. A review of drought concepts. *J. Hydrol.* **2010**, *391*, 202–216. [CrossRef]
6. Brando, P.M.; Nepstad, D.C.; Davidson, E.A.; Trumbore, S.E.; Ray, D.; Camargo, P. Drought effects on litterfall, wood production and belowground carbon cycling in an Amazon forest: Results of a throughfall reduction experiment. *Philos. Trans. R. Soc. B Biol. Sci.* **2008**, *1498*, 1839–1848. [CrossRef]
7. Leskinen, P.; Cardellini, G.; González García, S.; Hurmekoski, E.; Sathre, R.; Seppälä, J.; Smyth, C.E.; Stern, T.; Verkerk, H. Substitution effects of wood-based products in climate change mitigation. *From Sci. Policy* **2018**, *7*, 28.
8. Mitchard, E.T.A.; Feldpausch, T.R.; Brienen, R.J.W.; Lopez-Gonzalez, G.; Monteagudo, A.; Baker, T.R.; Lewis, S.L.; Lloyd, J.; Quesada, C.A.; Gloor, M.; et al. Markedly divergent estimates of Amazon forest carbon density from ground plots and satellites. *Glob. Ecol. Biogeogr.* **2014**, *23*, 935–946. [CrossRef]
9. Brienen, R.J.W.; Phillips, O.L.; Feldpausch, T.R.; Gloor, E.; Baker, T.R.; Lloyd, J.; Lopez-Gonzalez, G.; Monteagudo-Mendoza, A.; Malhi, Y.; Lewis, S.L.; et al. Long-term decline of the Amazon carbon sink. *Nature* **2015**, *519*, 344–348. [CrossRef]
10. Saatchi, S.; Houghton, R.A.; Dos Santos Alvalá, R.C.; Soares, J.V.; Yu, Y. Distribution of aboveground live biomass in the Amazon basin. *Glob. Chang. Biol.* **2007**, *13*, 816–837. [CrossRef]
11. Gardner, T.A.; Barlow, J.; Chazdon, R.; Ewers, R.M.; Harvey, C.A.; Peres, C.A.; Sodhi, N.S. Prospects for tropical forest biodiversity in a human-modified world. *Ecol. Lett.* **2009**, *12*, 561–582. [CrossRef] [PubMed]
12. Lewis, S.L.; Edwards, D.P.; Galbraith, D. Increasing human dominance of tropical forests. *Science* **2015**, *349*, 827–832. [CrossRef]
13. Seidl, R.; Schelhaas, M.J.; Lexer, M.J. Unraveling the drivers of intensifying forest disturbance regimes in Europe. *Glob. Chang. Biol.* **2011**, *17*, 2842–2852. [CrossRef]
14. Malhi, Y.; Aragão, L.E.O.C.; Galbraith, D.; Huntingford, C.; Fisher, R.; Zelazowski, P.; Sitch, S.; McSweeney, C.; Meir, P. Exploring the likelihood and mechanism of a climate-change-induced dieback of the Amazon rainforest. *Proc. Natl. Acad. Sci. USA* **2009**, *106*, 20610–20615. [CrossRef] [PubMed]
15. Exbrayat, J.F.; Liu, Y.Y.; Williams, M. Impact of deforestation and climate on the Amazon Basin's above-ground biomass during. *Sci. Rep.* **2017**, *7*, 1–7. [CrossRef] [PubMed]
16. Morton, D.C.; Le Page, Y.; DeFries, R.; Collatz, G.J.; Hurtt, G.C. Understorey fire frequency and the fate of burned forests in southern Amazonia. *Philos. Trans. R. Soc. B Biol. Sci.* **2013**, *368*, 20120163. [CrossRef] [PubMed]
17. Asner, G.P.; Knapp, D.E.; Broadbent, E.N.; Oliveira, P.J.C.; Keller, M.; Silva, J.N. Ecology: Selective logging in the Brazilian Amazon. *Science* **2005**, *310*, 480–482. [CrossRef] [PubMed]
18. Stocker, B.D.; Zscheischler, J.; Keenan, T.F.; Prentice, I.C.; Seneviratne, S.I.; Peñuelas, J. Drought impacts on terrestrial primary production underestimated by satellite monitoring. *Nat. Geosci.* **2019**, *12*, 264–270. [CrossRef]
19. Brodribb, T.J.; Powers, J.; Cochard, H.; Choat, B. Hanging by a thread? Forests and drought. *Science* **2020**, *368*, 261–266. [CrossRef] [PubMed]
20. Jiménez-Muñoz, J.C.; Mattar, C.; Barichivich, J.; Santamaría-Artigas, A.; Takahashi, K.; Malhi, Y.; Sobrino, J.A.; Schrier, G. Van Der Record-breaking warming and extreme drought in the Amazon rainforest during the course of El Niño 2015–2016. *Sci. Rep.* **2016**, *6*, 33130. [CrossRef] [PubMed]

21. Marengo, J.A.; Espinoza, J.C. Extreme seasonal droughts and floods in Amazonia: Causes, trends and impacts. *Int. J. Climatol.* **2016**, *36*, 1033–1050. [CrossRef]

22. Nepstad, D.C.; Stickler, C.M.; Soares-Filho, B.; Merry, F. Interactions among Amazon land use, forests and climate: Prospects for a near-term forest tipping point. *Philos. Trans. R. Soc. B Biol. Sci.* **2008**, *363*, 1737–1746. [CrossRef] [PubMed]

23. Marengo, J.A.; Souza, C.M.; Thonicke, K.; Burton, C.; Halladay, K.; Betts, R.A.; Alves, L.M.; Soares, W.R. Changes in Climate and Land Use Over the Amazon Region: Current and Future Variability and Trends. *Front. Earth Sci.* **2018**, *6*, 228. [CrossRef]

24. Brando, P.M.; Balch, J.K.; Nepstad, D.C.; Morton, D.C.; Putz, F.E.; Coe, M.T.; Silvério, D.; Macedo, M.N.; Davidson, E.A.; Nóbrega, C.C.; et al. Abrupt increases in Amazonian tree mortality due to drought-fire interactions. *Proc. Natl. Acad. Sci. USA* **2014**, *111*, 6347–6352. [CrossRef]

25. Rappaport, D.I.; Morton, D.C.; Longo, M.; Keller, M.; Dubayah, R.; Dos-Santos, M.N. Quantifying long-term changes in carbon stocks and forest structure from Amazon forest degradation. *Environ. Res. Lett.* **2018**, *13*, 065013. [CrossRef]

26. Walker, X.J.; Baltzer, J.L.; Cumming, S.G.; Day, N.J.; Ebert, C.; Goetz, S.; Johnstone, J.F.; Potter, S.; Rogers, B.M.; Schuur, E.A.G.; et al. Increasing wildfires threaten historic carbon sink of boreal forest soils. *Nature* **2019**, *572*, 520–523. [CrossRef] [PubMed]

27. Laurance, W.F.; Nascimento, H.E.M.; Laurance, S.G.; Andrade, A.; Ribeiro, J.E.L.S.; Giraldo, J.P.; Lovejoy, T.E.; Condit, R.; Chave, J.; Harms, K.E.; et al. Rapid decay of tree-community composition in Amazonian forest fragments. *Proc. Natl. Acad. Sci. USA* **2006**, *103*, 19010–19014. [CrossRef]

28. Barlow, J.; Gardner, T.A.; Lees, A.C.; Parry, L.; Peres, C.A. How pristine are tropical forests? An ecological perspective on the pre-Columbian human footprint in Amazonia and implications for contemporary conservation. *Biol. Conserv.* **2012**, *151*, 45–49. [CrossRef]

29. Pickell, P.D.; Hermosilla, T.; Frazier, R.J.; Coops, N.C.; Wulder, M.A. Forest recovery trends derived from Landsat time series for North American boreal forests. *Int. J. Remote Sens.* **2016**, *37*, 138–149. [CrossRef]

30. White, J.C.; Wulder, M.A.; Hermosilla, T.; Coops, N.C.; Hobart, G.W. A nationwide annual characterization of 25 years of forest disturbance and recovery for Canada using Landsat time series. *Remote Sens. Environ.* **2017**, *194*, 303–321. [CrossRef]

31. Andrade, R.B.; Balch, J.K.; Parsons, A.L.; Armenteras, D.; Roman-Cuesta, R.M.; Bulkan, J. Scenarios in tropical forest degradation: Carbon stock trajectories for REDD+. *Carbon Balance Manag.* **2017**, *12*, 1–7. [CrossRef]

32. Sato, L.Y.; Gomes, V.C.F.; Shimabukuro, Y.E.; Keller, M.; Arai, E.; dos-Santos, M.N.; Brown, I.F. Post-fire changes in forest biomass retrieved by airborne LiDAR in Amazonia. *Remote Sens.* **2016**, *8*, 839. [CrossRef]

33. Barlow, J.; Peres, C.A.; Lagan, B.O.; Haugaasen, T. Large tree mortality and the decline of forest biomass following Amazonian wildfires. *Ecol. Lett.* **2003**, *6*, 6–8. [CrossRef]

34. Balch, J.K.; Nepstad, D.C.; Curran, L.M.; Brando, P.M.; Portela, O.; Guilherme, P.; Reuning-Scherer, J.D.; de Carvalho, O. Size, species, and fire behavior predict tree and liana mortality from experimental burns in the Brazilian Amazon. *For. Ecol. Manag.* **2011**, *261*, 68–77. [CrossRef]

35. Feldpausch, T.R.; Jirka, S.; Passos, C.A.M.; Jasper, F.; Riha, S.J. When big trees fall: Damage and carbon export by reduced impact logging in southern Amazonia. *For. Ecol. Manag.* **2005**, *219*, 199–215. [CrossRef]

36. Marano, G.; Langella, G.; Basile, A.; Cona, F.; Michele, C.D.; Manna, P.; Teobaldelli, M.; Saracino, A.; Terribile, F. A geospatial decision support system tool for supporting integrated forest knowledge at the landscape scale. *Forests* **2019**, *10*, 690. [CrossRef]

37. Vacchiano, G.; Magnani, F.; Collalti, A. Modeling Italian forests: State of the art and future challenges. *IForest* **2012**, *5*, 113–120. [CrossRef]

38. Kumar, L.; Sinha, P.; Taylor, S.; Alqurashi, A.F. Review of the use of remote sensing for biomass estimation to support renewable energy generation. *J. Appl. Remote Sens.* **2015**, *9*, 097696. [CrossRef]

39. Keeling, H.; Phillips, O. The global relationship between forest productivity and biomass. *Glob. Ecol. Biogeogr.* **2007**, *16*, 618–631. [CrossRef]

40. Rödig, E.; Cuntz, M.; Rammig, A.; Fischer, R.; Taubert, F.; Huth, A. The importance of forest structure for carbon fluxes of the Amazon rainforest. *Environ. Res. Lett.* **2018**, *13*, 054013. [CrossRef]

41. Fauset, S.; Gloor, M.; Fyllas, N.M.; Phillips, O.L.; Asner, G.P.; Baker, T.R.; Patrick Bentley, L.; Brienen, R.J.W.; Christoffersen, B.O.; del Aguila-Pasquel, J.; et al. Individual-Based Modeling of Amazon Forests Suggests That Climate Controls Productivity While Traits Control Demography. *Front. Earth Sci.* **2019**, *7*, 83. [CrossRef]

42. Lewis, S.L.; Brando, P.M.; Phillips, O.L.; Van Der Heijden, G.M.F.; Nepstad, D. The 2010 Amazon drought. *Science* **2011**, *331*, 554. [CrossRef] [PubMed]

43. Berenguer, E.; Ferreira, J.; Gardner, T.A.; Aragão, L.E.O.C.; De Camargo, P.B.; Cerri, C.E.; Durigan, M.; De Oliveira, R.C.; Vieira, I.C.G.; Barlow, J. A large-scale field assessment of carbon stocks in human-modified tropical forests. *Glob. Chang. Biol.* **2014**, *20*, 3713–3726. [CrossRef] [PubMed]

44. De Faria, B.L.; Brando, P.M.; Macedo, M.N.; Panday, P.K.; Soares-Filho, B.S.; Coe, M.T. Current and future patterns of fire-induced forest degradation in amazonia. *Environ. Res. Lett.* **2017**, *12*, 095005. [CrossRef]

45. Hansen, M.C.; Potapov, P.V.; Moore, R.; Hancher, M.; Turubanova, S.A.; Tyukavina, A.; Thau, D.; Stehman, S.V.; Goetz, S.J.; Loveland, T.R.; et al. High-resolution global maps of 21st-century forest cover change. *Science* **2013**, *342*, 850–853. [CrossRef] [PubMed]

46. Wheeler, D.; Guzder-Williams, B.; Petersen, R.; Thau, D. Rapid MODIS-based detection of tree cover loss. *Int. J. Appl. Earth Obs. Geoinf.* **2018**, *69*, 78–87. [CrossRef]

47. Landsberg, J.J.; Waring, R.H. A generalised model of forest productivity using simplified concepts of radiation-use efficiency, carbon balance and partitioning. *For. Ecol. Manag.* **1997**, *95*, 209–228. [CrossRef]

48. Hirsch, A.I.; Little, W.S.; Houghton, R.A.; Scott, N.A.; White, J.D. The net carbon flux due to deforestation and forest re-growth in the Brazilian Amazon: Analysis using a process-based model. *Glob. Chang. Biol.* **2004**, *10*, 908–924. [CrossRef]

49. Coops, N.C.; Waring, R.H.; Landsberg, J.J. Assessing forest productivity in Australia and New Zealand using a physiologically-based model driven with averaged monthly weather data and satellite-derived estimates of canopy photosynthetic capacity. *For. Ecol. Manag.* **1998**, *104*, 113–127. [CrossRef]

50. Collalti, A.; Prentice, I.C. Is NPP proportional to GPP? Waring's hypothesis 20 years on. *Tree Physiol.* **2019**, *39*, 1473–1483. [CrossRef]

51. Collalti, A.; Ibrom, A.; Stockmarr, A.; Cescatti, A.; Alkama, R.; Fernández-Martínez, M.; Matteucci, G.; Sitch, S.; Friedlingstein, P.; Ciais, P.; et al. Forest production efficiency increases with growth temperature. *Nat. Commun.* **2020**. [CrossRef]

52. Waring, R.H.; Landsberg, J.J.; Williams, M. Net primary production of forests: A constant fraction of gross primary production? *Tree Physiol.* **1998**, *18*, 129–134. [CrossRef]

53. Collalti, A.; Perugini, L.; Santini, M.; Chiti, T.; Nolè, A.; Matteucci, G.; Valentini, R. A process-based model to simulate growth in forests with complex structure: Evaluation and use of 3D-CMCC Forest Ecosystem Model in a deciduous forest in Central Italy. *Ecol. Model.* **2014**, *272*, 362–378. [CrossRef]

54. Merganičová, K.; Merganič, J.; Lehtonen, A.; Vacchiano, G.; Sever, M.Z.O.; Augustynczik, A.L.D.; Grote, R.; Kyselová, I.; Mäkelä, A.; Yousefpour, R.; et al. Forest carbon allocation modelling under climate change. *Tree Physiol.* **2019**, *39*, 1937–1960. [CrossRef]

55. Baccini, A.; Goetz, S.J.; Walker, W.S.; Laporte, N.T.; Sun, M.; Sulla-Menashe, D.; Hackler, J.; Beck, P.S.A.; Dubayah, R.; Friedl, M.A.; et al. Estimated carbon dioxide emissions from tropical deforestation improved by carbon-density maps. *Nat. Clim. Chang.* **2012**, *2*, 182–185. [CrossRef]

56. Poorter, L.; Bongers, F.; Aide, T.M.; Almeyda Zambrano, A.M.; Balvanera, P.; Becknell, J.M.; Boukili, V.; Brancalion, P.H.S.; Broadbent, E.N.; Chazdon, R.L.; et al. Biomass resilience of Neotropical secondary forests. *Nature* **2016**, *530*, 211–214. [CrossRef]

57. Chambers, J.Q.; Tribuzy, E.S.; Toledo, L.C.; Crispim, B.F.; Higuchi, N.; Dos Santos, J.; Araújo, A.C.; Kruijt, B.; Nobre, A.D.; Trumbore, S.E. Respiration from a tropical forest ecosystem: Partitioning of sources and low carbon use efficiency. *Ecol. Appl.* **2004**, *14*, 72–88. [CrossRef]

58. Collalti, A.; Marconi, S.; Ibrom, A.; Trotta, C.; Anav, A.; D'Andrea, E.; Matteucci, G.; Montagnani, L.; Gielen, B.; Mammarella, I. validation of 3D-CMCC Forest Ecosystem Model (v.5.1) against eddy covariance data for ten European forest sites. *Geosci. Model Dev.* **2016**, *9*, 479–504. [CrossRef]

59. Almeida, A.C.; Landsberg, J.J.; Sands, P.J. Parameterisation of 3-PG model for fast-growing Eucalyptus grandis plantations. *For. Ecol. Manag.* **2004**, *193*, 179–195. [CrossRef]

60. Avitabile, V.; Herold, M.; Heuvelink, G.B.M.; Lewis, S.L.; Phillips, O.L.; Asner, G.P.; Armston, J.; Ashton, P.S.; Banin, L.; Bayol, N.; et al. An integrated pan-tropical biomass map using multiple reference datasets. *Glob. Chang. Biol.* **2016**, *22*, 1406–1420. [CrossRef]

61. Saatchi, S.S.; Harris, N.L.; Brown, S.; Lefsky, M.; Mitchard, E.T.A.; Salas, W.; Zutta, B.R.; Buermann, W.; Lewis, S.L.; Hagen, S.; et al. Benchmark map of forest carbon stocks in tropical regions across three continents. *Proc. Natl. Acad. Sci. USA* **2011**, *108*, 9899–9904. [CrossRef]

62. Phillips, O.L.; Aragão, L.E.O.C.; Lewis, S.L.; Fisher, J.B.; Lloyd, J.; López-González, G.; Malhi, Y.; Monteagudo, A.; Peacock, J.; Quesada, C.A.; et al. Drought sensitivity of the amazon rainforest. *Science* **2009**, *323*, 1344–1347. [CrossRef] [PubMed]

63. Zemp, D.C.; Schleussner, C.F.; Barbosa, H.M.J.; Hirota, M.; Montade, V.; Sampaio, G.; Staal, A.; Wang-Erlandsson, L.; Rammig, A. Self-amplified Amazon forest loss due to vegetation-atmosphere feedbacks. *Nat. Commun.* **2017**, *8*, 1–10. [CrossRef]

64. Aragão, L.E.O.C.; Malhi, Y.; Roman-Cuesta, R.M.; Saatchi, S.; Anderson, L.O.; Shimabukuro, Y.E. Spatial patterns and fire response of recent Amazonian droughts. *Geophys. Res. Lett.* **2007**, *34*. [CrossRef]

65. Andela, N.; Morton, D.C.; Giglio, L.; Paugam, R.; Chen, Y.; Hantson, S.; Van Der Werf, G.R.; Anderson, J.T. The Global Fire Atlas of individual fire size, duration, speed and direction. *Earth Syst. Sci. Data* **2019**, *11*, 529–552. [CrossRef]

66. Broadbent, E.N.; Asner, G.P.; Keller, M.; Knapp, D.E.; Oliveira, P.J.C.; Silva, J.N. Forest fragmentation and edge effects from deforestation and selective logging in the Brazilian Amazon. *Biol. Conserv.* **2008**, *141*, 1745–1757. [CrossRef]

67. Harris, I.; Jones, P.D.; Osborn, T.J.; Lister, D.H. Updated high-resolution grids of monthly climatic observations-the CRU TS3.10 Dataset. *Int. J. Climatol.* **2014**, *34*, 623–642. [CrossRef]

68. Lee, H. *Climate Algorithm Theoretical Basis Document (C-ATBD): Outgoing Longwave Radiation (OLR)-Daily. NOAA's Climate Data Record (CDR) Program, CDRP-ATBD-0526*; Broadway: New York, NY, USA, 2014.

69. Hunter, M.O.; Keller, M.; Victoria, D.; Morton, D.C. Tree height and tropical forest biomass estimation. *Biogeosciences* **2013**, *10*, 8385–8399. [CrossRef]

70. Shao, G.; Stark, S.C.; de Almeida, D.R.A.; Smith, M.N. Towards high throughput assessment of canopy dynamics: The estimation of leaf area structure in Amazonian forests with multitemporal multi-sensor airborne lidar. *Remote Sens. Environ.* **2019**, *221*, 1–13. [CrossRef]

71. Rex, F.E.; Silva, C.A.; Corte, A.P.D.; Klauberg, C.; Mohan, M.; Cardil, A.; da Silva, V.S.; de Almeida, D.R.A.; Garcia, M.; Broadbent, E.N.; et al. Comparison of statistical modelling approaches for estimating tropical forest aboveground biomass stock and reporting their changes in low-intensity logging areas using multi-temporal LiDAR data. *Remote Sens.* **2020**, *12*, 1498. [CrossRef]

72. Asner, G.P.; Powell, G.V.N.; Mascaro, J.; Knapp, D.E.; Clark, J.K.; Jacobson, J.; Kennedy-Bowdoin, T.; Balaji, A.; Paez-Acosta, G.; Victoria, E.; et al. High-resolution forest carbon stocks and emissions in the Amazon. *Proc. Natl. Acad. Sci. USA* **2010**, *107*, 16738–16742. [CrossRef]

73. Meyer, V.; Saatchi, S.; Ferraz, A.; Xu, L.; Duque, A.; García, M.; Chave, J. Forest degradation and biomass loss along the Chocó region of Colombia. *Carbon Balance Manag.* **2019**, *14*, 2. [CrossRef] [PubMed]

74. Aguiar, A.P.D.; Vieira, I.C.G.; Assis, T.O.; Dalla-Nora, E.L.; Toledo, P.M.; Oliveira Santos-Junior, R.A.; Batistella, M.; Coelho, A.S.; Savaget, E.K.; Aragão, L.E.O.C.; et al. Land use change emission scenarios: Anticipating a forest transition process in the Brazilian Amazon. *Glob. Chang. Biol.* **2016**, *22*, 1821–1840. [CrossRef] [PubMed]

75. Taylor, K.E.; Stouffer, R.J.; Meehl, G.A. An overview of CMIP5 and the experiment design. *Bull. Am. Meteorol. Soc.* **2012**, *93*, 485–498. [CrossRef]

76. Fonseca, M.G.; Alves, L.M.; Aguiar, A.P.D.; Arai, E.; Anderson, L.O.; Rosan, T.M.; Shimabukuro, Y.E.; de Aragão, L.E.O.E.C. Effects of climate and land-use change scenarios on fire probability during the 21st century in the Brazilian Amazon. *Glob. Chang. Biol.* **2019**, *25*, 2931–2946. [CrossRef] [PubMed]

77. Elias, F.; Ferreira, J.; Lennox, G.D.; Berenguer, E.; Ferreira, S.; Schwartz, G.; de Oliveira Melo, L.; Reis Júnior, D.N.; Nascimento, R.O.; Ferreira, F.N.; et al. Assessing the growth and climate sensitivity of secondary forests in highly deforested Amazonian landscapes. *Ecology* **2020**, *101*, e02954. [CrossRef]

78. Schwalm, C.R.; Anderegg, W.R.L.; Michalak, A.M.; Fisher, J.B.; Biondi, F.; Koch, G.; Litvak, M.; Ogle, K.; Shaw, J.D.; Wolf, A.; et al. Global patterns of drought recovery. *Nature* **2017**, *548*, 202–205. [CrossRef]

79. Engelbrecht, B.M.J.; Comita, L.S.; Condit, R.; Kursar, T.A.; Tyree, M.T.; Turner, B.L.; Hubbell, S.P. Drought sensitivity shapes species distribution patterns in tropical forests. *Nature* **2007**, *447*, 80–82. [CrossRef]

80. Saatchi, S.; Asefi-Najafabady, S.; Malhi, Y.; Aragão, L.E.O.C.; Anderson, L.O.; Myneni, R.B.; Nemani, R. Persistent effects of a severe drought on Amazonian forest canopy. *Proc. Natl. Acad. Sci. USA* **2013**, *110*, 565–570. [CrossRef]

81. Malhi, Y.; Roberts, J.T.; Betts, R.A.; Killeen, T.J.; Li, W.; Nobre, C.A. Climate change, deforestation, and the fate of the Amazon. *Science* **2008**, *319*, 169–172. [CrossRef]

82. Malhi, Y.; Doughty, C.E.; Goldsmith, G.R.; Metcalfe, D.B.; Girardin, C.A.J.; Marthews, T.R.; del Aguila-Pasquel, J.; Aragão, L.E.O.C.; Araujo-Murakami, A.; Brando, P.; et al. The linkages between photosynthesis, productivity, growth and biomass in lowland Amazonian forests. *Glob. Chang. Biol.* **2015**, *21*, 2283–2295. [CrossRef]

83. Collalti, A.; Trotta, C.; Keenan, T.F.; Ibrom, A.; Bond-Lamberty, B.; Grote, R.; Vicca, S.; Reyer, C.P.O.; Migliavacca, M.; Veroustraete, F.; et al. Thinning Can Reduce Losses in Carbon Use Efficiency and Carbon Stocks in Managed Forests Under Warmer Climate. *J. Adv. Model. Earth Syst.* **2018**, *10*, 2427–2452. [CrossRef]

84. Joetzjer, E.; Douville, H.; Delire, C.; Ciais, P. Present-day and future Amazonian precipitation in global climate models: CMIP5 versus CMIP3. *Clim. Dyn.* **2013**, *41*, 2921–2936. [CrossRef]

85. Boisier, J.P.; Ciais, P.; Ducharne, A.; Guimberteau, M. Projected strengthening of Amazonian dry season by constrained climate model simulations. *Nat. Clim. Chang.* **2015**, *5*, 656–660. [CrossRef]

86. Duffy, P.B.; Brando, P.; Asner, G.P.; Field, C.B. Projections of future meteorological drought and wet periods in the Amazon. *Proc. Natl. Acad. Sci. USA* **2015**, *112*, 13172–13177. [CrossRef]

87. Cai, W.; Borlace, S.; Lengaigne, M.; Van Rensch, P.; Collins, M.; Vecchi, G.; Timmermann, A.; Santoso, A.; Mcphaden, M.J.; Wu, L.; et al. Increasing frequency of extreme El Niño events due to greenhouse warming. *Nat. Clim. Chang.* **2014**, *4*, 111–116. [CrossRef]

88. Lau, W.K.M.; Kim, K.-M. Robust Hadley Circulation changes and increasing global dryness due to CO_2 warming from CMIP5 model projections. *Proc. Natl. Acad. Sci. USA* **2015**, *112*, 3630–3635. [CrossRef]

89. Nepstad, D.C.; Tohver, I.M.; David, R.; Moutinho, P.; Cardinot, G. Mortality of large trees and lianas following experimental drought in an amazon forest. *Ecology* **2007**, *88*, 2259–2269. [CrossRef]

90. Poulter, B.; Hattermann, F.; Hawkins, E.; Zaehle, S.; Sitch, S.; Restrepo-Coupe, N.; Heyder, U.; Cramer, W. Robust dynamics of Amazon dieback to climate change with perturbed ecosystem model parameters. *Glob. Chang. Biol.* **2010**, *16*, 2476–2495. [CrossRef]

91. Nepstad, D.; Lefebvre, P.; Da Silva, U.L.; Tomasella, J.; Schlesinger, P.; Solórzano, L.; Moutinho, P.; Ray, D.; Benito, J.G. Amazon drought and its implications for forest flammability and tree growth: A basin-wide analysis. *Glob. Chang. Biol.* **2004**, *10*, 704–717. [CrossRef]

92. Liu, J.; Vogelmann, J.E.; Zhu, Z.; Key, C.H.; Sleeter, B.M.; Price, D.T.; Chen, J.M.; Cochrane, M.A.; Eidenshink, J.C.; Howard, S.M.; et al. Estimating California ecosystem carbon change using process model and land cover disturbance data: 1951–2000. *Ecol. Model.* **2011**, *222*, 2333–2341. [CrossRef]

93. Doughty, C.E.; Metcalfe, D.B.; Girardin, C.A.J.; Amézquita, F.F.; Cabrera, D.G.; Huasco, W.H.; Silva-Espejo, J.E.; Araujo-Murakami, A.; Da Costa, M.C.; Rocha, W.; et al. Drought impact on forest carbon dynamics and fluxes in Amazonia. *Nature* **2015**, *519*, 78–82. [CrossRef] [PubMed]

94. Yang, Y.; Saatchi, S.S.; Xu, L.; Yu, Y.; Choi, S.; Phillips, N.; Kennedy, R.; Keller, M.; Knyazikhin, Y.; Myneni, R.B. Post-drought decline of the Amazon carbon sink. *Nat. Commun.* **2018**, *9*, 1–9. [CrossRef] [PubMed]

95. De Faria, B.L.; Staal, A.; Martin, P.A.; Panday, P.K.; Castanho, A.D.; Dantas, V.L. Climate change and deforestation boost post-fire grass invasion of Amazonian forests. *bioRxiv* **2019**. [CrossRef]

96. Jolly, W.M.; Cochrane, M.A.; Freeborn, P.H.; Holden, Z.A.; Brown, T.J.; Williamson, G.J.; Bowman, D.M.J.S. Climate-induced variations in global wildfire danger from 1979 to 2013. *Nat. Commun.* **2015**, *6*, 1–11. [CrossRef] [PubMed]

97. Cardil, A.; de-Miguel, S.; Silva, C.A.; Reich, P.B.; Calkin, D.E.; Brancalion, P.H.S.; Vibrans, A.C.; Gamarra, J.G.P.; Zhou, M.; Pijanowski, B.C.; et al. Recent deforestation drove the spike in Amazonian fires. *Environ. Res. Lett.* **2020**, *15*, 121003. [CrossRef]

98. Vanderwel, M.C.; Coomes, D.A.; Purves, D.W. Quantifying variation in forest disturbance, and its effects on aboveground biomass dynamics, across the eastern United States. *Glob. Chang. Biol.* **2013**, *19*, 1504–1517. [CrossRef]

99. Jin, W.; He, H.S.; Thompson, F.R. Are more complex physiological models of forest ecosystems better choices for plot and regional predictions? *Environ. Model. Softw.* **2016**, *75*, 1–14. [CrossRef]

100. Collalti, A.; Thornton, P.E.; Cescatti, A.; Rita, A.; Borghetti, M.; Nolè, A.; Trotta, C.; Ciais, P.; Matteucci, G. The sensitivity of the forest carbon budget shifts across processes along with stand development and climate change. *Ecol. Appl.* **2019**, *29*, 1–18. [CrossRef]

101. Collalti, A.; Tjoelker, M.G.; Hoch, G.; Mäkelä, A.; Guidolotti, G.; Heskel, M.; Petit, G.; Ryan, M.G.; Battipaglia, G.; Matteucci, G.; et al. Plant respiration: Controlled by photosynthesis or biomass? *Glob. Chang. Biol.* **2020**, *26*, 1739–1753. [CrossRef]

102. Swann, A.L.S.; Hoffman, F.M.; Koven, C.D.; Randerson, J.T. Plant responses to increasing CO_2 reduce estimates of climate impacts on drought severity. *Proc. Natl. Acad. Sci. USA* **2016**, *113*, 10019–10024. [CrossRef]

103. Holtum, J.A.M.; Winter, K. Elevated [CO_2] and forest vegetation: More a water issue than a carbon issue? *Funct. Plant Biol.* **2010**, *37*, 694–702. [CrossRef]

104. Jiang, M.; Medlyn, B.E.; Drake, J.E.; Duursma, R.A.; Anderson, I.C.; Barton, C.V.M.; Boer, M.M.; Carrillo, Y.; Castañeda-Gómez, L.; Collins, L.; et al. The fate of carbon in a mature forest under carbon dioxide enrichment. *Nature* **2020**, *580*, 227–231. [CrossRef] [PubMed]

105. Hofhansl, F.; Andersen, K.M.; Fleischer, K.; Fuchslueger, L.; Rammig, A.; Schaap, K.J.; Valverde-Barrantes, O.J.; Lapola, D.M. Amazon forest ecosystem responses to elevated atmospheric CO_2 and alterations in nutrient availability: Filling the gaps with model-experiment integration. *Front. Earth Sci.* **2016**, *4*, 19. [CrossRef]

106. Hubau, W.; Lewis, S.L.; Phillips, O.L.; Affum-Baffoe, K.; Beeckman, H.; Cuní-Sanchez, A.; Daniels, A.K.; Ewango, C.E.N.; Fauset, S.; Mukinzi, J.M.; et al. Asynchronous carbon sink saturation in African and Amazonian tropical forests. *Nature* **2020**, *579*, 80–87. [CrossRef]

107. Longo, M.; Keller, M.; dos-Santos, M.N.; Leitold, V.; Pinagé, E.R.; Baccini, A.; Saatchi, S.; Nogueira, E.M.; Batistella, M.; Morton, D.C. Aboveground biomass variability across intact and degraded forests in the Brazilian Amazon. *Glob. Biogeochem. Cycles* **2016**, *30*, 1639–1660. [CrossRef]

108. Longo, M.; Knox, R.G.; Medvigy, D.M.; Levine, N.M.; Dietze, M.C.; Kim, Y.; Swann, A.L.S.; Zhang, K.; Rollinson, C.R.; Bras, R.L.; et al. The biophysics, ecology, and biogeochemistry of functionally diverse, vertically and horizontally heterogeneous ecosystems: The Ecosystem Demography model, version 2.2-Part 1: Model description. *Geosci. Model. Dev. Discuss.* **2019**, *12*, 4309–4346. [CrossRef]

109. Longo, M.; Saatchi, S.; Keller, M.; Bowman, K.; Ferraz, A.; Moorcroft, P.R.; Morton, D.C.; Bonal, D.; Brando, P.; Burban, B.; et al. Impacts of Degradation on Water, Energy, and Carbon Cycling of the Amazon Tropical Forests. *J. Geophys. Res. Biogeosci.* **2020**, *125*, e2020JG005677. [CrossRef]

110. Dubayah, R.; Blair, J.B.; Goetz, S.; Fatoyinbo, L.; Hansen, M.; Healey, S.; Hofton, M.; Hurtt, G.; Kellner, J.; Luthcke, S.; et al. The Global Ecosystem Dynamics Investigation: High-resolution laser ranging of the Earth's forests and topography. *Sci. Remote Sens.* **2020**, *1*, 100002. [CrossRef]

111. Potapov, P.; Li, X.; Hernandez-Serna, A.; Tyukavina, A.; Hansen, M.C.; Kommareddy, A.; Pickens, A.; Turubanova, S.; Tang, H.; Silva, C.E.; et al. Mapping global forest canopy height through integration of GEDI and Landsat data. *Remote Sens. Environ.* **2020**, 112165. [CrossRef]

112. Duncanson, L.; Neuenschwander, A.; Hancock, S.; Thomas, N.; Fatoyinbo, T.; Simard, M.; Silva, C.A.; Armston, J.; Luthcke, S.B.; Hofton, M.; et al. Biomass estimation from simulated GEDI, ICESat-2 and NISAR across environmental gradients in Sonoma County, California. *Remote Sens. Environ.* **2020**, *242*, 111779. [CrossRef]

Publisher's Note: MDPI stays neutral with regard to jurisdictional claims in published maps and institutional affiliations.

 *forests*

Article

Stable Allometric Trajectories in *Picea abies* (L.) Karst. Trees along an Elevational Gradient

Claudio Mura [1],*, Christian Bianchi Strømme [2] and Tommaso Anfodillo [1]

[1] Department of Land, Environment, Agriculture and Forestry (TeSAF), University of Padua, 35020 Legnaro (PD), Italy; tommaso.anfodillo@unipd.it
[2] Department of Forestry and Wildlife Management, Faculty of Applied Ecology, Agricultural Sciences and Biotechnology, Inland Norway University of Applied Sciences, 2480 Evenstad, Norway; christian.stromme@uib.no
* Correspondence: cmura31@gmail.com

Received: 26 October 2020; Accepted: 17 November 2020; Published: 23 November 2020

Abstract: The effect of temperature on tree phenology and growth has gained particular attention in relation to climate change. While a number of reports indicate that warming can extend the length of the growing season and enhance tree growth rates, it is still debated whether temperature also affects biomass partitioning. Addressing the question of whether trees grown at different elevations invest similarly in various organs, we established four sites along an elevational gradient (320 to 595 m a.s.l.) in managed Norway spruce (*Picea abies* (L.) Karts) stands regenerating after clearcuts in central Norway. There, differences in temperature, bud break, tree growth, and allometric scaling were measured in small spruce trees (up to 3 m height). The results showed that bud break and shoot growth are affected by temperature, as lower sites completed the bud break process 5 days earlier than the higher sites did. There was some evidence indicating that the summer drought of 2018 affected tree growth during the season, and the implications of this are discussed. The allometric scaling coefficients did not change for the crown volume (slope value range 2.66–2.84), crown radius (0.77–0.89), and tree diameter (0.89–0.96) against tree height. A slight difference was found in the scaling coefficients of crown length against tree height (slope value range 1.04–1.12), but this did not affect the general scaling of the crown volume with tree height. Our results showed that different local environmental conditions affect both the growth rate and phenology in Norway spruce trees but, on the contrary, that the biomass partitioning among different parts of the tree remains essentially unchanged. This demonstrates that the allometric approach is an important tool for unraveling true vs. apparent plant plasticity, which in turn is an essential awareness for predicting plant responses to environmental changes.

Keywords: scaling; altitude; biomass partitioning; Norway spruce

---

1. Introduction

In temperate and boreal ecosystems, trees shift between growth and dormancy in cycles that are synchronized with seasonal environmental change, allowing survival in unfavourable seasons. These cycles are defined by physiological processes occurring in different plant tissues in response to local variations in temperature, day length, and light in terms of quantity and quality [1–4]. In addition, the magnitude and direction of phenological response to changes in temperature can be species-specific and/or site-specific [1,5]. Moreover, the uncertainty pertaining to climate change in terrestrial ecosystems calls for a better understanding of the thermal regulation of tree phenology and growth [5–8].

It is well known that growing seasons are shorter in colder climates—e.g., at higher latitudes and elevations [9–11]. Still, there is need for more species-specific research on the direct effect of temperature on phenology in the field, since most studies take place in controlled- and semi-controlled settings and therefore do not account for natural fluctuations in environmental conditions [3]. Temperature is the main environmental factor influencing bud break and apical growth in temperate and boreal climates [1,12,13], so we expect a later start of the growing season and lower growth rates at higher elevations. In these environments, temperature represents one main limiting factor for tree growth. The increase in temperatures due to anthropogenic climate change will therefore affect the distribution of many boreal tree species, both because of a change in environmental conditions [14,15] and because of an increased frequency and intensity of disturbance events such as fire [16,17] and windstorms [18].

Norway spruce is one tree species which will be affected by climate change. Its vast distribution, which spans from sea level in Northern Europe to 2400 m a.s.l. in the Italian Alps [19], makes it susceptible both to expansion and disturbance dynamics in the future [20,21]. The rapidity of climate change, opposed to the slow rate of expansion associated with seed dispersal, may cause the species distribution to shrink significantly [22,23] Additionally, the increased frequency of primary and secondary disturbances will decrease vitality in spruce forests [20,24]. Studying the effect of temperature on spruce growth and phenology in northern climates under field conditions can therefore provide significant contributions to the ecological knowledge of a tree species of primary ecological and economic importance in Northern Europe [19,25]. In particular, we ask whether spruce acclimates to colder, less favourable growth conditions by modifying its biomass partitioning. To answer this question, we adopt an allometric perspective and study how the scaling relationships of different functional traits vary during ontogenesis in a natural field setting.

Allometry is based on the observation that many traits (morphological, physiological, and ecological traits alike) in living organisms are integrated at the genetic, physiological, functional, and developmental levels [26]. Allometric relationships have been observed extensively in plants and are well known in forestry and forest ecology [27], and allometric models have long been used to infer valuable information, such as timber volume or global C stocks, from easily measurable field data [28–31].

At the tree level, allometric relationships define the balance that must exist between different functional parts in order to maintain functionality and increase fitness [32]. The fact that these scaling relationships are commonly observed within a very limited set of possible exponents suggests a strong driver of natural selection against individuals that deviate from "the least-bad" structure [33,34] and underlies existing models [35–37] and theories [38,39]. However, there is still uncertainty regarding the stability of allometric relationships between certain structural parts [30,40–42].

Investigating allometric scaling trajectories and their continuity or variations under different conditions (tree species, elevation, latitude, forest structure, stage of recovery from previous disturbance, etc.) can therefore provide significant contribution to the interpretation of forest structure and dynamics. In particular, studying allometric trajectories can give valuable insight into plant allocation patterns [43–46]. Reich et al. [47] found that trees allocate more biomass to leaves in warmer climates, while root development is favored in colder, nutrient-limited environments. This trend is more pronounced in conifers, therefore we could expect Norway spruce to grow comparatively fewer leaves at higher elevations. However, other studies have found little variations in the relationship between crown volume and tree height in forests with different compositions, disturbance histories, and elevations, but within limited latitudinal ranges [35–37]. This highlights the need for more field-based research on the drivers of tree structural balance and allocation patterns.

In this study, we aim to address gaps in existing knowledge on the climatic regulation of allocation patterns in *Picea abies*. Our objective is to contribute to the understanding of basic drivers of biomass partitioning. We focus on temperature, which is a particularly critical factor in the context of global warming. By adopting an allometric approach to quantify differences in the structural balance of trees, we propose a simple and easily replicable method to discern between true and apparent plasticity. Furthermore, by studying the variability of allometric trajectories within juvenile stands, we aim to unravel the variability of allocation patterns in young trees, an under-represented aspect in the literature since most studies compare the allometric scaling of juvenile stands to mature forests [31,48–51]. Therefore, we provide observational data that can improve our current understanding of environmental influence over a critical phase of stand development in a key European tree species.

In order to investigate possible variations in growth and allometric scaling trajectories due to temperature, we measured Norway spruce saplings along an elevational gradient in a northern continental climate. To quantify the differences in growing conditions, we observed the bud break process and tree growth throughout the season in relation to measured temperatures in selected plots. Further, we investigated whether the trees modified their structural balance in response to environmental variations by comparing allometric scaling trajectories.

Therefore, we aimed to test:

- The influence of temperature on bud break and tree growth in order to quantify differences in the growing conditions along the gradient;
- The stability of allometric trajectories with ontogenesis along the gradient.

We expect a positive effect of temperature on bud break and tree growth. The evidence behind this expectation comes from existing literature on the subject, which agrees on identifying temperature as the main environmental cue influencing bud break [1,2,12,13].

We also expect that allometric relationships, given their fundamental importance in determining tree fitness, are constant along the gradient. This hypothesis is based on the general quantitative theory of forest structure and dynamics [38,39] and is backed by existing studies providing evidence of the stability of allometric relationships despite environmental variations [33,35–37].

2. Materials and Methods

2.1. Experimental Set

The study area (61°25′29.9″ N, 11°04′46.6″ E) is located on a hillslope in the Glomma river valley, Inland county, Norway. The climate is continental, with a precipitation peak during the summer and generally low temperatures, resulting in a boreal biome (Figure 1). The conditions during data collection in the summer of 2018 were exceptionally warm and dry due to a severe drought that happened in Central and Northern Europe at the time. The event was quantified as "unprecedented" by Buras et al. [52]. In Norway, the season was described as "the hottest and driest summer since registrations started back in year 1900" by the Norwegian Meteorological Institute [53]. Available data suggest a strong influence of the drought on the study area, with the mean temperatures in May and July being more than 5 °C higher than average and the precipitation levels in July and August being more than 50 and 30 mm, respectively, lower than usual (Figure 1).

The field study was set up in mono-layered stands, dominated by Norway spruce and managed with a production purpose. Study sites were distributed among stands that underwent clearcutting between 2010 and 2011 and were planted with saplings in 2012: site A (320 m a.s.l.), B (420 m a.s.l.), C (500 m a.s.l.), and D (595 m a.s.l.).

The varieties of Norway spruce planted in the sites were "Opsahl" in sites A, B, and C and "Kaupanger" in site D. Opsahl was selected to maximize stand yield in the climatic conditions of the area. Seeds belonging to this variety were collected from stands with an average latitude of 61.11° N [54]. Kaupanger, on the other hand, was selected for its resistance to frost, particularly because of the anticipated dormancy in autumn. The average latitude of origin was 61.23° N [55]. In the field, we were not able to distinguish between individuals of natural and artificial origins. We accounted for this possible source of variation by including tree variety in our models, which allowed us to test its effect on growth and phenology at the site level.

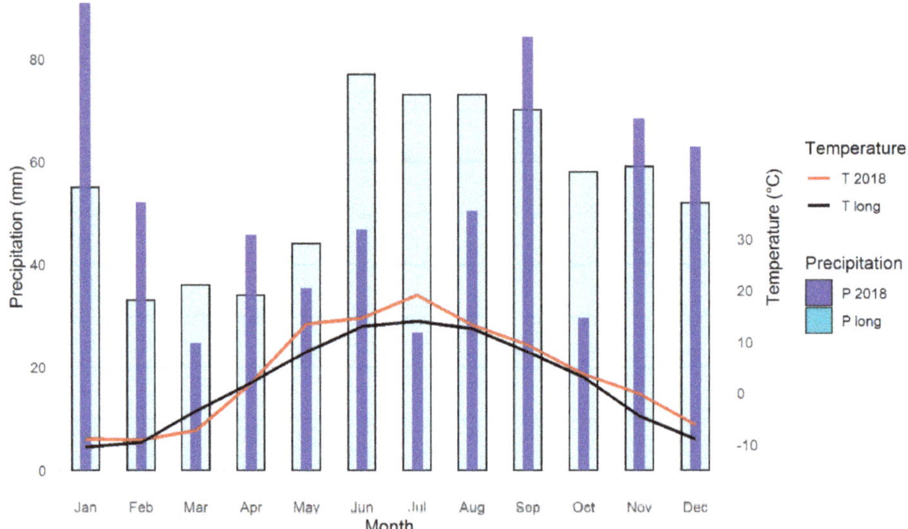

Figure 1. Mean temperature (T) and cumulative precipitation (P) monthly values for the study area. Long-term average values for the period 1961–1990 ("long") are shown together with the mean values recorded in 2018 ("2018"). Temperature data were measured at Evenstad's weather station (4 km from study area), precipitation data were measured at Rena's weather station (~30 km south of study area). Source of the data: Norwegian Meteorological Institute (www.yr.no).

2.2. Local Performance

To measure phenology and growth during the growing season, we established five plots for each site. The plots were established sparsely to account for possible local variations within the stand, while following these criteria: (i) areas with rocky or steeper soil were avoided, trying to limit the difference in soil water availability; (ii) each plot was more than 15 m from the border with the nearest forest stand. This was done to limit the edge effect of the nearby mature forest ecosystem due to shading and belowground competition.

After the determination of each plot center, we selected 10 juvenile trees among healthy individuals (showing no signs of desiccation or missing apical bud) in the vicinity of the center. Five plots were selected for each site, yielding 50 trees per site and a total of 200 trees across the sites. These individuals were marked with red tape and given IDs. This was done to perform repeated measurements of the following growth processes:

- Apical bud break process, measured every two days from 18 May to 5 June 2018;
- Apical shoot elongation, measured every three weeks from 23 May to 4 September 2018;
- Diameter at 20 cm from root collar, measured every three weeks from 23 May to 4 September 2018.

Our study design limited us to certain tree sizes and prevented us from performing a random sampling of trees, therefore sampled individuals should not be considered representative of the whole site in terms of tree size.

The values of apical shoot growth and diameter increment were normalized over tree height and diameter, respectively, as recorded at the start of the season. This allowed us to evaluate and compare the growth rate of trees of different dimensions during the season. The height and diameter values used for the normalization were measured at the start of the growing season.

Apical stages of bud break were classified using a phenological scale dedicated to spruce, adapted from Fløistad and Granhus [56]: (0) dormant buds; (1) buds slightly swollen; (2) buds swollen, bud scales still covering the new needles; (3) bud scales diverging, no elongation of needles; (4) elongation of needles, needles not yet spread; (5) needles spread. Compared to the original scale [56], ours is shorter and relies only on qualitative aspects. We removed stages relying exclusively on quantitative criteria, as they are inaccurate in evaluating bud break under field conditions.

At each plot, temperature loggers were installed, yielding a total of 20 loggers. We used HOBO 8K pendant® waterproof temperature loggers (Onset Computer Corporation, 470 MacArthur Blvd., Bourne, MA, USA), each placed inside a white solar shield case specifically built to shelter the logger from the influence of direct sunlight while also granting aeration. Temperature data were logged every 10 min. Due to logistic problems, the loggers were installed in the field as late as 3 July, well into the growing season. We recorded data from the 3 July until 24 September. We obtained the missing data (15 May to 3 July) by imputation, modelling the relationship between each logger and the Evenstad weather station, which was located 4 km away from the sites. The specific coefficients defining these relationships are available in Table S1 of the Supplementary Materials.

2.3. Allometric Relationships

The collection of allometric data occurred in the same sites (i.e., the same clearcut stands) as the measurement of bud break and growth.

We sampled trees varying from 10 to 300 cm in height and divided them into six groups on the basis of height: 0–50, 50–100, 100–150, 150–200, 200–250, and 250–300 cm. We sampled 10 trees per height interval, resulting in 60 trees per site and a total of 240 trees. Each tree was only sampled once; we sampled all sites between 19 and 24 September.

Because of this height range constraint, the trees sampled for allometric data included, but were not limited to, individuals on which we performed phenological and growth analyses. Plants were sampled based on their height and health status (no visible sign of desiccation, no broken branches or missing tips). We took the following measurements: (1) tree height; (2) tree base diameter; (3) crown radius (R_{cro})—i.e., the distance between the tip of the longest branch and the stem; (4) crown insertion height—i.e., the distance between the lowest active branch in the canopy and the ground. From these, we obtained also (5) crown length (L_{cro})—i.e., the distance between the lowest active branch and the apical tip—and (6) crown volume (V_{cro}), obtained by multiplying the crown length by the square of the crown radius ($V_{cro} = L_{cro} \times R_{cro}^2$). As in previous studies [35,36], this simple formula is preferred over other approaches to estimate crown volume because it easily allows one to investigate the relative change in crown volume—namely, the scaling exponent of V_{cro}—with tree height. The use of more complicated formulas to approach crown shapes (e.g., cone) does not improve the estimation of scaling parameters.

2.4. Statistical Analyses

All the statistical analyses were performed using the R software, version 3.6.1 [57].

We first tested for significant differences in the growing conditions between different sites. We applied analysis of variance (ANOVA) to test for differences in temperature, phenology, and growth. For temperature, we compared the mean of hourly temperatures. For phenology, we considered the day (in progressive "day of the year" DOY values) on which the final stage of bud break (stage 5) was reached. For growth, we considered the total apical shoot elongation and total diameter increment values, normalized over tree height and tree diameter, respectively.

We applied either the parametric ANOVA or Kruskal–Wallis tests depending on the data distribution. Normality in the data and residual distribution was tested with the Shapiro–Wilk test. The homogeneity of variance was tested with the Bartlett test.

Since the ANOVA only reveals that "at least" one site is different from the others, in the case of significant ($p < 0.05$) response we applied the Tukey (for parametric ANOVA) or Dunn (for Kruskal–Wallis) post hoc tests to determine specific differences between sites (complete statistical outputs available in Table S2 of the Supplementary Materials).

Additionally, we fitted mixed-effect models on phenology and growth to test the influence of temperature, daylength, date, and variety over the processes of bud break (including all observed stages) and apical shoot elongation (including all elongation measurements). In that respect, the temperature values from each plot were linked to the response variables measured from the saplings in the plots. For the mixed-effect models, we included plot (i.e., each plot's ID) as a random term when it improved the model and thus accounted for non-measured local environmental variations. We applied cumulative link mixed-effect models (R package "ordinal" [58]) to phenological-scale values, testing for the effect of temperature, daylength, and variety. We applied linear mixed-effect models (R packages "lme4" [59] and "lmerTest" [60]) to apical growth values, testing for the effect of temperature, date, and variety. We tested whether the inclusion of a random variable improved the models by comparing the AIC (Akaike Information Criterion) values of the mixed-effect models with that of models which did not include random variables (glm for continuous variables and clm for ordinal variables). Finally, we performed an AIC-based model selection using the dredge automated model selection function in the "MuMIn" package [61].

Repeated measurements of tree diameter resulted in being very prone to measurement errors because of the field conditions and small size scale of our trees. In order to reduce the error, we only used the first and last measurements to obtain the total diameter increment for the whole season. This allowed us to test for differences between the sites using ANOVA, but the full data series was not reliable for mixed-effect modelling.

Finally, we investigated allometric scaling to assess whether differences in growth are reflected in differences in the structural balance of the trees. We considered the scaling of crown volume, crown radius, crown length, and tree base diameter over tree height. All the scaling parameters were transformed by a base 10 logarithmic function, which allowed us to model the allometric relationship as a linear regression (Equation (1)) [62].

$$(\log)y = b(\log)x + (\log)a, \tag{1}$$

where y and x are the structural traits being considered (e.g., crown volume and tree height) and b is the scaling exponent and a is the intercept—i.e., the coefficients that define the relationship.

This approach allowed us to the quantify site-specific values for the allometric scaling exponent (b). Comparing b values, we were able to investigate the variation in allometric trajectories along the elevational gradient—i.e., the structural balance between traits—in different sites. Similarity or differences between sites are expressed by the confidence intervals (C.I.) of the b parameter; overlapping C.I. values for the b parameters indicate similarity.

3. Results

3.1. Local Performance

The ANOVA revealed significant ($p < 0.001$, $\chi^2 = 21.158$, $df = 3$) temperature differences between the sites. The post hoc Dunn test identified the two lower (A and B) and the two higher (C and D) sites as statistically different (Tables 1 and 2), with higher mean temperatures at lower elevations. More specifically, site A was warmer than sites C ($p < 0.001$) and D ($p = 0.002$). Similarly, site B was warmer than sites C ($p = 0.005$) and D ($p = 0.018$).

Table 1. Results of the post hoc test used to assess the differences in hourly temperature (°C) between sites for the growing season of 2018 (15 May to 4 September). "Group" letters identify significantly ($p <$ 0.05) different groups, where "a" is the group with the lower value. Corresponding mean temperature (°C) values for the growing season are also shown for each site. Values in brackets represent the standard error (S.E.).

Site	Group	Mean T (°C)
A (320 m a.s.l.)	b	17.16 (±0.1)
B (420 m a.s.l.)	b	17.09 (±0.1)
C (500 m a.s.l.)	a	16.39 (±0.1)
D (595 m a.s.l.)	a	16.49 (±0.1)

Table 2. Means of the hourly temperature differences between the sites, calculated for the growing season of 2018 (15 May to 4 September). Each value is the mean difference between the site in the first column and the one in the first row of the table. Values are shown in °C; values in brackets represent the standard error (S.E.).

	A	B	C	D
A	/	0.17 (±0.01)	0.80 (±0.02)	0.75 (±0.03)
B		/	0.64 (±0.01)	0.59 (±0.02)
C			/	−0.05 (±0.02)

The tree size variables varied in a similar fashion, as the ANOVA results identify significant differences between sites also for height ($p < 0.001$, $F = 13.4$, $df = 3$) and diameter ($p < 0.001$, $\chi^2 = 26.232$, $df = 3$) values, measured at the start of the season (23 May). Post hoc tests revealed differences between the two lower and two higher sites, with trees being smaller at higher elevations (Table 3). The trees in sites A and B resulted in being higher than the trees in sites C and D, with p-values < 0.001. The trees in site A also had larger diameters those in than sites C ($p < 0.001$) and D ($p = 0.009$). Similarly, individuals in site B had larger diameters than those in sites C ($p < 0.001$) and D ($p = 0.004$).

Table 3. Results of the post hoc tests used to assess the differences in tree height and diameter in all four sites. Values were measured at the start of the growing season (23 May). "Group" letters identify significantly ($p < 0.05$) different groups, where "a" is the group with the lower value. Corresponding mean height (cm) and diameter (cm) values are also shown for each site. Values in brackets represent the standard error (S.E.).

Site	Group	Mean Height (cm)
A (320 m a.s.l.)	b	169.97 (±5.2)
B (420 m a.s.l.)	b	166.44 (±5.3)
C (500 m a.s.l.)	a	136.35 (±4.7)
D (595 m a.s.l.)	a	137.47 (±4.7)

Site	Group	Mean Diameter (cm)
A (320 m a.s.l.)	b	3.16 (±0.11)
B (420 m a.s.l.)	b	3.19 (±0.10)
C (500 m a.s.l.)	a	2.55 (±0.10)
D (595 m a.s.l.)	a	2.75 (±0.11)

To compare the timing of growth onset across sites, we considered the day of completion of the bud break process—namely, the date in which each individual reached stage 5 of the phenological scale. It took 13 days, from 18 May to 31 May, for all individuals in sites A and B to complete the bud break process. This happened 5 days earlier than in sites C and D, where the process was completed on 5 June (Figure 2).

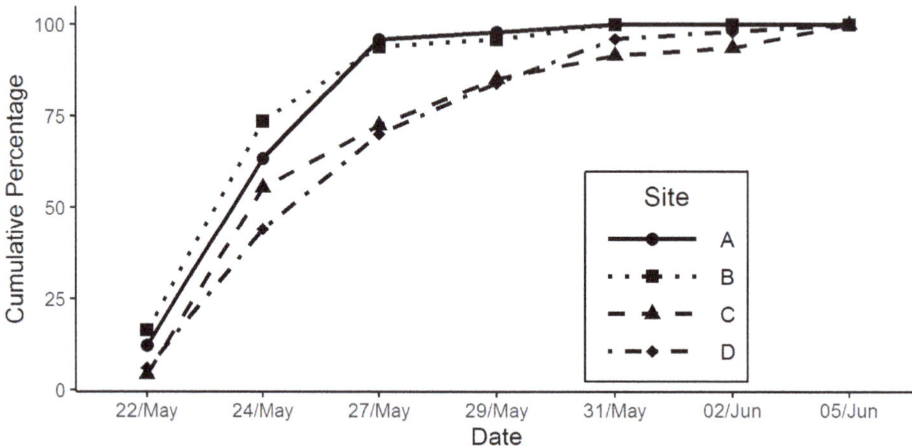

Figure 2. Cumulative percentage of trees reaching the final stage of bud break by site.

The mean date of completion of the bud break process is only one to two days earlier in lower sites A and B (Table 4). Nevertheless, applying ANOVA revealed significant ($p = 0.0015$, $\chi^2 = 15.395$, $df = 3$) differences between the sites. The post hoc analysis results indicate an earlier mean date of bud break in site A than in site D ($p = 0.017$) and in site B than sites C ($p = 0.021$) and D ($p = 0.003$) (Table 4).

Table 4. Results of the post hoc test used to assess differences in the bud break process in the study sites. The parameter being tested is the mean date of completion of the bud break process (in day of the year values). "Group" letters identify statistically different groups, where "a" is the group with the lower value. Dates are expressed as "day of the year" (DOY) progressive values, the corresponding dd/mm dates are also shown. Values in brackets represent the standard error (S.E.).

Site	Group	Mean Date (DOY)	Approximate Date
A (320 m a.s.l.)	ab	144.98 (±0.29)	25 May
B (420 m a.s.l.)	a	144.67 (±0.31)	25 May
C (500 m a.s.l.)	bc	146.47 (±0.53)	26 May
D (595 m a.s.l.)	c	146.62 (±0.4)	27 May

The CLMM that was applied on the apical stage data included the random variable "plot", and indicated a significant ($p < 0.001$, $Z = 28.15$, S.E. $= 0.03002$) positive (0.845) correlation between temperature and bud break. Tree variety and daylength were discarded during model selection, indicating no influence of these variables over the bud break process.

Tree growth differed significantly between sites according to ANOVA, in terms of both normalized apical shoot elongation ($p = 0.0318$, $F = 3$, $df = 3$) and normalized diameter increment ($p = 0.0014$, $\chi^2 = 15.51$, $df = 3$). The differences in apical growth are limited, as the only significant difference is between sites A and C ($p = 0.048$), with site A growing comparatively less than site C. The differences in diameter increment are stronger, as the normalized radial growth in site A was lower than in sites C ($p = 0.0138$) and D ($p = 0.0026$), and lower in site B than in site D ($p = 0.03$) (Table 5). These differences do not follow the same elevational pattern observed for temperature, tree size, and phenology; when looking at normalized growth values, trees in lower sites grew similarly or even less than those at higher elevations.

Table 5. Results of the post hoc tests used to assess the differences in tree growth during the season. Parameters being tested were the mean normalized values of apical shoot elongation and diameter increment. Mean values, both normalized and absolute (cm), are also shown for each site. "Group" letters identify statistically different groups, where "a" is the group with the lower value. Values in brackets represent the standard error (S.E.).

Site	Group	Mean Normalized Shoot Elongation	Mean Shoot Elongation (cm)
A	a	0.22 (±0.01)	37.48 (±2.2)
B	ab	0.25 (±0.01)	42.12 (±1.9)
C	b	0.26 (±0.01)	35.77 (±1.7)
D	ab	0.23 (±0.01)	32.32 (± 2)

Site	Group	Mean Normalized Diameter Increment	Mean Diameter Increment (cm)
A	a	0.105 (±0.007)	0.334 (±0.02)
B	ab	0.112 (±0.008)	0.351 (±0.02)
C	bc	0.139 (±0.009)	0.348 (±0.02)
D	c	0.138 (±0.007)	0.374 (±0.02)

The LME (linear mixed effect) model applied to apical shoot growth for the whole growing season showed a significant positive influence of temperature on growth (coeff = 14.1238, p = 0.00904, t = 2.619, S.E. = 5.39236). The inclusion of "date" improved the model but did not have a significant effect (p = 0.16104, t = 1.403, S.E. = 0.52593). Date, however, had a significant negative interaction with temperature (coeff = −0.07035, p = 0.01252, t = −2.505, S.E. = 0.02809), indicating that the influence of temperature on apical shoot growth decreased over time. Including "plot" as a random variable accounting for environmental variation improved the model, while the "variety" variable was discarded during model selection.

Unexpectedly, the diameter increment during the season displayed a peaks-and-trough pattern. This pattern was found in all four sites, but the trough was deeper in lower sites (Figure 3). In the three-week period from 13 June to 4 July, the average normalized diameter increment values were lower in all sites if compared with the previous and following three-week periods. This means that the growth either slowed down or stopped altogether in the middle of the season. The normalized apical shoot growth values do not show any interruption and simply decrease to a stop, with most of the elongation being over by 4 July (Figure 3).

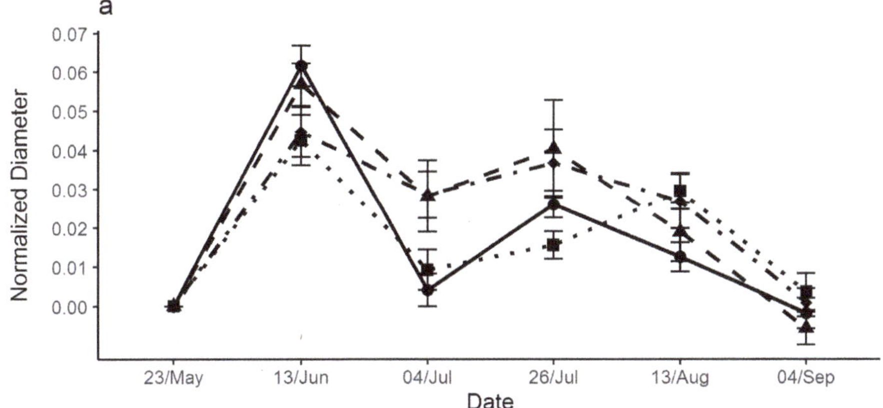

Figure 3. *Cont.*

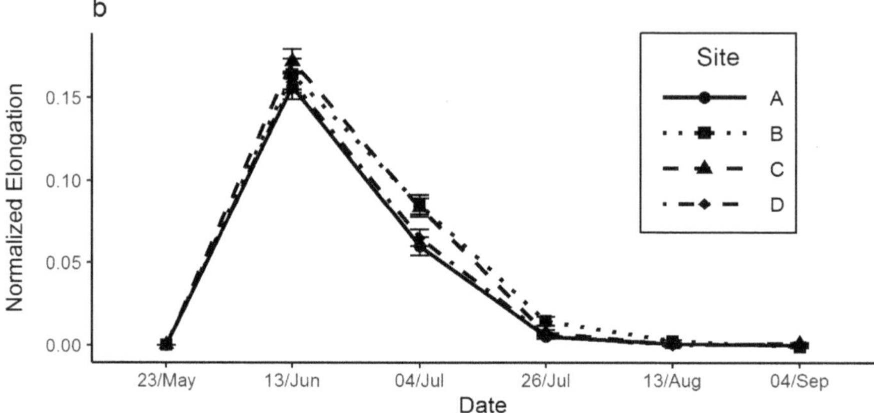

Figure 3. Normalized values of diameter increment (**a**) and shoot elongation (**b**) measured throughout the growing season of 2018. Notice the peaks-and-trough pattern in (**a**), which points to a decrease in the diameter increment during the season. Error bars represent the standard error.

3.2. Allometric Relationships

The slope coefficients in the linear regression function, modelling the allometric scaling of the crown volume vs. tree height, did not show significant differences among sites (Figure 4).

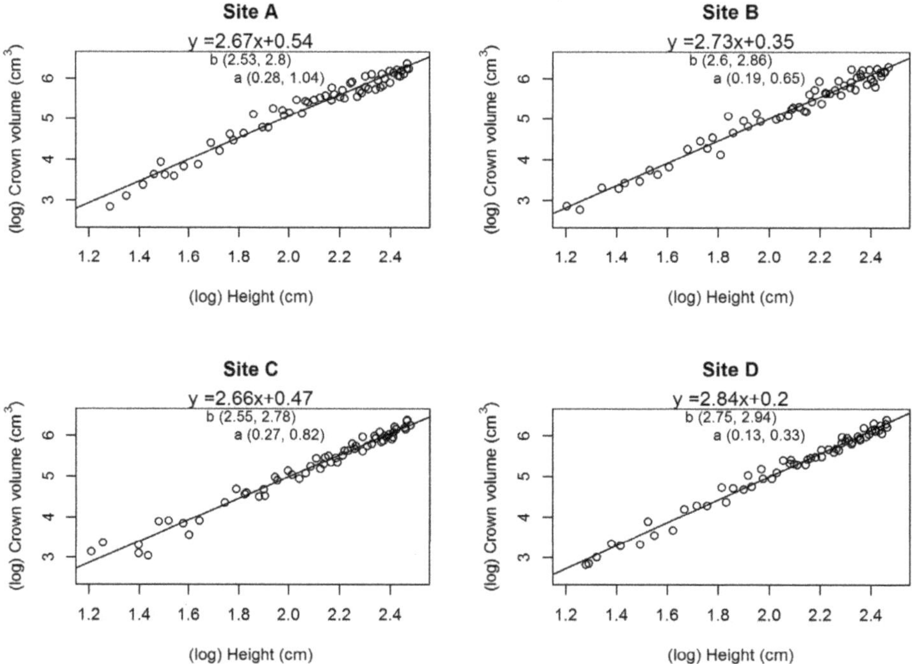

Figure 4. Linear regression models for the allometric scaling of crown volume over height. Regression equations are shown above each plot. Black lines represent the regression model. Dots represent observations. Values in brackets are the 95% confidence intervals (C.I.) of the slope (b) and intercept (a) coefficients.

Further modelling of the allometric scaling of structural parameters other than the crown volume supports this result, with the allometric trajectories being consistently similar in all sites for the values of tree diameter (D), crown length (L_{cro}), and crown radius (R_{cro}) over tree height (H) (Table 6). One single exception is the L_{cro} vs. H relationship in site A (Table 6), whose *b* value is significantly higher than in other sites.

Table 6. Summary of the resulting a and b coefficients and relative C.I. (confidence intervals, 95%) that describe the different allometric relationships in the four sites.

Allometric Relationship	Site	Slope (b)	C.I. (2.5%)	C.I. (97.5%)	Intercept (a)	C.I. (2.5%)	C.I. (97.5%)
Diameter Vs. Height (D vs. H)	A	0.96	0.91	0.99	0.03	0.02	0.03
	B	0.89	0.85	0.94	0.03	0.03	0.04
	C	0.91	0.86	0.96	0.03	0.02	0.04
	D	0.93	0.88	0.98	0.03	0.02	0.04
Crown Length Vs. Height (L_{cro} vs. H)	A	1.12	1.1	1.15	0.48	0.43	0.54
	B	1.04	1.03	1.06	0.74	0.68	0.79
	C	1.04	1.02	1.06	0.71	0.66	0.79
	D	1.06	1.03	1.09	0.63	0.55	0.72
Crown radius Vs. Height (R_{cro} vs. H)	A	0.77	0.7	0.84	1.05	0.78	1.48
	B	0.84	0.78	0.91	0.69	0.5	0.95
	C	0.81	0.75	0.86	0.81	0.62	1.05
	D	0.89	0.84	0.94	0.58	0.45	0.72

4. Discussion

Our results are in line with the present literature concerning the role of temperature in regulating bud break in northern tree species [1,2] and provide detailed modelling of this process under field conditions. This increases the strength of the current projections, associating an earlier growth onset with the increasing temperatures linked to climate change [5–7].

Similarly, we expected to see higher growth rates in warmer sites, but the normalized values of apical and radial growth appear irregular and do not follow the same pattern as the elevational temperature gradient. When looking at the radial increment values (Figure 3), our data point to a slowing in the tree growth rate during the season, which contrasts with existing knowledge [63] and possibly indicates the influence of the year's drought.

Although we lack the necessary data to determine the drivers behind these patterns, we argue that the observed significant differences in growth rates point to differences in the growing conditions between sites. Despite these variations, we find that the exponents of allometric trajectories are similar in the four sites. This provides evidence that spruce saplings maintain the same trajectories of structural balance during growth, regardless of the local growing conditions.

The measured temperatures along the elevational gradient indicate two distinct local climates: a "lower" area, comprising sites A and B and ranging from 320 to 420 m a.s.l., and a "higher" area comprising sites C and D and ranging from 500 to 595 m a.s.l. (Tables 1 and 2). The sampled trees in these sites seem to mirror these conditions, with trees in low-elevation sites (A–B) being significantly bigger than those at higher elevations (C–D) both in height and diameter (Table 3). Similarly, the positive influence of temperature on bud break caused saplings at the lower, warmer sites A and B to experience a faster growth onset. Lower sites showed a faster rate of completion (perc. of trees reaching the final stage) during the whole period (Figure 2), and completed the process 5 days earlier than the higher sites did. Differences in the mean day of completion of the bud break process are limited, amounting only to one to two days, but still indicate a later start of the season at higher elevations (Table 4). This is consistent with existing knowledge, as temperature is the main environmental cue influencing bud break in northern tree species [1,2] and temperatures decrease with increasing elevation, resulting in delayed bud break [9–11].

Despite the variation in phenology across elevation, the measured values of tree growth did not follow such a pattern, as the growth rate variables were generally of smaller value at lower elevation, and the differences between the sites were much less pronounced. The values of normalized apical shoot elongation appear to be significantly different only between sites A and C, with trees in the lowest site A producing comparatively shorter shoots. Differences in the normalized diameter increment are slightly more evident (Table 5), but again indicate that trees in lower sites had a comparatively lower radial increment than trees in higher sites. Overall, it appears that bigger trees in lower sites (A and B) grew comparatively less than smaller trees in higher sites (C and D) during the season. This appears to be in contrast with the well-known direct correlation between temperature and growth rate, as confirmed by our modelling. This inconsistency could be explained, at least for radial growth values, by the pattern of measured values during the season (Figure 3). Rossi et al. [63] found that conifers in cold climates increase their radial growth rate with day length, with a peak at the day of maximum day length. This would be the summer solstice, 21 June 2018—i.e., at the time of the season when we recorded the lowest radial growth rates (Figure 3). This discrepancy is likely due to the summer drought of 2018 [52,53]. Severe effects of the drought on forests were observed in Central Europe during the same event [64], and similar or worse effects likely happened in Scandinavia, where deviations from the norm where stronger [52].

The warm and dry conditions of the summer may have caused the peak-and-trough pattern either by slowing down the tree growth or by causing the shrinking of the xylem cells due to dehydration [65]. It appears that lower-elevation sites were more affected by this event (Figure 3). This would explain the comparatively higher radial growth values observed in the higher-elevation sites C and D (Table 5), where lower temperatures may have diminished the drought effect. This is an untested hypothesis, since assessing and quantifying the local impacts of the drought was not within the scopes of our study and would need further investigation. Furthermore, the fact that our measurements cover such a small area makes it difficult to generalize our results. However, given the existing knowledge on the effects of the 2018 summer drought [52,64] and the prospected increase in the frequency of extreme events due to ongoing clime change [66], we consider this aspect of our study relevant enough to be discussed.

The growth rate pattern for the normalized apical shoot elongation values is consistently similar in all sites; after an initial peak, it slows down and almost stops by 4 July (Figure 3). The significant negative effect of the interaction between the temperature and date variables shows a decrease in the positive effect of temperature on growth over time. This is due to apical growth cessation during the season, while temperatures are still high. Unlike radial growth, there is no recovery in the growth rate after it slows down. This may be caused by the fixed growth pattern of *P. abies*; at the start of the season, the bud already contains all the cell primordia that will develop during the season [2]. As soon as the pre-determined number of cells has developed, apical shoot growth stops and the apical bud enters dormancy. We have no way to establish whether the apical shoot growth ceased because all of the bud primordia properly developed or because the drought forced an early dormancy of the apical buds.

Variety does not seem to have a significant effect on bud break or apical shoot elongation. This is unexpected, as the characteristics of the two varieties are very different. Trees planted in sites A, B, and C belong to the "Opsahl" variety and are selected to maximize stand yield [54,55]. Therefore, we expected to see some differences—e.g., higher growth rates in these sites. Our sampling involved both natural and artificial regeneration, so it is possible that we sampled few individuals of non-local origin and the influence of genotypes was not represented in the dataset. Another possibility is that the drought offset the effect of different varieties. If lower-elevation sites really grew less because of the drought, this may have counterbalanced the effect of the "Opsahl" variety. Again, it should be noted that this hypothesis is untested, but this should be discussed and investigated by future studies. The possible management implications for Scandinavian forestry in the face of climate change are huge, given the importance of Norway spruce in the area [19]. As the tree species distribution is projected to change dramatically [14,15], spruce forests will experience more frequent primary and secondary disturbances [22,24]. Observed and predicted increases in forest growth with global warming [67,68] could be disrupted by these events, as noted by other studies [69,70]. Our results seem to point in this direction, and highlight the need for more field-based studies to adapt future silviculture to the changing climate.

Although the variations in tree growth rates during the season do not reflect the elevational temperature gradient, they were significant nonetheless. This points to local differences in growing conditions, with factors other than temperature influencing growth onset and growth rates during the same season. Despite these differences, the scaling exponents that define the allometric relationships of structural parts of the trees do not vary significantly (Figure 4, Table 6). This means that crown volume, stem diameter, crown length, and crown radius grow similarly, in relation to tree height, in all four sites. The only exception appears to be the scaling relationship L_{cro} vs. H (Table 6), as the scaling value C.I.s of site A do not overlap with other sites, being slightly higher. A higher scaling value of L_{cro} vs. H means that the trees in site A tend to favor vertical growth over time. This is offset by the relatively lower scaling rate in site A for R_{cro} vs. H (Table 6). As a result, the scaling of V_{cro} vs. H is the same in all sites (Figure 4). Therefore, it would appear that trees in site A develop comparatively narrower crowns over time.

The difference between the slope values of L_{cro} vs. H is minimal, yet this is a noteworthy discrepancy since the common knowledge is that Norway spruce tends to grow narrower crowns in colder, harsher environments typically associated with higher elevations and latitudes [71,72]. The fact that this applies to the lowest and warmest site A is surprising. It was observed in the field that competition with *Betula* spp. is stronger at lower elevations, likely because of the higher temperatures. It is possible that this drives spruce saplings in site A to favor vertical growth in early development phases. Despite this, the general balance between productive organs (in our case, crown volume) and tree dimension (tree height) follows the same trajectory in all sites. An important aspect of our study is that the regeneration of both natural and artificial origin comes from a restricted range of latitudes (61.11° N to 61.25° N). Another study conducted by Anfodillo et al. [35] found that the scaling exponent of crown volume versus tree height does not vary significantly (2.22 to 2.31) between four temperate mountain forests with different composition and disturbance levels but within a restricted range of latitudes (46.06° N to 47.27° N). Sellan et al. [36] obtained a similar outcome when comparing the scaling exponents (3.30 to 3.38) of three tropical forest plots with different species richness and composition between 0° and 1° N. Therefore, latitude appears to be one major factor influencing the slope of the locally optimal allometric relationships between these traits.

The continuity in the measured scaling exponents supports the hypothesis that allometric relationships are fundamental to tree functioning and, therefore, that trees acclimate to local conditions by changing their growth rate while maintaining an optimal structural balance during onthogenesis [33]. From this perspective, allometric relationships are a major constraint that governs tree growth, and can be a powerful tool in modelling tree and forest dynamics. This is consistent with the general quantitative theory of allometry proposed by West, Brown, and Enquist [38,39]. It is fundamental to keep in mind that this is observed for the scaling of structural parts versus tree height. The scaling of traits versus diameter has shown more variability in response to environmental variation [30,40,73].

5. Conclusions

In this study, we quantified the influence of temperature on bud break and growth under field conditions. Temperatures follow an elevational gradient, being higher in lower sites and positively influencing the bud break process. Consequently, the season started earlier at lower elevation sites. Differences in growth appear less consistent with temperature, as the bigger trees in the lowest, warmest site A grew comparatively less than in the higher sites C and D. The values of diameter increment during the season point to a possible negative influence of the exceptional 2018 summer drought, which would explain this discrepancy.

The measured differences in temperatures and growth point to significant differences in local growing conditions. Despite this, the allometric relationships of crown volume and tree diameter versus tree height do not vary along the elevational gradient. This indicates that young spruce trees do not appear to modify their structural balance and biomass partitioning to acclimate to local conditions. The allometries of crown length and crown radius versus tree height are stable as well, with the exception of a minor difference in crown length in site A, which did not have an impact on the overall crown volume–height balance.

Our data support the hypothesis that maintaining specific allometric trajectories is fundamental for tree functioning, even at such a young age of tree and stand development. By comparing the growing conditions and allometric scaling in Norway spruce saplings, we offer insight into the ecological factors regulating its growth and allocation patterns. We show that the simple application of an allometric approach can provide valuable information on true vs. apparent plant plasticity, thereby increasing our understanding of tree and forest dynamics and our predictive ability in the face of climate change.

We recommend that future studies wishing to understand variations in tree biomass partitioning adopt a similar allometric-based perspective. On a broader scale, further investigations are needed to fully understand the drivers of allometric scaling in trees. Working along gradients is one efficient way of testing for the variations and drivers of tree and forest growth in the ongoing effort to formulate a comprehensive theory. Additionally, our results highlight the importance of field-based studies that allow for unforeseen changes in environmental conditions. As climate change intensifies, extreme events such as droughts are projected to increase in a diffused manner [66], and field studies can provide useful data on tree responses to such disturbances.

Supplementary Materials: The following are available online at http://www.mdpi.com/1999-4907/11/11/1231/s1: Table S1: Intercept and slope values defining the relationship between the temperature in each plot and the Evenstad weather station. These values were used to compute missing temperature values for the period ranging from 15 May to 3 July, Table S2: Results for post hoc tests performed to identify statistical differences between sites. "Comparison" columns indicate the two sites being compared. Significance levels for p-values: "***" 0.001, "**" 0.01, "*" 0.05. Z = Z value for Dunn test; p.unadj = unadjusted *p*-value for Dunn test; p.adj = *p*-value adjusted to multiple comparisons; diff = difference in average between compared sites; lwr = lower value for 95% C.I.; upr = upper value for 95% C.I.

Author Contributions: Conceptualization and methodology, C.M., C.B.S. and T.A.; formal analysis, C.M., C.B.S. and T.A.; investigation, C.M., C.B.S.; resources, C.B.S.; data curation, C.M., C.B.S.; visualization, C.M.; writing—original draft preparation, C.M.; writing—review and editing, C.M., C.B.S. and T.A.; supervision, C.B.S., T.A. All authors have read and agreed to the published version of the manuscript.

Funding: This research received no external funding. The APC (Article Processing Charges) were partially funded by the Department of Land, Environment, Agriculture and Forestry (TeSAF) of the University of Padua, 35020 Legnaro (PD), Italy.

Acknowledgments: We would like to thank J. Olsen from NMBU, Ås (Norway) for sharing her knowledge of boreal tree species phenology; R. Aamold from the Norwegian Forest Management Company (Statskog), for allowing us to work in the stands and providing useful information about forestry practices and tree provenances; A. Poleo from the Dpt. Of Forestry, Wildlife and Management, INN, Evenstad (Norway) for providing the temperature loggers for our study. This study was conceived and fieldwork carried out during an Erasmus+ study exchange program financed by the European Union.

Conflicts of Interest: The authors declare no conflict of interest.

References

1. Hänninen, H.; Tanino, K. Tree seasonality in a warming climate. *Trends Plant Sci.* **2011**, *16*, 412–416. [CrossRef] [PubMed]

2. Junttila, O. Regulation of annual shoot growth cycle in northern tree species. In *Physiology of Northern Plants under Changing Environment*; Research Signpost: Kerala, India, 2007; pp. 177–210.

3. Olsen, J.E. Light and temperature sensing and signaling in induction of bud dormancy in woody plants. *Plant Mol. Biol.* **2010**, *73*, 37–47. [CrossRef] [PubMed]

4. Strømme, C.B.; Julkunen-Tiitto, R.; Krishna, U.; Lavola, A.; Olsen, J.E.; Nybakken, L. UV-B and temperature enhancement affect spring and autumn phenology in *Populus tremula*: Climate change effects on tree phenology. *Plant Cell Environ.* **2015**, *38*, 867–877. [CrossRef] [PubMed]

5. Körner, C.; Basler, D. Phenology under Global Warming. *Science* **2010**, *327*, 1461–1462. [CrossRef] [PubMed]

6.   Cleland, E.E.; Chuine, I.; Menzel, A.; Mooney, H.A.; Schwartz, M.D. Shifting plant phenology in response to global change. *Trends Ecol. Evol.* **2007**, *22*, 357–365. [CrossRef]

7.   Khanduri, V.P.; Sharma, C.M.; Singh, S.P. The effects of climate change on plant phenology. *Environmentalist* **2008**, *28*, 143–147. [CrossRef]

8.   Morin, X.; Lechowicz, M.J.; Augspurger, C.; O'keefe, J.; Viner, D.; Chuine, I. Leaf phenology in 22 North American tree species during the 21st century. *Glob. Chang. Biol.* **2009**, *15*, 961–975. [CrossRef]

9.   Jyske, T.; Mäkinen, H.; Kalliokoski, T.; Nöjd, P. Intra-annual tracheid production of Norway spruce and Scots pine across a latitudinal gradient in Finland. *Agric. For. Meteorol.* **2014**, *194*, 241–254. [CrossRef]

10.   Moser, L.; Fonti, P.; Buntgen, U.; Esper, J.; Luterbacher, J.; Franzen, J.; Frank, D. Timing and duration of European larch growing season along altitudinal gradients in the Swiss Alps. *Tree Physiol.* **2010**, *30*, 225–233. [CrossRef]

11.   Rossi, S.; Anfodillo, T.; Čufar, K.; Cuny, H.E.; Deslauriers, A.; Fonti, P.; Frank, D.; Gričar, J.; Gruber, A.; Huang, J.-G.; et al. Pattern of xylem phenology in conifers of cold ecosystems at the Northern Hemisphere. *Glob. Chang. Biol.* **2016**, *22*, 3804–3813. [CrossRef]

12.   Hänninen, H. Effects of Climatic Change on Overwintering of Forest Trees in Temperate and Boreal Zones. In Proceedings of the International Conference on Impacts of Global Change on Tree Physiology and Forest Ecosystems, Wageningen, The Netherlands, 26–29 November 1996; Mohren, G.M.J., Kramer, K., Sabaté, S., Eds.; Forestry Sciences. Springer: Dordrecht, The Netherlands, 1997; pp. 149–158, ISBN 978-94-015-8949-9.

13.   Sarvas, R. *Investigations on the Annual Cycle of Development of Forest Trees. Active Period*; Communicationes Instituti Forestalis Fenniae: Helsinki, Finland, 1972; Volume 76.

14.   Dyderski, M.K.; Paź, S.; Frelich, L.E.; Jagodziński, A.M. How much does climate change threaten European forest tree species distributions? *Glob. Chang. Biol.* **2018**, *24*, 1150–1163. [CrossRef] [PubMed]

15.   Sykes, M.T.; Prentice, I.C. Climate change, tree species distributions and forest dynamics: A case study in the mixed conifer/northern hardwoods zone of northern Europe. *Clim. Chang.* **1996**, *34*, 161–177. [CrossRef]

16.   Amiro, B.D.; Stocks, B.J.; Alexander, M.E.; Flannigan, M.D.; Wotton, B.M. Fire, climate change, carbon and fuel management in the Canadian boreal forest. *Int. J. Wildland Fire* **2001**, *10*, 405–413. [CrossRef]

17.   Walker, X.J.; Baltzer, J.L.; Cumming, S.G.; Day, N.J.; Ebert, C.; Goetz, S.; Johnstone, J.F.; Potter, S.; Rogers, B.M.; Schuur, E.A.G.; et al. Increasing wildfires threaten historic carbon sink of boreal forest soils. *Nature* **2019**, *572*, 520–523. [CrossRef]

18.   Gregow, H.; Laaksonen, A.; Alper, M.E. Increasing large scale windstorm damage in Western, Central and Northern European forests, 1951–2010. *Sci. Rep.* **2017**, *7*, 46397. [CrossRef]

19.   San Miguel Ayanz, J.; de Rigo, D.; Caudullo, G.; Durrant, T.H.; Mauri, A. *European Atlas of Forest Tree Species*; Publication Office of the European Union: Luxembourg, 2016; ISBN 978-92-79-36740-3.

20.   Jönsson, A.M.; Linderson, M.-L.; Stjernquist, I.; Schlyter, P.; Bärring, L. Climate change and the effect of temperature backlashes causing frost damage in Picea abies. *Glob. Planet. Chang.* **2004**, *44*, 195–207. [CrossRef]

21.   Prentice, I.C.; Sykes, M.T.; Cramer, W. A simulation model for the transient effects of climate change on forest landscapes. *Ecol. Model.* **1993**, *65*, 51–70. [CrossRef]

22.   Bradshaw, R.H.; Holmqvist, B.H.; Cowling, S.A.; Sykes, M.T. The effects of climate change on the distribution and management of *Picea abies* in southern Scandinavia. *Can. J. For. Res.* **2000**, *30*, 1992–1998. [CrossRef]

23.   Pitelka, L.; Ash, J.; Berry, S.; Bradshaw, R.; Brubaker, L.B.; Clark, J.; Davis, M.; Dyer, J.; Gardner, R.; Gitay, H.; et al. Plant migration and climate change. *Am. Sci.* **1997**, *85*, 464–473.

24.   Schlyter, P.; Stjernquist, I.; Bärring, L.; Jönsson, A.; Nilsson, C. Assessment of the impacts of climate change and weather extremes on boreal forests in northern Europe, focusing on Norway spruce. *Clim. Res.* **2006**, *31*, 75–84. [CrossRef]

25.   Jansson, G.; Danusevičius, D.; Grotehusman, H.; Kowalczyk, J.; Krajmerova, D.; Skrøppa, T.; Wolf, H. Norway Spruce (*Picea abies* (L.) H.Karst.). In *Forest Tree Breeding in Europe: Current State-of-the-Art and Perspectives*; Pâques, L.E., Ed.; Managing Forest Ecosystems; Springer: Dordrecht, The Netherlands, 2013; pp. 123–176. ISBN 978-94-007-6146-9.

26.   Shingleton, A.W. Allometry: The Study of Biological Scaling. *Nat. Educ. Knowl.* **2010**, *3*, 2.

27.   Niklas, K.J. *Plant Allometry: The Scaling of Form and Process*; University of Chicago Press: Chicago, IL, USA, 1994; ISBN 0-226-58080-6.

28. Chave, J.; Réjou-Méchain, M.; Búrquez, A.; Chidumayo, E.; Colgan, M.S.; Delitti, W.B.C.; Duque, A.; Eid, T.; Fearnside, P.M.; Goodman, R.C.; et al. Improved allometric models to estimate the aboveground biomass of tropical trees. *Glob. Chang. Biol.* **2014**, *20*, 3177–3190. [CrossRef] [PubMed]

29. Chave, J.; Andalo, C.; Brown, S.; Cairns, M.A.; Chambers, J.Q.; Eamus, D.; Fölster, H.; Fromard, F.; Higuchi, N.; Kira, T.; et al. Tree allometry and improved estimation of carbon stocks and balance in tropical forests. *Oecologia* **2005**, *145*, 87–99. [CrossRef] [PubMed]

30. Duncanson, L.I.; Dubayah, R.O.; Enquist, B.J. Assessing the general patterns of forest structure: Quantifying tree and forest allometric scaling relationships in the United States: Forest allometric variability in the United States. *Glob. Ecol. Biogeogr.* **2015**, *24*, 1465–1475. [CrossRef]

31. Pilli, R.; Anfodillo, T.; Carrer, M. Towards a functional and simplified allometry for estimating forest biomass. *For. Ecol. Manag.* **2006**, *237*, 583–593. [CrossRef]

32. West, G.B. A General Model for the Origin of Allometric Scaling Laws in Biology. *Science* **1997**, *276*, 122–126. [CrossRef]

33. Anfodillo, T.; Petit, G.; Sterck, F.; Lechthaler, S.; Olson, M.E. Allometric Trajectories and "Stress": A Quantitative Approach. *Front. Plant Sci.* **2016**, *7*. [CrossRef]

34. Enquist, B.J. Universal scaling in tree and vascular plant allometry: Toward a general quantitative theory linking plant form and function from cells to ecosystems. *Tree Physiol.* **2002**, *22*, 1045–1064. [CrossRef]

35. Anfodillo, T.; Carrer, M.; Simini, F.; Popa, I.; Banavar, J.R.; Maritan, A. An allometry-based approach for understanding forest structure, predicting tree-size distribution and assessing the degree of disturbance. *Proc. R. Soc. B* **2013**, *280*, 20122375. [CrossRef]

36. Sellan, G.; Simini, F.; Maritan, A.; Banavar, J.R.; de Haulleville, T.; Bauters, M.; Doucet, J.-L.; Beeckman, H.; Anfodillo, T. Testing a general approach to assess the degree of disturbance in tropical forests. *J. Veg. Sci.* **2017**, *28*, 659–668. [CrossRef]

37. Simini, F.; Anfodillo, T.; Carrer, M.; Banavar, J.R.; Maritan, A. Self-similarity and scaling in forest communities. *Proc. Natl. Acad. Sci. USA* **2010**, *107*, 7658–7662. [CrossRef]

38. Enquist, B.J.; West, G.B.; Brown, J.H. Extensions and evaluations of a general quantitative theory of forest structure and dynamics. *Proc. Natl. Acad. Sci. USA* **2009**, *106*, 7046–7051. [CrossRef]

39. West, G.B.; Enquist, B.J.; Brown, J.H. A general quantitative theory of forest structure and dynamics. *Proc. Natl. Acad. Sci. USA* **2009**, *106*, 7040–7045. [CrossRef] [PubMed]

40. Anderson-Teixeira, K.J.; McGarvey, J.C.; Muller-Landau, H.C.; Park, J.Y.; Gonzalez-Akre, E.B.; Herrmann, V.; Bennett, A.C.; So, C.V.; Bourg, N.A.; Thompson, J.R.; et al. Size-related scaling of tree form and function in a mixed-age forest. *Funct. Ecol.* **2015**, *29*, 1587–1602. [CrossRef]

41. Muller-Landau, H.C.; Condit, R.S.; Chave, J.; Thomas, S.C.; Bohlman, S.A.; Bunyavejchewin, S.; Davies, S.; Foster, R.; Gunatilleke, S.; Gunatilleke, N.; et al. Testing metabolic ecology theory for allometric scaling of tree size, growth and mortality in tropical forests. *Ecol. Lett.* **2006**, *9*, 575–588. [CrossRef]

42. Russo, S.E.; Wiser, S.K.; Coomes, D.A. Growth-size scaling relationships of woody plant species differ from predictions of the Metabolic Ecology Model. *Ecol. Lett.* **2007**, *10*, 889–901. [CrossRef] [PubMed]

43. Weiner, J. Allocation, plasticity and allometry in plants. *Perspect. Plant Ecol. Evol. Syst.* **2004**, *6*, 207–215. [CrossRef]

44. Xie, J.-B.; Xu, G.-Q.; Jenerette, G.D.; Bai, Y.; Wang, Z.-Y.; Li, Y. Apparent plasticity in functional traits determining competitive ability and spatial distribution: A case from desert. *Sci. Rep.* **2015**, *5*, 12174. [CrossRef]

45. Cheng, D.-L.; Niklas, K.J. Above- and Below-ground Biomass Relationships across 1534 Forested Communities. *Ann. Bot.* **2007**, *99*, 95–102. [CrossRef]

46. Poorter, H.; Niklas, K.J.; Reich, P.B.; Oleksyn, J.; Poot, P.; Mommer, L. Biomass allocation to leaves, stems and roots: Meta-analyses of interspecific variation and environmental control: Tansley review. *New Phytol.* **2012**, *193*, 30–50. [CrossRef]

47. Reich, P.B.; Luo, Y.; Bradford, J.B.; Poorter, H.; Perry, C.H.; Oleksyn, J. Temperature drives global patterns in forest biomass distribution in leaves, stems, and roots. *Proc. Natl. Acad. Sci. USA* **2014**, *111*, 13721–13726. [CrossRef] [PubMed]

48. Fatemi, F.R.; Yanai, R.D.; Hamburg, S.P.; Vadeboncoeur, M.A.; Arthur, M.A.; Briggs, R.D.; Levine, C.R. Allometric equations for young northern hardwoods: The importance of age-specific equations for estimating aboveground biomass. *Can. J. For. Res.* **2011**, *41*, 881–891. [CrossRef]

49. Peichl, M.; Arain, M.A. Allometry and partitioning of above- and belowground tree biomass in an age-sequence of white pine forests. *For. Ecol. Manag.* **2007**, *253*, 68–80. [CrossRef]

50. Poorter, H.; Jagodzinski, A.M.; Ruiz-Peinado, R.; Kuyah, S.; Luo, Y.; Oleksyn, J.; Usoltsev, V.A.; Buckley, T.N.; Reich, P.B.; Sack, L. How does biomass distribution change with size and differ among species? An analysis for 1200 plant species from five continents. *New Phytol.* **2015**, *208*, 736–749. [CrossRef] [PubMed]

51. Seo, Y.O.; Lumbres, R.I.C.; Lee, Y.J. Partitioning of above and belowground biomass and allometry in the two stand age classes of Pinus rigida in South Korea. *Life Sci. J.* **2012**, *9*, 3553–3559.

52. Buras, A.; Rammig, A.; Zang, C.S. Quantifying impacts of the 2018 drought on European ecosystems in comparison to 2003. *Biogeosciences* **2020**, *17*, 1655–1672. [CrossRef]

53. Lippestad, H. Cooperation Is a Must for Adaptation to and Mitigation of Climate Change. Available online: https://www.met.no/en/archive/cooperation-is-a-must-for-adaptation-to-and-mitigation-of-climate-change (accessed on 25 May 2020).

54. Skogfrøverket Frøplantasje nr. 1122 Opsahl. Available online: http://www.skogfroverket.no/userfiles/files/Fr%C3%B8plantasjeveiledning/Fr%C3%B8kildebeskrivelser_april2018/1122_Opsahl.pdf (accessed on 1 April 2020).

55. Skogfrøverket Frøplantasje, nr. 1221 Kaupanger. Available online: http://www.skogfroverket.no/userfiles/files/Fr%C3%B8plantasjeveiledning/Fr%C3%B8kildebeskrivelser_april2018/1221_Kaupanger-Frost.pdf (accessed on 1 April 2020).

56. Fløistad, I.S.; Granhus, A. Bud break and spring frost hardiness in Picea abies seedlings in response to photoperiod and temperature treatments. *Can. J. For. Res.* **2010**, *40*, 968–976. [CrossRef]

57. R Core Team. *R: A Language and Environment for Statistical Computing*; R Core Team: Vienna, Austria, 2019.

58. Christensen, R.H.B. Ordinal—Regression Models for Ordinal Data. 2019. Available online: https://rdrr.io/cran/ordinal/ (accessed on 17 November 2020).

59. Bates, D.; Mächler, M.; Bolker, B.; Walker, S. Fitting Linear Mixed-Effects Models Using lme4. *J. Stat. Softw.* **2015**, *67*, 1–48. [CrossRef]

60. Kuznetsova, A.; Brockhoff, P.B.; Christensen, R.H.B. lmerTest Package: Tests in Linear Mixed Effects Models. *J. Stat. Softw.* **2017**, *82*, 1–26. [CrossRef]

61. Barton, K. MuMIn: Multi-Model Inference. 2019. Available online: https://rdrr.io/cran/MuMIn/ (accessed on 17 November 2020).

62. Niklas, K.J. Plant allometry: Is there a grand unifying theory? *Biol. Rev.* **2004**, *79*, 871–889. [CrossRef]

63. Rossi, S.; Deslauriers, A.; Anfodillo, T.; Morin, H.; Saracino, A.; Motta, R.; Borghetti, M. Conifers in cold environments synchronize maximum growth rate of tree-ring formation with day length. *New Phytol.* **2006**, *170*, 301–310. [CrossRef] [PubMed]

64. Schuldt, B.; Buras, A.; Arend, M.; Vitasse, Y.; Beierkuhnlein, C.; Damm, A.; Gharun, M.; Grams, T.E.E.; Hauck, M.; Hajek, P.; et al. A first assessment of the impact of the extreme 2018 summer drought on Central European forests. *Basic Appl. Ecol.* **2020**, *45*, 86–103. [CrossRef]

65. Mäkinen, H.; Nojd, P.; Saranpaa, P. Seasonal changes in stem radius and production of new tracheids in Norway spruce. *Tree Physiol.* **2003**, *23*, 959–968. [CrossRef] [PubMed]

66. IPCC. *Climate Change 2014: Synthesis Report. Contribution of Working Groups I, II and III to the Fifth Assessment Report of the Intergovernmental Panel on Climate Change*; Core Writing Team, Pachauri, R.K., Meyer, L.A., Eds.; IPCC: Geneva, Switzerland, 2014; p. 151.

67. Kellomäki, S.; Peltola, H.; Nuutinen, T.; Korhonen, K.T.; Strandman, H. Sensitivity of managed boreal forests in Finland to climate change, with implications for adaptive management. *Philos. Trans. R. Soc. B Biol. Sci.* **2008**, *363*, 2339–2349. [CrossRef]

68. Kauppi, P.E.; Posch, M.; Pirinen, P. Large Impacts of Climatic Warming on Growth of Boreal Forests since 1960. *PLoS ONE* **2014**, *9*, e111340. [CrossRef]

69. Kurz, W.A.; Stinson, G.; Rampley, G. Could increased boreal forest ecosystem productivity offset carbon losses from increased disturbances? *Phil. Trans. R. Soc. B* **2008**, *363*, 2259–2268. [CrossRef]

70. D'Orangeville, L.; Houle, D.; Duchesne, L.; Phillips, R.P.; Bergeron, Y.; Kneeshaw, D. Beneficial effects of climate warming on boreal tree growth may be transitory. *Nat. Commun.* **2018**, *9*, 3213. [CrossRef]

71. Caré, O.; Müller, M.; Vornam, B.; Höltken, A.; Kahlert, K.; Krutovsky, K.; Gailing, O.; Leinemann, L. High Morphological Differentiation in Crown Architecture Contrasts with Low Population Genetic Structure of German Norway Spruce Stands. *Forests* **2018**, *9*, 752. [CrossRef]

72. Geburek, T.; Robitschek, K.; Milasowszky, N. A tree of many faces: Why are there different crown types in Norway spruce (*Picea abies* [L.] Karst.)? *Flora Morphol. Distrib. Funct. Ecol. Plants* **2008**, *203*, 126–133. [CrossRef]

73. Lines, E.R.; Zavala, M.A.; Purves, D.W.; Coomes, D.A. Predictable changes in aboveground allometry of trees along gradients of temperature, aridity and competition. *Glob. Ecol. Biogeogr.* **2012**, *21*, 1017–1028. [CrossRef]

Publisher's Note: MDPI stays neutral with regard to jurisdictional claims in published maps and institutional affiliations.

Article

Wood vs. Canopy Allocation of Aboveground Net Primary Productivity in a Mediterranean Forest during 21 Years of Experimental Rainfall Exclusion

Romà Ogaya [1,2,*] and Josep Peñuelas [1,2]

[1]   CSIC, Global Ecology Unit CREAF-CSIC-UAB, Bellaterra, 08913 Catalonia, Spain; Josep.Penuelas@uab.cat
[2]   CREAF, Cerdanyola del Vallès, 08913 Catalonia, Spain
*   Correspondence: r.ogaya@creaf.uab.cat; Tel.: +34-9358-140-36

Received: 18 September 2020; Accepted: 9 October 2020; Published: 14 October 2020

Abstract: A Mediterranean holm oak forest was subjected to experimental partial rainfall exclusion during 21 consecutive years to study the effects of the expected decrease in water availability for Mediterranean vegetation in the coming decades. Allocation in woody structures and total aboveground allocation were correlated with annual rainfall, whereas canopy allocation and the ratio of wood/canopy allocation were not dependent on rainfall. Fruit productivity was also correlated with annual rainfall, but only in *Quercus ilex*. In the studied site, there were two types of forest structure: high canopy stand clearly dominated by *Quercus ilex*, and low canopy stand with more abundance of a tall shrub species, *Phillyrea latifolia*. In the tall canopy stand, the allocation to woody structures decreased in the experimental rainfall exclusion, but not the allocation to canopy. In the low canopy stand, wood allocation in *Quercus ilex* was very small in both control and plots with rainfall exclusion, but wood allocation in *Phillyrea latifolia* was even higher than that obtained in tall canopy plots, especially in the plots receiving the experimental rainfall exclusion. These results highlight likely future changes in the structure and functioning of this ecosystem induced by the decrease in water availability. A serious drop in the capacity to mitigate climate change for this Mediterranean forest can be expected, and the ability of *Phillyrea latifolia* to take advantage of the limited capacity to cope with drought conditions detected in *Quercus ilex* makes likely a forthcoming change in species dominance, especially in the low canopy stands.

Keywords: carbon sink; climate change; forest dieback; holm oak; Mediterranean forest; tree growth; wood/canopy allocation

---

1. Introduction

The net primary productivity (NPP) allocation in different tissues is one of the best descriptors of ecosystem functioning [1]. Every woody plant has a trade-off between resource allocation in canopy structures (leaves, flowers, and fruits) or in woody structures (branches and stem). The fraction invested in canopy structures will affect the production (and the consumption) of flowers and fruits, canopy leaf area, their photosynthetic capacity, litter through fall, and its decomposition by soil organisms [1]. The fraction invested in woody structures will be more related to the growth of vegetation and its capacity to ameliorate climate change due to atmospheric CO_2 uptake [2]. Variations in wood/canopy allocation could drive important changes in ecosystem functioning and services. Several factors contribute to the variation in resource allocation in trees: tree age [3], stand structure, environment in which the tree develops [4–6], silvicultural treatments [7,8], ontogeny, and many other factors [6,9].

Water availability is the main limiting factor for plant productivity in some semiarid ecosystems such as Mediterranean forests, where dry season coincides with the hot summer. For these Mediterranean areas, higher rates of evapotranspiration are induced by higher air temperatures with no increase in precipitation [10]. Furthermore, an increase in the frequency of extreme droughts [11] and heat waves [12] is forecasted for these areas. The Mediterranean forest, seasonally exposed to water stress, is often particularly vulnerable to a decrease in water availability, which can induce reductions in stem growth [13,14], defoliation [15,16], tree mortality, and forest dieback [17–22].

The effect of climate change on ecosystem functioning is quite uncertain, but changes in woody and canopy allocation could be a key factor determining some variations of ecosystem structure and functioning because how carbon is allocated in different tissues determines how long carbon remains in plant biomass and thus remains a central challenge for understanding the global carbon cycle [23]. For example, if carbon allocation in tissues with long turnover time (such as stem and branches) decreases and carbon allocation in tissues with short turnover time (leaves, flowers, and fruits) proportionally increases, the function of atmospheric CO_2 sinks and its capacity to ameliorate climate change will strongly decrease.

Climate manipulating experiments have the potential to verify ecosystem responses when it is subjected to continuous climate change [24–26]. Furthermore, long-term experiments could highlight the usual dampening effect size of treatments reported in global-change experiments [27]. As a result, the use of experiments simulating the environmental conditions projected by models of global change as long as possible is the best tool to study long-term climate change effects on natural ecosystems. The Mediterranean forest studied here has been subjected to rainfall exclusion since 1999, and is one of the longest ongoing climate manipulation experiments conducted in natural ecosystems in the world [26].

Plant development in Mediterranean forests is mainly limited by summer drought, and a future decrease in water availability is projected in these ecosystems (about 15%) [13,14,21]. For these reasons, we aimed to study the effects on carbon allocation in woody or canopy structures, derived from an experimental 15% decrease in soil moisture. We also aimed to discuss the possible variations in the structure and functioning of the Mediterranean forest as a consequence of these effects on carbon wood/canopy allocation.

2. Materials and Methods

2.1. Study Site

The study was carried out on a south-facing slope (25%) in the Prades holm oak forest, in Catalonia, NE Spain (41°21′ N, 1°2′ E; 930 m asl). This forest has not been perturbed during the last 70 years, but as a result of ancient coppicing, the vegetation is a dense multi-stem forest (15.433 stems ha^{-1}) dominated by *Quercus ilex* L., with a high presence of evergreen tall shrub species such as *Phillyrea latifolia* L (Appendix A Figure A1). In the forest distribution, there are two types of forest structure: a tall canopy forest (8–10 m tall) clearly dominated by *Q. ilex*, and a low canopy forest (4–6 m tall) with more abundance of *P. latifolia*. Different soil depths determined different water and nutrient availability and this different forest structure [28]. Eight 15 × 10 m plots were delimited at the same altitude along the slope (four in the tall canopy area and another four in the low canopy area). In four of these plots (two in the tall canopy area and another two in the low canopy area) a partial rainfall exclusion system was installed, whereas the other four plots were control. The partial rainfall exclusion treatment consisted of transparent PVC strips installed 1 m above ground level and covering ca. 30% of the plot surface, as shown in Figure A1; these PVC strips also covered a buffer of 2 m above and below the plot area. Soil moisture in these drought plots was on average 15% smaller than in control plots during the overall studied period [21].

2.2. Carbon Allocation Measurements

Before the start of the experiment and outside the plots, different sizes of 10 *Q. ilex* and 10 *P. latifolia* trees were cut (covering the range of stem sizes of trees from the experimental plots). The circumference of each stem at 50 cm height was measured with a metric tape, and after that, each stem was cut from the base and all the aboveground biomass was dried in an oven at 105 °C for enough time to reach constant weight. For each species, allometric relationships between the stem circumference and the weight of dry leaves and total aboveground dry biomass were determined, and these allometric relationships were used to estimate total aboveground and leaf biomass of all stems from the plots (Table A1). Every winter, the circumference at 50 cm height of all stems from the experimental plots was measured, and the aboveground and leaf biomass of all species per plot was estimated using the allometric relationships. We estimated annual wood net primary productivity (Wood NPP) from the difference of the total aboveground biomass increment and the leaf biomass increment during the overall studied period (1999–2019).

In each plot, 20 baskets (27 cm diameter, and with a 1.5 mm mesh inside them) were randomly placed on the ground (Figure A1). The litter that fell in the baskets was collected every two months and separated per species and litter fraction (leaves, flowers, and fruits). Finally, the litter was weighed after drying in an oven at 70 °C. Litter collected in the baskets for one year was used to estimate total flower, fruit, and leaf production per species and plot; it was assumed that all fallen leaves were replaced by new ones. The canopy net primary productivity (canopy NPP) was estimated as the sum of all fallen leaves, fruits, and flowers, and the aboveground net primary productivity (ANPP) was estimated as the sum of annual wood NPP and canopy NPP.

2.3. Statistics

General linear models were performed to test the effect of climatic conditions and rainfall exclusion treatment on ANPP, wood NPP, canopy NPP (flower and fruit production and leaf replacement), and the proportion of wood allocation in relation to canopy allocation (wood/canopy allocation). For each type of canopy plot, several analyses of variance (ANOVAs) were conducted with the percentage of ANPP, wood NPP, canopy NPP (related to the initial biomass per species and plot at the start of the experiment), and wood/canopy allocation as dependent variables, and species (only the two dominant species, *Q. ilex* and *P. latifolia*, were considered separately) and rainfall exclusion treatment as independent factors. Other ANOVAs were conducted with each fraction of canopy NPP as a dependent variable, and species, rainfall exclusion, and type of canopy as independent factors. Analyses of covariance (ANCOVAs) were performed with ANPP, wood NPP, canopy NPP, and wood/canopy allocation as dependent variables; species, canopy type, and rainfall exclusion treatment as independent factors; and mean annual temperature and total annual rainfall as covariates. Finally, linear regressions with the same variables were conducted to determine the effect of annual climatic conditions on each dependent variable. The proportions of ANPP, wood NPP, and canopy NPP (p) were arcsin $p^{0.5}$ transformed to reach the assumptions of a normal distribution. All statistical analyses were performed with the Statistica 12 software package (StatSoft Inc., Tulsa, OK, USA).

3. Results

3.1. Climate and Productivity

3.1.1. Climatic Data

Mean annual temperature was on average 12.38 °C and mean annual precipitation 619 mm during these 21 years of experiments at the studied site (Figure A1). Both temperature and rainfall fluctuated a lot depending on the year. For example in 2015, the highest mean temperature (13.25 °C) and also the lowest precipitation (355 mm) were recorded. On the contrary, the coldest year, 2010 (10.98 °C), coincided with the second wettest year (902 mm, just below the 1021 mm corresponding to year 2018) (Figure A2).

3.1.2. Relationships between Productivity and Climatic Conditions

Forest wood NPP and total ANPP were positively correlated with annual rainfall ($p < 0.01$ and $p = 0.05$ for wood NPP and ANPP, respectively), but there was no significant relationship between canopy NPP or wood/canopy allocation and annual rainfall (Figure 1), and none of the wood NPP, canopy NPP, wood/canopy, or ANPP variables were significantly correlated with mean annual temperature. There was a trend of wood NPP inversely correlating with temperature ($p = 0.1$) (Figure 1). When both dominant species were analyzed separately, the wood NPP of both species was correlated with annual rainfall ($p < 0.05$), and *Q. ilex* ANPP was positively correlated with annual rainfall ($p < 0.05$) (Figure 2) but not with mean annual temperature. However, neither the canopy NPP nor the wood/canopy allocation ratio were correlated with meteorological data. Finally, ANPP was correlated with annual rainfall ($p < 0.05$) in *Q. ilex*, but not in *P. latifolia* (Figure 2). The meteorological data also determined the fruit allocation, which was correlated with annual rainfall only in *Q. ilex* ($p < 0.05$) (Figure 3).

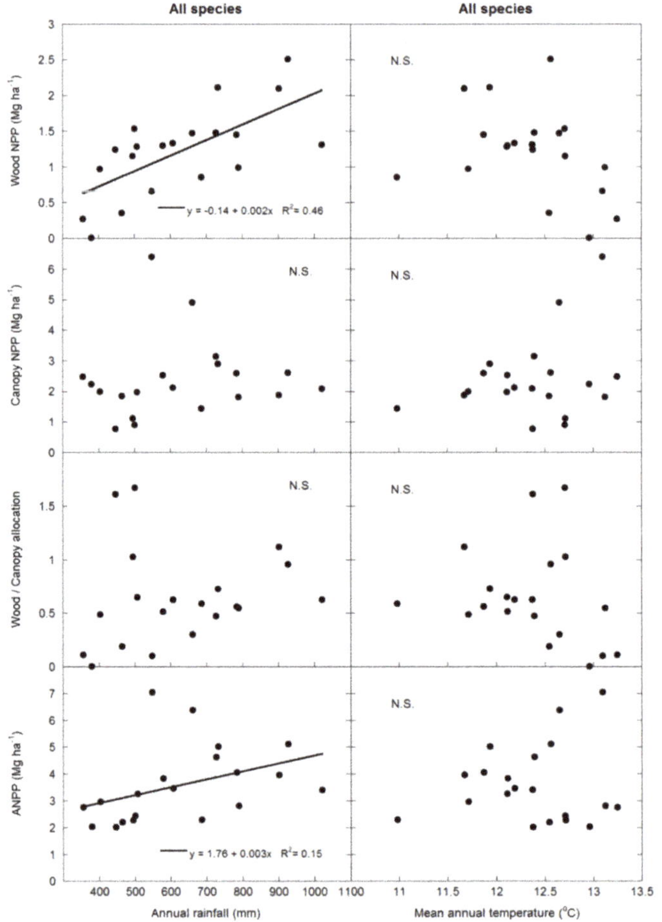

Figure 1. Linear relationships between average net primary productivity (NPP) values (wood NPP, canopy NPP, aboveground net primary productivity (ANPP), and wood/canopy allocation) and meteorological data (annual rainfall and mean annual temperature) of all species analyzed together.

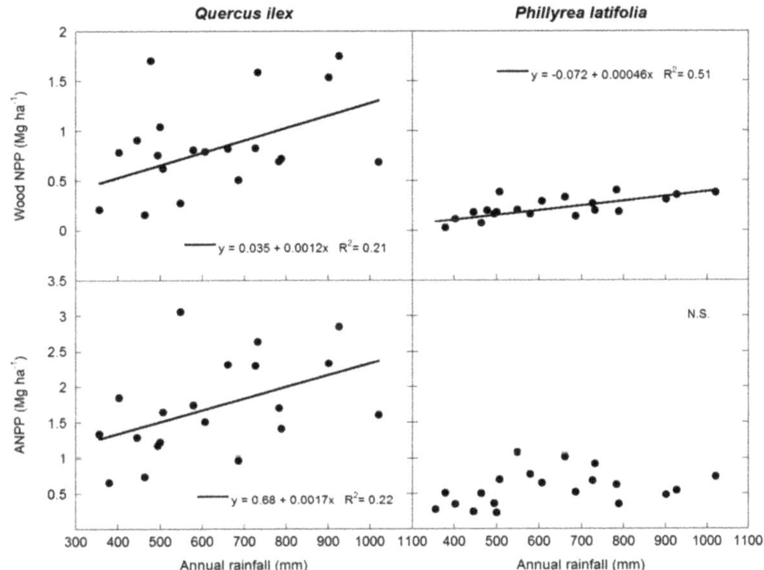

Figure 2. Linear relationships between wood NPP and ANPP values, and annual rainfall in the two dominant species of the studied forest.

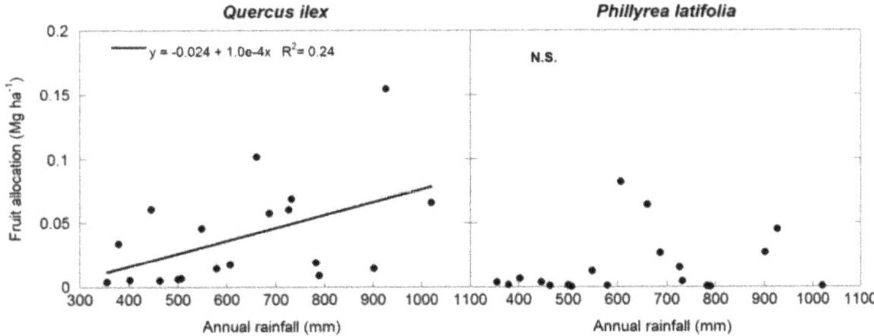

Figure 3. Linear relationships between fruit allocation and annual rainfall in the two dominant species of the studied forest.

The effect of annual rainfall on the studied variables was similar in plots located in tall and low canopy areas, and also in control and plots with the partial rainfall exclusion system.

3.2. Productivity Allocation in Different Forest Stands

3.2.1. Net Primary Productivity Allocation

The average ANPP of the studied forest during the entire studied period was 3.6 Mg ha^{-1} y^{-1}. About one third of this ANPP corresponded to wood NPP (1.2 Mg ha^{-1} y^{-1}), and about two thirds of the total ANPP corresponded to canopy NPP (2.4 Mg ha^{-1} y^{-1}), so wood/canopy allocation was on average 0.56 Mg ha^{-1} y^{-1}. ANPP was higher in tall canopy plots than in low canopy plots (4.0 and 3.2 Mg ha^{-1} y^{-1}, respectively) ($p < 0.01$), but the wood/canopy allocation ratio was similar in both canopy-type plots. When ANPP was examined in relation to the previous aboveground biomass in each plot, *Q. ilex* experienced a similar percentage of ANPP in both tall and low canopy plots, but ANPP in *P. latifolia* was relatively higher in low canopy plots ($p < 0.05$).

3.2.2. Net Primary Productivity Allocation in Tall Canopy Forest Stand

In tall canopy plots, the percentage of wood NPP was similar in both species, but the partial rainfall exclusion reduced wood NPP in both species ($p < 0.05$) (Figure 4). The percentage of canopy NPP was higher in *P. latifolia* than in *Q. ilex* ($p < 0.01$). As a result of this, the percentage of ANPP was also higher in *P. latifolia* ($p < 0.05$) (Figure 4). In these tall canopy plots, wood/canopy allocation was higher in *Q. ilex* than in *P. latifolia* ($p < 0.01$), and this wood/canopy allocation ratio decreased in both species when submitted to rainfall exclusion treatment ($p = 0.01$) (Figure 4).

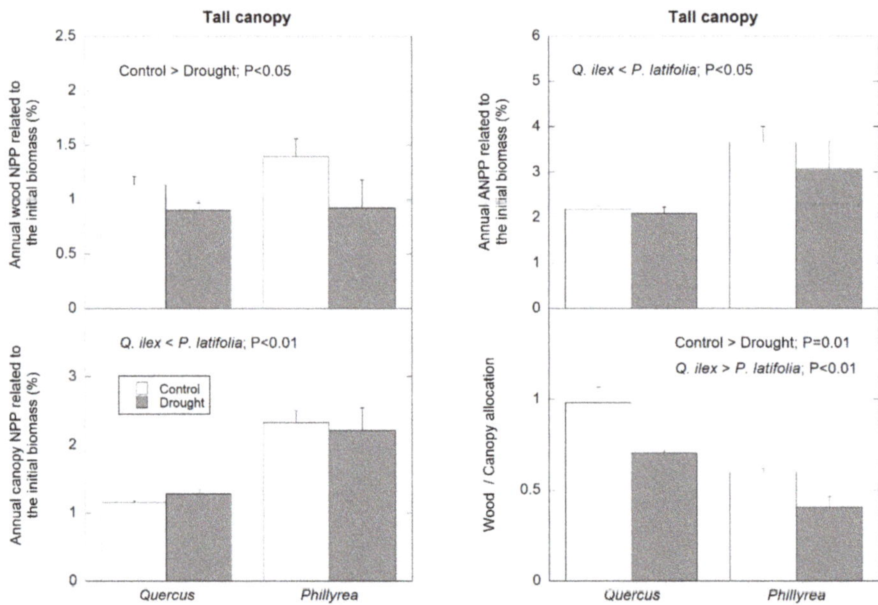

Figure 4. Average percentage of wood NPP, canopy NPP, ANPP, and wood/canopy allocation of the two dominant species in control and plots with partial rainfall exclusion (drought plots), located in the tall canopy stand. Depicted data correspond to the NPP during the overall studied period (1999–2019).

3.2.3. Net Primary Productivity Allocation in Low Canopy Forest Stand

In low canopy plots, the percentage of wood NPP and the percentage of canopy NPP were higher in *P. latifolia* than in *Q. ilex* ($p < 0.05$ and $p < 0.01$ for wood NPP and canopy NPP, respectively), and as a result of this, the percentage of ANPP was also higher in *P. latifolia* than in *Q. ilex* ($p < 0.01$) (Figure 5). In low canopy plots, there was no significant difference between both species and plots with different treatment for the wood/canopy allocation (Figure 5). Surprisingly, the only significant effect of rainfall exclusion was an increase of ANPP in *P. latifolia* (Figure 5).

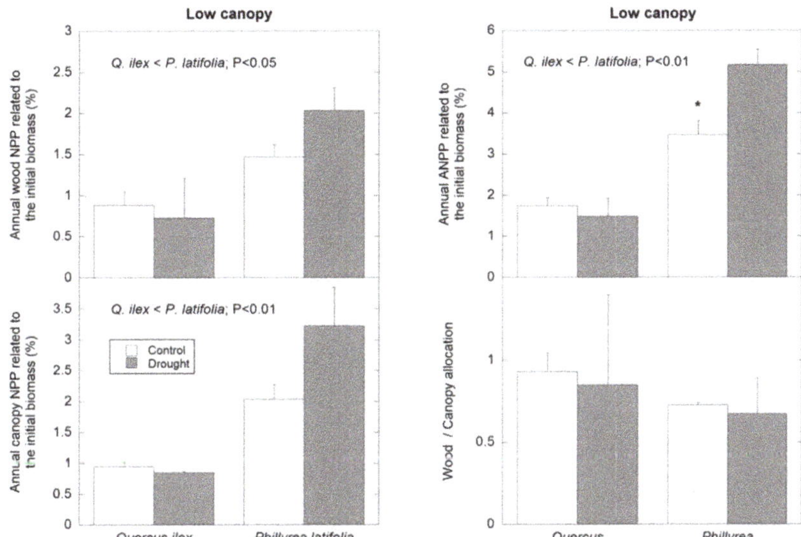

Figure 5. Average percentage of wood NPP, canopy NPP, ANPP, and wood/canopy allocation of the two dominant species, in control and plots with partial rainfall exclusion (drought plots), located in the low canopy stand. Depicted data correspond to the NPP during the overall studied period (1999–2019). *: significant.

4. Discussion

A decrease in water availability usually drives lower carbon allocation in stem and branches compared to the allocation in roots and canopy [29]. In our experiment, wood allocation decreased during the drier years, especially in *Q. ilex*. In plots located in the tall forest stand, where water availability is not as limited as in low canopy stand, rainfall exclusion decreased wood allocation in both studied species, as expected, whereas no effect was observed in canopy allocation; as a result, wood/canopy allocation decreased in both species. In the low forest stand, wood and canopy allocation were very low in *Q. ilex* in both control and rainfall exclusion plots, showing that growth in this species is seriously constrained by the low water availability of this low canopy stand. In contrast, *P. latifolia* seemed to be more able to obtain an advantage from the decay of *Q. ilex*, because *P. latifolia* experienced larger wood and canopy allocation in the low canopy stand and even larger allocation in plots with rainfall exclusion as a result of the decrease in stem density induced by the tree mortality observed in *Q. ilex* [20,21]. As observed in previous studies conducted at the same experimental site, *P. latifolia* was better able to cope with drought conditions than *Q. ilex* [13,14,21].

Analyzing the fruit production, once again *Q. ilex* was more dependent than *P. latifolia* on water availability, because fruit allocation decreased when annual rainfall decreased in *Q. ilex*, but not in *P. latifolia*. Despite a slight decrease of *Q. ilex* fruit production in rainfall exclusion plots detected during the first years of study [30], this effect disappeared with time, and globally there was no effect of rainfall exclusion on fruit allocation during the overall studied period [31], in contrast with what was observed in another rainfall exclusion experiment where *Q. ilex* acorn production decreased in plots with rainfall exclusion [32]. However, rainfall exclusion exerted a decrease in wood growth of *Q. ilex*, but not in *P. latifolia*, so the more drought-sensitive species, *Q. ilex*, was able to maintain fecundity by shifting allocation of resources away from growth [31].

Our results highlight a strong decrease in the capacity of Mediterranean forest to mitigate climate change induced by future limitations in water availability, because of the high decrease in wood growth. However, this forest seems to be able to maintain leaf turnover and the production of flowers and fruits. Moreover, our estimation of wood allocation was provided from stem radial growth, but sometimes water scarcity could decrease wood density in tree rings instead of radial growth [33], as observed in *Q. ilex* [34]. If water scarcity also drives lower wood density, the decrease in wood resource allocation could be even greater than that detected in our experiment. The allocation to fine roots is another fraction of resource allocation competing with wood and canopy allocation, and fine root allocation could be more variable depending on water availability than wood and canopy allocation [1]. We reported the effects of rainfall exclusion in aboveground biomass allocation, but there is a lack of any estimation about the allocation to fine roots. Usually, water scarcity enhanced the growth of fine roots [1]. Another experiment conducted near the study site also reported shorter fine-root longevity in *Q. ilex* when water availability decreased [35], so more resource allocation is needed to restore dead roots. We did not measure how much is lost in the form of volatiles, root exudates, mycorrhizal associations, consumption by animals, and dissociated dead parts [36]. Previous measurements conducted at the same experimental site highlighted a decrease in volatile organic compound (VOC) emission under heavy drought in *Q. ilex*, whereas no VOC emission decrease was observed in *P. latifolia* [37]. It was also observed in *Q. ilex*, that 21 consecutive days of rainfall exclusion increased 21% root exudates compared to well-watered conditions [38]. Despite these changes in resource allocation described before, other authors reported higher plant ability to alter organ morphology, such as specific leaf area (SLA), to cope with climatic variations than to adjust allocation [6].

The small values of wood allocation described in *Q. ilex* trees from low canopy plots and the high wood and total aboveground allocation in *P. latifolia* in the same plots are in agreement with the defoliation and tree mortality observed in *Q. ilex* in the same study area when it is subjected to severe drought conditions [20,21], and highlight the ability of *P. latifolia* to take competitive advantage of *Q. ilex* decay. In the low canopy stand, *Q. ilex* cannot actually adjust the allocation of resources more to wood structures to cope with the drier conditions posed by climate change. The dominant tree species of this Mediterranean forest, *Q. ilex*, is currently starting to be replaced especially in the low canopy stand by *P. latifolia*, which is more adapted to the current and forthcoming drier and hotter conditions [13,14,21].

5. Conclusions

Increasing water deficits characterized by lower soil moisture and higher atmospheric aridity are leading to several changes in the ecosystem functioning of the Mediterranean forest, such as a dramatic decrease in the capacity to mitigate climate change, but it seems that this forest will be able to maintain leaf turnover and the production of reproductive structures. However, if current climate evolution leads to forest dieback, as is occurring in *Q. ilex* in the low canopy stand, the maintenance of the production and dispersion of seeds will be not enough to compensate for *Q. ilex* decay. A progressive substitution of the species most sensitive to water scarcity (*Q. ilex*) by other species more adapted to drought (such as *P. latifolia*) is expected, and it may transform the current Mediterranean forest to a tall shrubland, with different species composition and different ecosystem behavior and services [39].

Author Contributions: Conceptualization, design of the experiment, statistical analyses, and discussion of the results—R.O. and J.P.; writing—R.O.; supervision—J.P. All authors have read and agreed to the published version of the manuscript.

Funding: This research was financially supported by the European Research Council Synergy grant ERC-2013-SyG-2013-610028 IMBALANCE-P, the Spanish government project PID2019-110521GB-I00 ELEMENTAL SHIFT, and the Catalan government grant SGR-2017–1005.

Acknowledgments: This study was undertaken with the approval of the staff of the PNIN de Poblet Natural Park.

Conflicts of Interest: The authors declare no conflict of interest.

Appendix A

Table A1. Allometric relationships between total aboveground biomass (AB) and total biomass of leaves (LB) from stem diameter at 50 cm height (D50), in *Q. ilex* and *P. latifolia*.

Species	Total Aboveground Biomass	Total Biomass of Leaves
Quercus ilex	ln AB = 4.900 + 2.277 ln D50	ln LB = 3.481 + 1.695 ln D50
Phillyrea latifolia	ln AB = 4.251 + 2.463 ln D50	ln LB = 1.433 + 2.426 ln D50

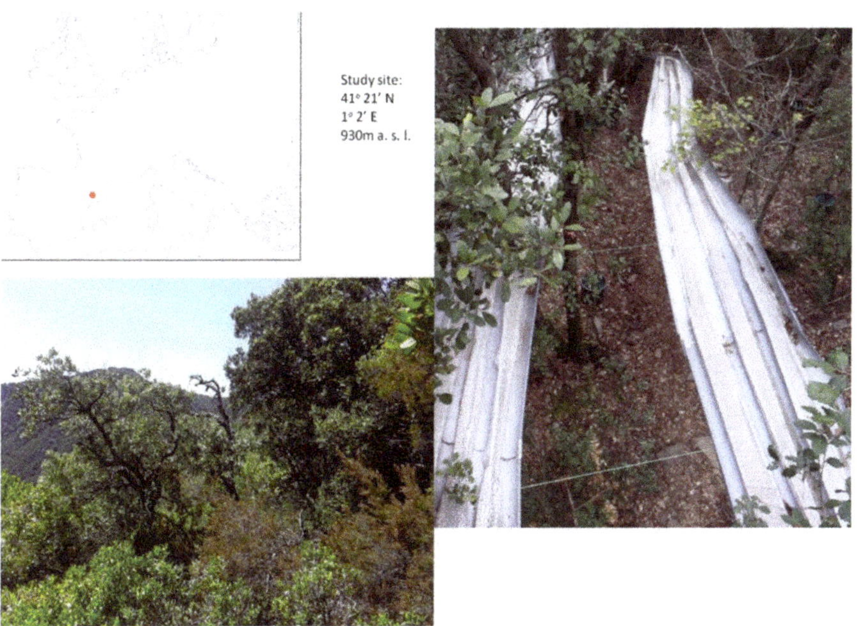

Figure A1. View of the study site and its localization, and view of the partial rainfall exclusion treatment installation and litter collectors. The PVC strips were only placed in plots subjected to partial rainfall exclusion, whereas litter collector baskets were placed in all plots.

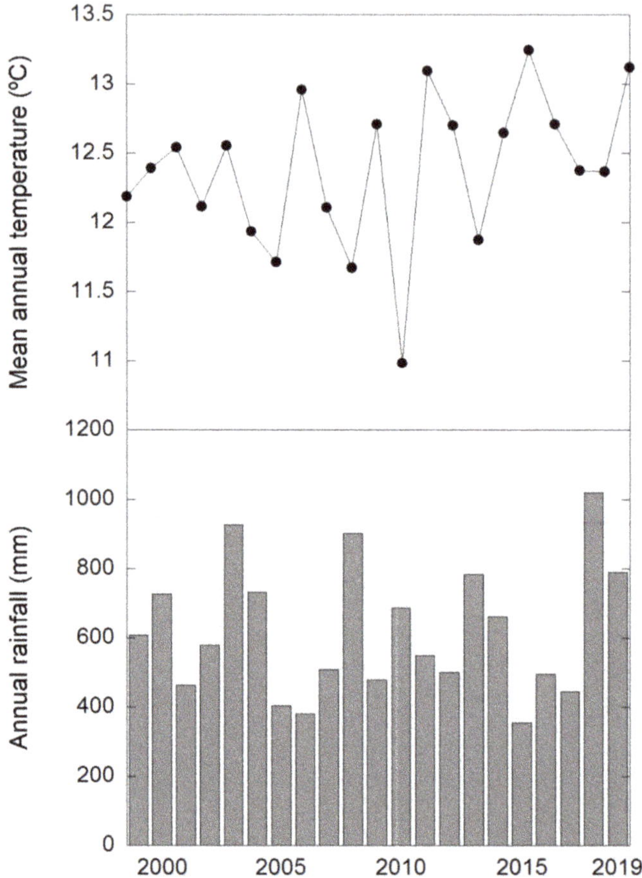

Figure A2. Mean annual temperature and annual rainfall values of the study site during the overall duration of the experiment.

References

1. Malhi, Y.; Doughty, C.; Galbraith, D. The allocation of ecosystem net primary productivity in tropical forests. *Philos. Trans. R. Soc. B Biol. Sci.* **2011**, *366*, 3225–3245. [CrossRef] [PubMed]
2. Hyvönen, R.; Ågren, G.I.; Linder, S.; Persson, T.; Cotrufo, M.F.; Ekblad, A.; Freeman, M.; Grelle, A.; Janssens, I.A.; Jarvis, P.G.; et al. The likely impact of elevated [CO_2], nitrogen deposition, increased temperature and management on carbon sequestration in temperate and boreal forest ecosystems: A literature review. *New Phytol.* **2007**, *173*, 463–480. [CrossRef] [PubMed]
3. Bartelink, H.H. A model of dry matter partitioning in trees. *Tree Physiol.* **1998**, *18*, 91–101. [CrossRef] [PubMed]
4. Iivonen, S.; Kaakinen, S.; Jolkkonen, A.; Vapaavuori, E.; Linder, S. Influence of long-term nutrient optimization on biomass, carbon, and nitrogen acquisition and allocation in Norway spruce. *Can. J. For. Res.* **2006**, *36*, 1563–1571. [CrossRef]
5. Poorter, H.; Nagel, O. The role of biomass allocation in the growth response of plants to different levels of light, CO_2, nutrients and water: A quantitative review. *Funct. Plant Biol.* **2000**, *27*, 595–607. [CrossRef]

6. Poorter, H.; Niklas, K.J.; Reich, P.B.; Oleksyn, J.; Poot, P.; Mommer, L. Biomass allocation to leaves, stems and roots: Meta-analyses of interspecific variation and environmental control. *New Phytol.* **2011**, *193*, 30–50. [CrossRef] [PubMed]

7. Litton, C.M.; Raich, J.W.; Ryan, M.G. Carbon allocation in forest ecosystems. *Glob. Chang. Biol.* **2007**, *13*, 2089–2109. [CrossRef]

8. Lopez, B.C.; Sabaté, S.; Gracia, C.A. Thinning effects on carbon allocation to fine roots in a *Quercus ilex* forest. *Tree Physiol.* **2003**, *23*, 1217–1224. [CrossRef]

9. Merganičová, K.; Merganič, J.; Lehtonen, A.; Vacchiano, G.; Sever, M.Z.O.; Augustynczik, A.L.D.; Grote, R.; Kyselová, I.; Mäkelä, A.; Yousefpour, R.; et al. Forest carbon allocation modelling under climate change. *Tree Physiol.* **2019**, *39*, 1937–1960. [CrossRef]

10. IPCC. "IPCC 2018: Summary of Policymakers" in Global warming of 1.5 °C. In *An IPCC Special Report on the Impacts of Global Warming of 1.5 °C Above Pre-Industrial Levels and Related Global Greenhouse Gas Emission Pathways, in the Context of Strengthening the Global Response to the Threat of Climate Change, Sustainable Development, and Efforts to Eradicate Poverty*; Masson-Delmotte, V., Zhai, P., Pörtner, H.O., Roberts, D., Skea, J., Shukla, P.R., Pirani, A., Moufouma-Okia, W., Péan, C., Pidcock, R., et al., Eds.; World Meteorological Organization: Geneva, Switzerland, 2018; p. 32.

11. Bates, B.C.; Kundzcwicz, Z.W.; Wu, S.; Palutikof, J.P. *Climate Change and Water. Technical Paper of the Intergovernmental Panel on Climate Change*; IPCC Secretariat: Geneva, Switzerland, 2008; p. 210.

12. Fischer, E.M.; Schär, C. Consistent geographical patterns of changes in high-impact European heatwaves. *Nat. Geosci.* **2010**, *3*, 398–403. [CrossRef]

13. Barbeta, A.; Ogaya, R.; Peñuelas, J. Dampening effects of long-term experimental drought on growth and mortality rates of a Holm oak forest. *Glob. Chang. Biol.* **2013**, *19*, 3133–3144. [CrossRef] [PubMed]

14. Liu, D.; Ogaya, R.; Barbeta, A.; Yang, X.; Peñuelas, J. Long-term experimental drought combined with natural extremes accelerate vegetation shift in a Mediterranean holm oak forest. *Environ. Exp. Bot.* **2018**, *151*, 1–11. [CrossRef]

15. Carnicer, J.; Coll, M.; Ninyerola, M.; Pons, X.; Sanchez, G.; Peñuelas, J. Widespread crown condition decline, food web disruption, and amplified tree mortality with increased climate change-type drought. *Proc. Natl. Acad. Sci. USA* **2011**, *108*, 1474–1478. [CrossRef] [PubMed]

16. Galiano, L.; Martinez-Vilalta, J.; Sabaté, S.; Lloret, F. Determinants of drought effects on crown condition and their relationship with depletion of carbon reserves in a Mediterranean holm oak forest. *Tree Physiol.* **2012**, *32*, 478–489. [CrossRef]

17. Allen, C.D.; Macalady, A.K.; Chenchouni, H.; Bachelet, D.; McDowell, N.; Vennetier, M.; Kitzberger, T.; Rigling, A.; Breshears, D.D.; Hogg, E.; et al. A global overview of drought and heat-induced tree mortality reveals emerging climate change risks for forests. *For. Ecol. Manag.* **2010**, *259*, 660–684. [CrossRef]

18. Breshears, D.D.; Cobb, N.S.; Rich, P.M.; Price, K.P.; Allen, C.D.; Balice, R.G.; Romme, W.H.; Kastens, J.H.; Floyd, M.L.; Belnap, J.; et al. Regional vegetation die-off in response to global-change-type drought. *Proc. Natl. Acad. Sci. USA* **2005**, *102*, 15144–15148. [CrossRef]

19. Peng, C.; Ma, Z.; Lei, X.; Zhu, Q.; Chen, H.; Wang, W.; Liu, S.; Li, W.; Fang, X.; Zhou, X. A drought-induced pervasive increase in tree mortality across Canada's boreal forests. *Nat. Clim. Chang.* **2011**, *1*, 467–471. [CrossRef]

20. Ogaya, R.; Barbeta, A.; Başnou, C.; Peñuelas, J. Satellite data as indicators of tree biomass growth and forest dieback in a Mediterranean holm oak forest. *Ann. For. Sci.* **2014**, *72*, 135–144. [CrossRef]

21. Ogaya, R.; Liu, D.; Barbeta, A.; Peñuelas, J. Stem Mortality and Forest Dieback in a 20-Years Experimental Drought in a Mediterranean Holm Oak Forest. *Front. For. Glob. Chang.* **2020**, *2*, 89. [CrossRef]

22. Williams, A.P.; Allen, C.D.; Macalady, A.K.; Griffin, D.; Woodhouse, C.A.; Meko, D.M.; Swetnam, T.W.; Rauscher, S.A.; Seager, R.; Grissino-Mayer, H.D.; et al. Temperature as a potent driver of regional forest drought stress and tree mortality. *Nat. Clim. Chang.* **2012**, *3*, 292–297. [CrossRef]

23. Montané, F.; Fox, A.M.; Arellano, A.F.; MacBean, N.; Alexander, M.R.; Dye, A.W.; Bishop, D.A.; Trouet, V.; Babst, F.; Hessl, A.; et al. Evaluating the effect of alternative carbon allocation schemes in a land surface model (CLM4.5) on carbon fluxes, pools, and turnover in temperate forests. *Geosci. Model Dev.* **2017**, *10*, 3499–3517. [CrossRef]

24. Wu, Z.; Dijkstra, P.; Koch, G.W.; Peñuelas, J.; Hungate, B.A. Responses of terrestrial ecosystems to temperature and precipitation change: A meta-analysis of experimental manipulation. *Glob. Chang. Biol.* **2011**, *17*, 927–942. [CrossRef]

25. Beier, C.; Beierkuhnlein, C.; Wohlgemuth, T.; Peñuelas, J.; Emmet, B.; Körner, C.; de Boeck, H.; Christensen, J.H.; Leuzinger, S.; Janssens, I.A.; et al. Precipitation manipulation experiments—Challenges and recommendations for the future. *Ecol. Lett.* **2012**, *15*, 899–911. [CrossRef]

26. Peñuelas, J.; Sardans, J.; Filella, I.; Estiarte, M.; Llusià, J.; Ogaya, R.; Carnicer, J.; Bartrons, M.; Rivas-Ubach, A.; Grau, O.; et al. Assessment of the impacts of climate change on Mediterranean terrestrial ecosystems based on data from field experiments and long-term monitored field gradients in Catalonia. *Environ. Exp. Bot.* **2018**, *152*, 49–59. [CrossRef]

27. Leuzinger, S.; Luo, Y.; Beier, C.; Dieleman, W.; Vicca, S.; Körner, C. Do global change experiments overestimate impacts on terrestrial ecosystems? *Trends Ecol. Evol.* **2011**, *26*, 236–241. [CrossRef]

28. Rivas-Ubach, A.; Gargallo-Garriga, A.; Sardans, J.; Oravec, M.; Mateu-Castell, L.; Pérez-Trujillo, M.; Parella, T.; Ogaya, R.; Urban, O.; Peñuelas, J. Drought enhances folivory by shifting foliar metabolomes in *Quercus ilex* trees. *New Phytol.* **2014**, *202*, 874–885. [CrossRef]

29. Doughty, C.E.; Metcalfe, D.B.; Girardin, C.A.J.; Amezquita, F.F.; Durand, L.; Huasco, W.H.; Silva-Espejo, J.E.; Araujo-Murakami, A.; Da Costa, M.C.; Da Costa, A.C.L.; et al. Source and sink carbon dynamics and carbon allocation in the Amazon basin. *Glob. Biogeochem. Cycles* **2015**, *29*, 645–655. [CrossRef]

30. Ogaya, R.; Peñuelas, J. Species-specific drought effects on flower and fruit production in a Mediterranean holm oak forest. *Forestry* **2007**, *80*, 351–357. [CrossRef]

31. Bogdziewicz, M.; Fernández-Martínez, M.; Espelta, J.; Ogaya, R.; Peñuelas, J. Is forest fecundity resistant to drought? Results from an 18-yr rainfall-reduction experiment. *New Phytol.* **2020**, *227*, 1073–1080. [CrossRef]

32. Pérez-Ramos, I.M.; Ourcival, J.M.; Limousin, J.M.; Rambal, S. Mast seeding under increasing drought: Results from a long-term data set and from a rainfall exclusion experiment. *Ecology* **2010**, *91*, 3057–3068. [CrossRef] [PubMed]

33. Balducci, L.; DesLauriers, A.; Giovannelli, A.; Beaulieu, M.; Delzon, S.; Rossi, S.; Rathgeber, C.B.K. How do drought and warming influence survival and wood traits of Picea mariana saplings? *J. Exp. Bot.* **2014**, *66*, 377–389. [CrossRef] [PubMed]

34. Zalloni, E.; Battipaglia, G.; Cherubini, P.; Saurer, M.; De Micco, V. Wood Growth in Pure and Mixed *Quercus ilex* L. Forests: Drought Influence Depends on Site Conditions. *Front. Plant Sci.* **2019**, *10*, 397. [CrossRef] [PubMed]

35. López, B.; Sabaté, S.; Gracia, C. Fine-root longevity of *Quercus ilex*. *New Phytol.* **2001**, *151*, 437–441. [CrossRef]

36. Reich, P.B. Root–shoot relations: Optimality in acclimation and adaptation or the "Emperor's New Clothes"? In *Plant Roots: The Hidden Half*, 3rd ed.; Waisel, Y., Eshel, A., Kafkafi, U., Eds.; Marcel Dekker: New York, NY, USA, 2002; pp. 205–220.

37. Llusià, J.; Peñuelas, J.; Alessio, G.A.; Ogaya, R. Species-specific, seasonal, inter-annual, and historically-accumulated changes in foliar terpene emission rates in *Phillyrea latifolia* and *Quercus ilex* submitted to rain exclusion in the Prades Mountains (Catalonia). *Russ. J. Plant Physiol.* **2011**, *58*, 126–132. [CrossRef]

38. Preece, C.; Farré-Armengol, G.; Llusià, J.; Peñuelas, J. Thirsty tree roots exude more carbon. *Tree Physiol.* **2018**, *38*, 690–695. [CrossRef]

39. Peñuelas, J.; Sardans, J.; Filella, I.; Estiarte, M.; Llusià, J.; Ogaya, R.; Carnicer, J.; Bartrons, M.; Rivas-Ubach, A.; Grau, O.; et al. Impacts of Global Change on Mediterranean Forests and Their Services. *Forestry* **2017**, *8*, 463. [CrossRef]

Publisher's Note: MDPI stays neutral with regard to jurisdictional claims in published maps and institutional affiliations.

 *forests*

Article

Patterns for *Populus* spp. Stand Biomass in Gradients of Winter Temperature and Precipitation of Eurasia

Vladimir Andreevich Usoltsev [1,2,3], Baozhang Chen [1,4,*], Seyed Omid Reza Shobairi [1,2], Ivan Stepanovich Tsepordey [3], Viktor Petrovich Chasovskikh [2] and Shoaib Ahmad Anees [5]

[1] School of Remote Sensing and Geomatics Engineering, Nanjing University of Information Science and Technology, Nanjing 210000, China; usoltsev50@mail.ru (V.A.U.); general@usfeu.ru or Omidshobeyri214@gmail.com (S.O.R.S.)
[2] Ural State Forest Engineering University, Faculty of Forestry, Sibirskiy Trakt, 37, 620100 Yekaterinburg, Russia; u2007u@ya.ru
[3] Botanical Garden of Ural Branch of RAS, Department of Forest Productivity, ul. 8 Marta, 202a, 620144 Yekaterinburg, Russia; common@botgard.uran.ru
[4] LREIS, Institute of Geographic Sciences & Nature Resources Research Chinese Academy Sciences (CAS) 11A Datun Rd, Beijing 100101, China
[5] Beijing Key Laboratory of Precision Forestry, Forestry College, Beijing Forestry University, Beijing 100083, China; saanees@bjfu.edu.cn
[*] Correspondence: Baozhang.chen@nuist.edu.cn or Baozhang.Chen@igsnrr.ac.cn

Received: 7 July 2020; Accepted: 13 August 2020; Published: 19 August 2020

Abstract: Based on a generated database of 413 sample plots, with definitions of stand biomass of the genus *Populus* spp. in Eurasia, from France to Japan and southern China, statistically significant changes in the structure of forest stand biomass were found, with shifts in winter temperatures and average annual precipitation. When analyzing the reaction of the structure of the biomass of the genus *Populus* to temperature and precipitation in their transcontinental gradients, a clearly expressed positive relationship of all components of the biomass with the temperature in January is visible. Their relationship with precipitation is less clear; in warm climate zones, when precipitation increases, the biomass of all wood components decreases intensively, and in cold climate zones, this decrease is less pronounced. The foliage biomass does not increase when precipitation decreases, as is typical for wood components, but decreases. This can be explained by the specifics of the functioning of the assimilation apparatus, namely its transpiration activity when warming, and the corresponding increase in transpiration, which requires an increase in the influx of assimilates into the foliage, and the desiccation of the climate that reduces this influx of assimilates. Comparison of the obtained patterns with previously published results for other species from Eurasia showed partial or complete discrepancies, the causes of which require special physiological studies. The results obtained can be useful in the management of biosphere functions of forests, which is important in the implementation of climate stabilization measures, as well as in the validation of the results of simulation experiments to assess the carbon-deposition capacity of forests.

Keywords: genus *Populus* spp.; regression models; stand biomass; biomass structure; climate change; average January temperature; average annual precipitation

---

1. Introduction

Active human economic activity has led to significant global changes in the functioning of the biosphere, and the observed climate warming has had a significant impact on the vegetation cover of the planet [1,2]. If earlier, the problems of assessing climate impacts on vegetation had a regional character [3], then in recent decades it has become clear that the problem has reached a global, general planetary level, and largely determines the future fate of human civilization [4,5]. Mapping the distribution of net primary production (NPP) over the surface of the planet, by extrapolating empirical NPP data obtained from forest sample plots to large areas of biomes [6,7] or to latitude gradients [8,9] does not allow for making any predictions of changes in the climate-NPP system. The same can be said about the common planetary patterns of distribution of NPP harvest data by gradients of average temperatures and precipitation [10].

Due to current climate changes, priority is given to changing the biomass and NPP of forest ecosystems under the influence of average temperatures and precipitation. Similar studies are performed at both a regional [11–13] and transcontinental [14,15] levels. Their implementation, especially in the latter case, is one of the problems that constitute the subject of biogeography [16]. The forest, as we know, is a geographical phenomenon [17], and in view of the topic indicated in the title of this work, it is important for us to identify the geographical aspects of the biomass of forest ecosystems, i.e., to make a choice in favor of those geographical characteristics that determine the distribution of forest biomass on the territory of a particular continent.

However, the biomass of a stand represented by a particular tree species is primarily determined by its age and morphological (taxation) structure, i.e., a set of characteristics such as age, mean height, mean diameter at breast height, the basal area, and the volume stock, which are interrelated. The problem of multicollinearity arises in empirical modeling of biomass. One of the solutions to the problem is to harmonize the system by constructing recursive (recurrent, related) equations, in which the dependent variable of the previous equation is included as one of the independent variables of the subsequent one [18]. This approach provides a multivariate conditionality of factors that provide flexibility and universality of the regression system describing the dynamics of biomass of stands.

To account for the geographical effect in this recursive system, each equation of the system must be supplemented with corresponding regressors. One possible option is to introduce dummy variables [18] that encode the regional affiliation of the harvest data [11,18,19] as one of the methods for model harmonization [20]. The disadvantage of such equations is that they only take into account the geographical shifts of the desired variables by the value of the interception term. It is assumed that the regression coefficients in such cases are unchanged by region, which is not true. The second option is to include indices of natural zoning and continentality of climate in the equations of the system [21], using the basis that changes in vegetation cover occur both in the latitudinal direction due to changes in the PhAR [22], and in the meridional direction due to changes in the continentality of climate [23]. Therefore, models of the phytomass of trees and plantings have been developed, including their mass-forming indices as independent variables, as well as indices of natural zoning and climate continentality [21]. However, such models do not provide an answer to the question of in which direction the biomass structure of a particular tree species may change with the expected change in air temperature or annual precipitation. The use of evapotranspiration as a combined index in the assessment of tree production is futile, since it explains only 24% of its variability compared to 42%, which provides the relation to mean annual precipitation, and compared to 31%, which provides the relation to mean annual temperature [24]. It is assumed that orography, soil water balance, PhAR, and climate continentality are indirectly reflected in the territorial features of temperatures and precipitation.

Studies of forest stand biomass at the transcontinental level, performed for five species from Eurasia, showed that changes in their biomass due to temperatures and precipitation are species-specific, i.e., they differ between species in the total biomass [25]. If we adhere to the concept of species-specific responses of forest biomass to changes in the main climatic characteristics, then when we reach the transcontinental level, we are faced with the obvious fact that no species grows throughout the continent, precisely because of regional climate differences. Moving from refuges under the influence of geological processes and climate changes, a particular species adapted to changing environmental conditions, forming a series of vicariate species within the genus [26,27]. This gives grounds for analyzing the response of tree species to changes in climate characteristics, to combine them into one climate-dependent set within the entire genus, since differences in ecological and physiological properties of different species of the genus, for example, *Populus tremula* vs. *P. trichocarpa* vs. *P. pruinosa* are derived from regional climatic features.

Eurasia is the largest continental area on Earth, located primarily in the Northern and Eastern Hemispheres, it is bordered by the Atlantic Ocean to the west, the Pacific Ocean to the east, the Arctic Ocean to the north, and by Africa, the Mediterranean Sea, and the Indian Ocean to the south. Eurasia covers around 55,000,000 square kilometers (21,000,000 sq mi), or around 36.2% of the Earth's total land area. The landmass contains well over 5 billion people, equating to approximately 70% of the human population. The unique size and complexity of the natural conditions differentiate Eurasia from the rest of continents. No continent has such an original history of paleogeographic development. Structural differences are reflected in the features of the morphological structure. In the territory of Eurasia there are the highest mountain systems, vast highlands, plateaus, and plains. Climatic and landscape conditions are no less diverse. Here you can trace all the geographical zones that are characteristic of the land of the globe from the icy deserts in the North to the humid equatorial forests in the South [28].

In our work, we made the first attempt to study transcontinental trends in the structure of biomass of the genus *Populus* spp., formed under the influence of geographically distributed temperatures and precipitation in the territory of Eurasia. Across the Northern Hemisphere, this genus plays a disproportionately important role in promoting biodiversity and sequestering carbon. It is illustrative of efforts to move beyond single-species conservation worldwide. The genus *Populus* is valued for many reasons, but one highlights their potential as key contributors to regional and global biodiversity [29]. A tremendous need for paper, cardboard, and board materials open almost unlimited opportunities for the economic use of this genus' wood. By density and cellulose content, poplar wood does not come up short compared to the coniferous species. Despite the slightly shorter wood fiber of poplar in comparison to spruce, modern technologies make the first class production of paper, cardboard, and wood board materials out of this "disgraced" species possible. Today, however, the genus *Populus* is an example of a particularly evident disparity, between the potential organic matter production in plantations, and its actual implementation in the boreal natural forests [30].

2. Objects of Research

To analyze geographical patterns of biomass distribution in Eurasian forests formed by stands of the genus *Populus* spp., from the author's database of eight thousand sample plots [31], the materials of 413 determinations with the data of the biomass structure were used. These biomass data were presented in different components (stems, branches, foliage, and roots). The distribution of sample plots with biomass data of the genus *Populus* spp. on the map-scheme of Eurasia is shown in Figure 1, and according to tree species and countries, in Table 1.

Figure 1. Allocation of sample plots with biomass (t/ha) determinations of 413 *Populus* forest stands in the territory of Eurasia.

Table 1. Distribution of plots with determinations of Populus biomass (t/ha) by species and countries.

Species	Botanical Name	Country	Plot Quantity
Quaking aspen	*Populus tremula* L.	Russia, Ukraine, Kazakhstan, Estonia, Belarus	188
David's aspen	*P. davidiana* Dode	China, Japan	129
Californian poplar	*P. trichocarpa* Torr. & A.Gray ex Hook.	France, Austria, Belgium, Netherlands	37
Poplar larrity	*P. laurifolia* Ledeb.	Russia	12
White poplar	*P. alba* Ledeb.	Russia, Kazakhstan	10
Poplar «Robusta»	*Populus × euroamericana*	Ukraine	10
Asiatic poplar	*P. euphratica Olivier*	China	9
Hybrid	*Populus hybrid*	Japan	8
Poplar berry-bearing	*P. deltoids* W. Bartram ex Humphry Marshall	China	6
Black poplar	*P. nigra* L.	Russia	2
Bahala poplar	*Populus × bachelieri* Solemacher	Bulgaria	1
Ploomy poplar	*P. pruinosa* Schrenk	Tajikistan	1
	Total		413

3. Methods

As the plots for estimating biomass of forest stands were usually established in typical 'background' habitats, which were representative in relation to this type of plant communities, one can make on their basis a preliminary geographical analysis of biomass gradients of *Populus* forests. For an analytical description of the geographic distribution patterns of the biomass productivity of forest cover, one must choose the geographical characteristics of the territory of Eurasia that can be expressed by quantity and measure.

The actual values of the biomass of 413 stands of the genus *Populus* (see Figure 1), based on the known coordinates of the sample plots established, we superimposed on the maps of winter (January) temperatures and average annual precipitation distribution [32], and related them to the isolines of the mentioned indices on the maps. In our case, the schematic map of the isolines of mean January temperature, rather than that of the mean annual temperature, was used. With an inter-annual time step, the predominant influence of summer temperature is quite normal [33]. However, against the background of long-term climatic shifts for decades, the prevailing influence is acquired by winter temperatures [34,35]. For example, Toromani and Bojaxhi [36] write: "Earlier studies has shown that phytosynthesis is possible for *Abies alba* in winter, where high temperatures could play an important role in improving carbohydrate storage and growth at following year. For species grown under a Mediterranen climate high temperatures and low precipitation during growing season may cause water stress, which is the main limiting factor for tree growth".

We should keep in mind that winter temperatures in the Northern Hemisphere have increased faster than summer ones during the 20th century [37–39]. In terms of regression analysis, a weak temporal trend of summer temperatures compared to a steep trend of winter ones, means a smaller regression slope and a worse ratio of residual variance to the total variance explained by this regression. Obviously, taking the mean winter temperature as one of the independent variables, we get a more reliable dependence having the higher predictive ability.

Then, the compiled matrix of harvest data (Table 2) were subjected to the common regression analysis.

Table 2. A fragment of the original matrix of experimental data *.

A	N	V	P_i						Tm	PRm
			P_s	P_b	P_f	P_a	P_r	P_t		
40	0.790	208	89	5.5	2.40	98.0	21.6	119.6	−7	570
21	0.278	218	99.8	20.7	4.19	129.7	29.2	158.9	−3	570
12	12.54	62.5	34.8	4.45	1.91	41.2	15.0	56.2	−13	290
22	4.550	30	16.1	4.24	0.80	21.1	6.0	27.1	−13	290
49	0.650	284	113	22.5	3.07	138.6	57.0	195.6	−20	317
41	0.526	192	76.0	22.2	2.40	100.6	55.0	155.6	−18	250
78	0.518	200	88.83	28.37	4.99	127.7	38.79	166.5	−15	570
45	0.500	105	49.62	10.92	4.52	67.56	8.58	76.14	−26	570
78	0.666	185	103.1	45.44	8.75	163.9	37.8	201.7	−15	570
27	2.935	142	84.29	17.56	7.01	114.0	42.73	156.7	−9	820
68	1.244	223	102.4	22.44	7.43	138.5	51.47	190.0	−15	570
25	4.066	122	73.51	11.31	5.89	95.04	36.6	131.6	−15	570
40	1.062	224	99.81	32.82	7.69	146.6	45.47	192.1	−15	570
34	1.595	182	95.77	13.15	7.91	122.6	52.18	174.8	−10	444
50	1.510	163	75.11	22.44	6.21	108.4	34.26	142.7	−25	444
28	7.32	129	73.0	11.40	2.00	89.93	17.2	107.1	−15	570
37	2.913	153	86.65	18.56	6.27	116.7	41.79	158.5	−15	570
69	0.811	284	110.9	17.20	7.62	142.2	56.36	198.6	−26	444
58	1.188	124	61.47	16.79	3.66	85.63	27.24	112.9	−26	444
79	0.403	163	68.75	18.91	3.66	95.43	29.6	125.0	−26	444
38	4.255	121	73.56	11.79	6.19	95.95	38.21	134.2	−15	570
68	1.822	234	117.3	30.29	7.23	162.0	54.8	216.8	−26	444
29	2.000	61	34.42	16.31	2.77	55.88	16.57	72.45	−5	826
39	2.774	62	37.92	8.72	3.00	58.27	9.41	67.68	−5	826

* A = stand age, y; V = stem volume, m³/ha; N = tree density, 1000/ha; i = index of biomass component: total wood storey (*t*), aboveground wood storey (*a*), underground wood storey, or roots (*r*), stem over the bark (*s*), foliage (*f*), and branches (*b*); PRm = mean annual precipitation, mm; Tm = mean January temperature, °C.

The basic principles of modelling and the results obtained by means of regression analysis should have an ecologic-geographical interpretation. The biological productivity of forests is dependent on climatic factors, but only as a first approximation, since there are ontogenetic, cenotic, edaphic, and other levels of its variability. Therefore, we included in the regression equations the independent variables explaining the variability of the dependent variable, expressing not only with climatic parameters but also with forest age, tree density, and stem volume.

As the mean January temperature in the northern part of Eurasia has negative values, the corresponding independent variable was modified to the form (Tm + 50). Then, the technique of multiple regression analysis (http://www.statgraphics.com/for more information), according to three blocks of recursive equations, was used: two blocks of mass-forming indices, N and V, and a single block of biomass P_i (arrows show the sequence of calculations)

$$\ln N = a_0 + a_1(\ln A) + a_2[\ln (Tm + 50)] + a_3(\ln PRm); \tag{1}$$

$$\ln V = a_0 + a_1(\ln A) + a_2(\ln N) + a_3 [\ln (Tm + 50)] + a_4(\ln PRm); \tag{2}$$

$$\ln P_i = a_0 + a_1(\ln A) + a_2(\ln V) + a_3(\ln N) + a_4[\ln(Tm + 50)] + a_5(\ln PRm) \tag{3}$$

4. Results and Discussion

The results of the calculation of Equations (1)–(3) are listed in the Table 3. Only the variables that are significant at the level of probability of P_{95} and above are showed in this table. The equations were tabulated in the sequence illustrated by the arrows. The results of tabulating the models in the sequence of Equations (1)–(3) present the rather cumbersome table. We took from it the values of the component composition of the biomass of the *Populus* forests of the age of 50 years and built 3D-graphs of their dependence upon temperature and precipitation (Figure 2).

Table 3. Characteristics of biomass Equations (1) to (3).

Dependent Variables	Coefficients and Independent Variables						adjR^2 **	SE ***
	a_0 *	a_1(ln$A$)	a_2(ln$V$)	a_3(ln$N$)	a_5[ln(Tm + 50)]	a_6(lnPRm)		
ln(N)	10.7307	−1.2994	-	-	−2.4045	0.4747	0.623	0.74
ln(V)	5.9573	0.3617	-	−0.2589	0.5288	−0.6169	0.534	0.47
ln(P_s)	−1.8923	0.2068	0.9123	0.0646	0.0764	0.0811	0.963	0.17
ln(P_b)	−2.8796	0.2421	0.5520	−0.0537	0.3792	0.0678	0.675	0.44
ln(P_f)	−4.0545	0.0070	0.4127	0.1332	−0.0589	0.5272	0.522	0.38
ln(P_r)	−3.5174	0.0497	0.7113	0.0563	0.2306	0.3533	0.720	0.35
ln(P_a)	−1.2511	0.1782	0.8183	0.0463	0.0984	0.0960	0.955	0.17
ln(P_t)	−1.1022	0.0477	0.7969	0.0486	0.1467	0.1849	0.918	0.18

* The constant corrected for logarithmic retransformation by [40]; ** $adjR^2$ = determination coefficient adjusted for the number of variables; *** SE = standard error of the equations.

When analyzing the reaction of the biomass structure of the genus *Populus* to temperature and precipitation in their transcontinental gradients, a clearly expressed unambiguous positive relationship of all components of the biomass with the average temperature of January is seen. Their relationship with precipitation is less clear; in warm climate zones (Tm = 0 °C), when precipitation increases the biomass of all wood components decreases most intensively, and in cold climate zones (Tm = −40 °C) this decrease is expressed to a much lesser extent.

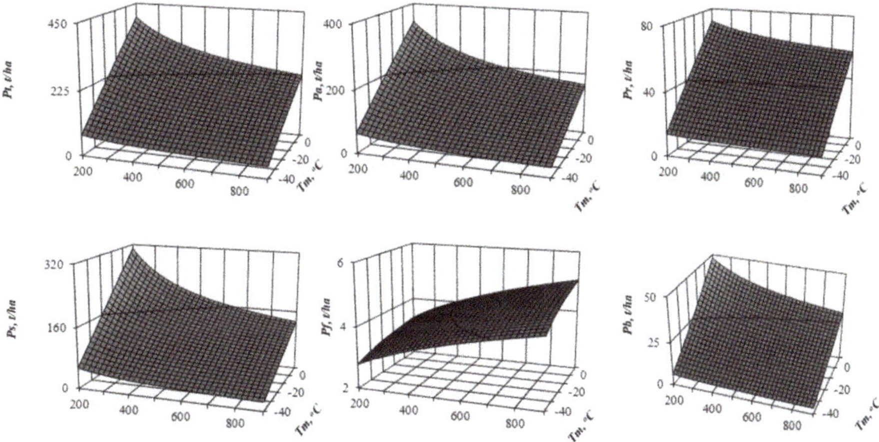

Figure 2. Dependence of *Populus* ecosystems of Eurasia upon the mean January temperature (*Tm*) and mean annual precipitation (*PRm*). Abbreviations: *Pt, Pa, Ps, Pr, Pf, Pb* are, respectively, biomass of: total wood storey, aboveground, stems (wood and bark), roots, foliage, and branches, t/ha.

It is interesting to compare the obtained patterns of changes in the total biomass of the genus *Populus* with previously published results for other forest-forming species of Eurasia, obtained using a similar methodology [25]. The increase in the total biomass of *Populus* during the transition from cold to warm regions was confirmed earlier for *Larix* spp., *Picea* spp., *Abies* spp., and *Betula* spp. However, the decrease in the total biomass as precipitation increases was confirmed only in larch, whereas in spruce, fir, and birch, the dependence is the opposite of that established for *Populus*. The specific pattern of the change in the total biomass was obtained for two-needled pines *Pinus* L.; its increase during the transition from cold to warm regions was recorded only in regions with heavy precipitation, and with the transition to water-deficit regions, the pattern changes to the opposite. If, in *Populus* and *Larix*, the decrease in total biomass with the transition from water-deficient to moisture-rich regions was observed in all thermal zones, then in two-needled pines (only in cold zones, and with the transition from cold to warm climatic zones) this negative trend changes to the opposite.

We can see that the reaction of foliage biomass with increasing precipitation does not decrease, as is typical for wood components, but increases (Figure 2). This is consistent with a similar situation observed in Russian Siberia with respect to forest cover [41], where with a warming climate and a simultaneous decrease in precipitation, the share of assimilation mass decreases, and the share of wood components increases. This is explained by the specifics of foliage functioning, namely, its transpiration activity when warming, and a corresponding increase in transpiration requires an increase in the influx of assimilates into the foliage, while the desiccation of the climate reduces this influx of assimilates due to a decrease in transpiration activity. Perhaps this phenomenon demonstrates the future scenario of acclimatization of trees to the ongoing warming and changes in the water balance of territories. However, in Canada's forests, a contradictory result was obtained. If the January temperature and humidity conditions of the growing season had a positive effect on the growth of *Betula papyrifera* Marsh. and the growth of *Picea mariana* Mill., then *Populus tremuloides* Michx. might be the least responsive species [42].

The patterns of biomass-amount change, under assumed changed climatic conditions (Figure 2), are hypothetical. They reflect long-term adaptive responses of forest stands to regional climatic conditions and do not take into account rapid trends of current environmental changes, which place serious constraints on the ability of forests to adapt to new climatic conditions [43–49]. Climate changes are manifested primarily in shifts in the phenology of a particular species, and are determined by the degree of species-specific phenotypic plasticity [50], which were not taken into account in our work, and require special study.

The law of limiting factors [51] works well in stationary conditions. With a rapid change in limiting factors (such as air temperature or precipitation), forest ecosystems are in a transitional (non-stationary) state, in which some factors that were not significant may come to the fore, and the end result may be determined by other limiting factors [52].

The main pool of biomass harvest data in Eurasia was obtained during the 1970s–1990s, and the climate maps used, cover the period of the late 1990s–early 2000s. Some discrepancy between the two time periods may cause some biases in the results obtained, but for such a small time difference in the used data, the inclusion of compensatory mechanisms or phenological shifts in forest communities is unlikely [47,48].

5. Conclusions

Based on a database of 413 sample plots, with definitions of forest biomass of the genus *Populus* spp. in Eurasia, in the territory from France to the South of China and Japan, a statistically significant increase in stem, aboveground, and underground biomass was found with an increase in winter temperatures and a decrease in precipitation, especially in warm climate regions. In contrast to the woody components of biomass, the mass of foliage, while increasing with warming, simultaneously decreases with a decrease in precipitation, which is due to the specifics of the functioning of foliage, namely, its transpiration activity.

Comparison of the results obtained for *Populus* showed that the regularities of *Populus* are repeated only in *Larix*, and in other species, only partially. In our work, we can only state the species-specificity of the reaction of various Eurasian species to changes in temperature and precipitation, but explaining this specificity at the level of physiological processes is the task of the future.

The results obtained can be useful in the management of biosphere functions of forests, which is important in the implementation of climate stabilization measures, as well as in the validation of the results of simulation experiments to assess the carbon-deposition capacity of forests. They also provide a preliminary idea of possible shifts in forest biological productivity indicators under the influence of climate change.

Author Contributions: V.A.U.—analyzed the data, mathematical processing, prepared the manuscript; B.C. contributed to the conceptualization of ideas, the methodology and the review of the manuscript; S.O.R.S.—designed the study and the review of the manuscript; I.S.T.—mathematical processing, design; V.P.C.—mathematical adjustment; S.A.A.—spelling edit. All authors have read and agreed to the published version of the manuscript.

Funding: This research was funded by the National Key R&D Program of China (2018YFA0606001,2017YFA0604302, 2017YFC0503904,and 2017YFA0604301), an international partnership program of Chinese Academy of Sciences (Grant 131A11KYSB20170025), a research project funded by the State Key Laboratory of Resources and Environmental Information System (O88RA901YA) and a project funded by the National Natural Science Foundation of China (41771114 & 41977404).

Conflicts of Interest: The authors declare no conflict of interest.

References

1. Halofsky, J.S.; Conklin, D.R.; Donato, D.C.; Halofsky, J.E.; Kim, J.B. Climate change, wildfire, and vegetation shifts in a high-inertia forest landscape: Western Washington, U.S.A. *PLoS ONE* **2018**, *13*, e0209490. [CrossRef] [PubMed]

2. Kosanic, A.; Anderson, K.; Harrison, S.; Turkington, T.; Bennie, J. Changes in the geographical distribution of plant species and climatic variables on the West Cornwall peninsula (South West UK). *PLoS ONE* **2018**, *13*, e0191021. [CrossRef] [PubMed]

3. Glebov, F.; Litvinenko, V. The dynamics of tree ring width in relation to meteorological indices in different types of wetland forests. *Lesovedenie* **1976**, *4*, 56–62.

4. Tarko, A. *Antropogennye Izmeneniya Global'nykh Biosfernykh Protsessov [Anthropogenic Changes of Global Biosphere Processes]*; Fizmatlit: Moscow, Russia, 2005.

5. Behrensmeyer, A.K. ATMOSPHERE: Climate Change and Human Evolution. *Science* **2006**, *311*, 476–478. [CrossRef]

6. Bazilevich, N.; Rodin, L. Schematic Maps of Productivity and Biological Turnover of Elements in the main Types of Land vegetation. *Izvestiya Vsesoyuznogo Geograficheskogo Obshchestva* **1967**, *99*, 190–194.

7. Rodin, L.E.; Bazilevich, N. *Production and Mineral Cycling in Terrestrial Vegetation*; Oliver & Boyd: Edinburgh/London, UK, 1967.

8. Anderson, K.J.; Allen, A.P.; Gillooly, J.F.; Brown, J.H. Temperature-dependence of biomass accumulation rates during secondary succession. *Ecol. Lett.* **2006**, *9*, 673–682. [CrossRef]

9. Huston, M.A.; Wolverton, S. The global distribution of net primary production: resolving the paradox. *Ecol. Monogr.* **2009**, *79*, 343–377. [CrossRef]

10. Lieth, H. Modeling the Primary Productivity of the World. *Ecol. Stud.* **1975**, *4*, 237–263.

11. Fu, L.; Sun, W.; Wang, G. A climate-sensitive aboveground biomass model for three larch species in northeastern and northern China. *Trees* **2016**, *31*, 557–573. [CrossRef]

12. Forrester, D.I.; Tachauer, I.; Annighoefer, P.; Barbeito, I.; Pretzsch, H.; Ruiz-Peinado, R.; Stark, H.; Vacchiano, G.; Zlatanov, T.; Chakraborty, T.; et al. Generalized biomass and leaf area allometric equations for European tree species incorporating stand structure, tree age and climate. *For. Ecol. Manag.* **2017**, *396*, 160–175. [CrossRef]

13. Zeng, W.; Duo, H.; Lei, X.; Chen, X.; Wang, X.; Pu, Y.; Zou, W. Individual tree biomass equations and growth models sensitive to climate variables for Larix spp. in China. *Eur. J. For. Res.* **2017**, *136*, 233–249. [CrossRef]

14. Usoltsev, V.A.; Merganičová, K.; Konôpka, B.; Osmirko, A.A.; Tsepordey, I.S.; Chasovskikh, V.P. Fir (*Abies* spp.) stand biomass additive model for Eurasia sensitive to winter temperature and annual precipitation. *Central Eur. For. J.* **2019**, *65*, 166–179. [CrossRef]

15. Hubau, W.; Lewis, S.L.; Phillips, O.L.; Affum-Baffoe, K.; Beeckman, H.; Cuní-Sanchez, A.; Daniels, A.K.; Ewango, C.E.N.; Fauset, S.; Mukinzi, J.M.; et al. Asynchronous carbon sink saturation in African and Amazonian tropical forests. *Nature* **2020**, *579*, 80–87. [CrossRef] [PubMed]

16. Lomolino, M.V.; Riddle, B.R.; Brown, J.H. *Biogeography*, 3rd ed.; Sinauer Associates Inc.: Sunderland, MA, USA, 2006.

17. Morozov, G.F. *The Theory of Forest Stand Types*; Selkolkhozgiz: Moscow-Leningrad, Russia, 1931.

18. Draper, N.; Smith, H. *Applied Regression Analysis*; Wiley: New York, NY, USA, 1966.

19. Zeng, W. Developing Tree Biomass Models for Eight Major Tree Species in China. In *Biomass Volume Estimation and Valorization for Energy*; IntechOpen: London, UK, 2017; pp. 3–21. [CrossRef]

20. Jacobs, M.W.; Cunia, T. Use of dummy variables to harmonize tree biomass tables. *Can. J. For. Res.* **1980**, *10*, 483–490. [CrossRef]

21. Usoltsev, V.A.; Shobairi, S.O.; Tsepordey, I.S.; Chasovskikh, V.P. Additive Allometric model of *Populus* sp. Single-Tree Biomass as a Basis of Regional Taxation Standards for Eurasia. *Indian For.* **2019**, *145*, 625–630.

22. Grigoriev, A.; Budyko, M. The periodicity law of geographic zonality. *Doklady Akademii Nauk SSSR* **1956**, *110*, 129–132.

23. Komarov, V. Meridional zonality of organisms. In Proceedings of the 1st All-Russian Congress of Russian Botanists in Petrograd, Leningrad, Russia, 16 September 1921.

24. Ni, J.; Zhang, X.-S.; Scurlock, J.M. Synthesis and analysis of biomass and net primary productivity in Chinese forests. *Ann. For. Sci.* **2001**, *58*, 351–384. [CrossRef]

25. Usoltsev, V.A.; Tsepordey, I.S.; Osmirko, A.A. Biological productivity of Eurasian forests due to temperature and precipitation. In Proceedings of the Forest Ecosystems of Boreal Zone: Biodiversity, Bioeconomy, Ecological Risks, All-Russian Conference with International Participation. Krasnoyarsk, Russia, 26–31 August 2019; pp. 458–460. (In Russian with English title, summary and contents)

26. Hultén, E.; Lehre, J. *Outline of the History of Arctic and Boreal Biota during the Quaternary Period*; Cramer: New York, NY, USA, 1937.

27. Tolmachev, A.I. *Fundamentals of Plant Habitat Theory: Introduction to Plant Community Chorology*; State University Publishing: Leningrad, Russia, 1962.

28. Lavrinovich, M.V. *Physical Geography of Eurasia (Regional Overview)*; Belarusian State University: Minsk, Belarusia, 2003. Available online: https://b-ok.cc/book/3128971/f9cd53 (accessed on July 2003). (In Russian)

29. Rogers, P.C.; Pinno, B.D.; Šebesta, J.; Albrectsen, B.R.; Li, Q.; Ivanova, N.; Kusbach, A.; Kuuluvainen, T.; Landhäusser, S.M.; Liu, H.; et al. A global view of aspen: Conservation science for widespread keystone systems. *Glob. Ecol. Conserv.* **2020**, *21*, 00828. [CrossRef]

30. Usoltsev, V. *Forest Arabesques, or Sketches of Our Trees' Life*, 3rd ed.; Radomska Szkoła Wyższa w Radomiu: Radom, Poland, 2019.

31. Usoltsev, V.A. *Forest Biomass and Primary Production Database for Eurasia: Digital Version*, 3rd ed.; Ural State Forest Engineering University: Yekaterinburg, Russia, 2020. [CrossRef]

32. World Weather Maps. 2007. Available online: https://www.mapsofworld.com/referrals/weather/ (accessed on 15 June 2007).

33. Zubairov, B.; Heußner, K.-U.; Schröder, H. Searching for the best correlation between climate and tree rings in the Trans-Ili Alatau, Kazakhstan. *Dendrobiology* **2018**, *79*, 119–130. [CrossRef]

34. Morley, J.W.; Batt, R.D.; Pinsky, M.L. Marine assemblages respond rapidly to winter climate variability. *Glob. Chang. Biol.* **2016**, *23*, 2590–2601. [CrossRef] [PubMed]

35. Bijak, S. Tree-ring chronology of silver fir and ist dependence on climate of the Kaszubskie Lakeland (Northern Poland). *Geochronometria* **2010**, *35*, 91–94. [CrossRef]

36. Toromani, E.; Bojaxhi, F. Growth response of silver fir and Bosnian pine from Kosovo. *South-East Eur. For.* **2010**, *1*, 20–28. [CrossRef]

37. Emanuel, W.R.; Shugart, H.H.; Stevenson, M.P. Climatic change and the broad-scale distribution of terrestrial ecosystem complexes. *Clim. Chang.* **1985**, *7*, 29–43. [CrossRef]

38. Laing, J.; Binyamin, J. Climate change effect on winter temperature and precipitation of Yellowknife, Northwest Territories, Canada from 1943 to 2011. *Am. J. Clim. Chang.* **2013**, *2*, 275–283. [CrossRef]

39. Felton, A.; Nilsson, U.; Sonesson, J.; Felton, A.M.; Roberge, J.-M.; Ranius, T.; Ahlström, M.; Bergh, J.; Björkman, C.; Boberg, J.; et al. Replacing monocultures with mixed-species stands: Ecosystem service implications of two production forest alternatives in Sweden. *Ambio* **2016**, *45*, 124–139. [CrossRef]

40. Baskerville, G.L. Use of logarithmic regression in the estimation of plant biomass. *Can. J. For. Res.* **1972**, *2*, 9–53. [CrossRef]

41. Lapenis, A.; Shvidenko, A.; Shepaschenko, D.; Nilsson, S.; Aiyyer, A.R. Acclimation of Russian forests to recent changes in climate. *Glob. Chang. Biol.* **2005**, *11*, 2090–2102. [CrossRef]

42. Huang, J.; Tardif, J.C.; Bergeron, Y.; Denneler, B.; Berninger, F.; Girardin, M.P. Radial growth response of four dominant boreal tree species to climate along a latitudinal gradient in the eastern Canadian boreal forest. *Glob. Chang. Biol.* **2010**, *16*, 711–731. [CrossRef]

43. Givnish, T.J. Adaptive significance of evergreen vs. deciduous leaves: solving the triple paradox. *Silva Fenn.* **2002**, *36*, 703–743. [CrossRef]

44. Schaphoff, S.; Reyer, C.P.; Schepaschenko, D.; Gerten, D.; Shvidenko, A. Tamm Review: Observed and projected climate change impacts on Russia's forests and its carbon balance. *For. Ecol. Manag.* **2016**, *361*, 432–444. [CrossRef]

45. Spathelf, P.; Stanturf, J.; Kleine, M.; Jandl, R.; Chiatante, D.; Bölte, A. Adaptive measures: integrating adaptive forest management and forest landscape restoration. *Ann. For. Sci.* **2018**, *75*, 55. [CrossRef]

46. Vasseur, F.; Exposito-Alonso, M.; Ayala-Garay, O.J.; Wang, G.; Enquist, B.J.; Vile, D.; Violle, C.; Weigel, D. Adaptive diversification of growth allometry in the plant *Arabidopsis thaliana*. *Proc. Natl. Acad. Sci. USA* **2018**, *115*, 3416–3421. [CrossRef] [PubMed]

47. Anderegg, W.R.; Anderegg, L.D.L.; Kerr, K.L.; Trugman, A.T. Widespread drought-induced tree mortality at dry range edges indicates that climate stress exceeds species' compensating mechanisms. *Glob. Chang. Biol.* **2019**, *25*, 3793–3802. [CrossRef] [PubMed]

48. DeLeo, V.L.; Menge, D.N.L.; Hanks, E.M.; Juenger, T.E.; Lasky, J.R. Effects of two centuries of global environmental variation on phenology and physiology of *Arabidopsis thaliana*. *Glob. Chang. Biol.* **2019**, *26*, 523–538. [CrossRef]

49. Denney, D.A.; Anderson, J.T. Natural history collections document biological responses to climate change: A commentary on DeLeo et al.(2019), Effects of two centuries of global environmental variation on phenology and physiology of *Arabidopsis thaliana*. *Glob. Chang. Biol.* **2020**, *26*, 340–342. [CrossRef]

50. Bigot, S.; Buges, J.; Gilly, L.; Jacques, C.; Le Boulch, P.; Berger, M.; Delcros, P.; Domergue, J.-B.; Koehl, A.; Ley-Ngardigal, B.; et al. Pivotal roles of environmental sensing and signaling mechanisms in plant responses to climate change. *Glob. Chang. Biol.* **2018**, *24*, 5573–5589. [CrossRef]

51. Shelford, V.E. *Animal Communities in Temperate America: As Illustrated in the Chicago region: A Study in Animal Ecology*; University of Chicago Press: Chicago, IL, USA, 1913.
52. Odum, E.P. *Fundamentals of Ecology*; Saunders: Philadelphia, PA, USA, 1971.

Article

Structural Carbon Allocation and Wood Growth Reflect Climate Variation in Stands of Hybrid White Spruce in Central Interior British Columbia, Canada

Anastasia Ivanusic, Lisa J. Wood * and Kathy Lewis

Ecosystem Science and Management, University of Northern British Columbia, Prince George, BC V2N 4Z9, Canada; Anastasia.Ivanusic@alumni.unbc.ca (A.I.); Kathy.Lewis@unbc.ca (K.L.)
* Correspondence: lisa.wood@unbc.ca; Tel.: +1-250-960-5352

Received: 11 July 2020; Accepted: 10 August 2020; Published: 12 August 2020

Abstract: *Research Highlights:* This research presents a novel approach for comparing structural carbon allocation to tree growth and to climate in a dendrochronological analysis. Increasing temperatures reduced the carbon proportion of wood in some cases. *Background and Objectives*: Our goal was to estimate the structural carbon content of wood within hybrid white spruce (*Picea glauca* (Moench) × *engelmannii* (Parry) grown in British Columbia, Canada, and compare the percent carbon content to wood properties and climate conditions of the region. Specific objectives included: (i) the determination of average incremental percent carbon, ring widths (RW), earlywood (EW) and latewood (LW) widths, cell wall thickness, and density over time; (ii) the determination of differences between percent carbon in individual forest stands and between regions; and (iii) the evaluation of the relationships between percent carbon and climate variation over time. *Methods:* Trees were sampled from twelve sites in northern British Columbia. Wood cores were analyzed with standard dendrochronology techniques and SilviScan analysis. Percent structural carbon was determined using acetone extraction and elemental analysis for 5 year increments. Individual chronologies of wood properties and percent carbon, and chronologies grouped by region were compared by difference of means. Temperature and precipitation values from the regions were compared to the carbon chronologies using correlation, regression, and visual interpretation. *Results:* Significant differences were found between the percent structural carbon of wood in individual natural and planted stands; none in regional aggregates. Some significant relationships were found between percent carbon, RW, EW, LW, and the cell wall thickness and density values. Percent carbon accumulation in planted stands and natural stands was found in some cases to correlate with increasing temperatures. Natural stand percent carbon values truncated to the last 30 years of growth was shown as more sensitive to climate variation compared to the entire time series. *Conclusions*: Differences between the stands in terms of structural carbon proportion vary by site-specific climate characteristics in areas of central interior British Columbia. Wood properties can be good indicators of variation in sequestered carbon in some stands. Carbon accumulation was reduced with increasing temperatures; however, warmer late-season conditions appear to enhance growth and carbon accumulation.

Keywords: carbon allocation; carbon; forest growth; hybrid white spruce; climate; natural and planted stands; wood density; wood cell wall thickness; tree rings

---

1. Introduction

Tree growth in both naturally occurring, and managed forests is a key process that influences carbon balance in terrestrial ecosystems, that is subject to the impacts of environmental change. The estimations of carbon content in tree stems are usually based on modelled data, calculated from measured variables such as tree height and diameter at breast height (DBH), but can be enhanced with knowledge of wood density, carbon concentration, and wood volume [1]. It has been suggested that wood density and cell wall thickness correlate with carbon sequestration; cellulose and lignin are components of xylem cell walls, and thicker, denser cell walls should have greater proportions of sequestered carbon [1–4]. However, the relationships between wood properties and sequestered carbon are not well understood because past research has focused mainly on biomass (or allometric biomass equations as determined from DBH and height measurements) instead of the direct measurements of volatile and structural carbon [5,6]. Expanding knowledge of the variation in naturally grown (hereafter referred to as 'natural stands') and managed plantation stands ('planted stands') and between wood properties, such as density and cell wall thickness, and carbon could improve the projections of carbon sequestration [1].

Forest growth, and the subsequent carbon accumulation, are strongly affected by changes in the climate. Changes in the climate are predicted to cause deviations in tree photosynthetic and respiration rates, increase disturbance, and increase tree mortality related to chronic drought [7–11]. Changes in the climate in British Columbia (BC), Canada, are predicted to include warmer and wetter conditions, with increased maximum and minimum temperatures and a decreased depth and water content of snowpack that will vary across the topographic landscape [12,13]. Over the next century, substantial changes in the temperature and precipitation in central interior BC, particularly in the spruce–willow–birch (SWB) and sub-boreal spruce (SBS) biogeoclimatic zones, are expected [12,14]. Increasing temperatures may push forests beyond growth sustainability thresholds, reducing the amount of carbon dioxide uptake and carbon accumulation [15].

Tree-ring analysis was used to determine forest growth dynamics and has provided climate variability information through radial growth and climate reconstructions [10,16]. Dendrochronological techniques may also be used to enhance the understandings of relationships between above ground carbon accumulation and the climate [17], as most carbon research relates to productivity based on climate [18], biomass equations [19], and changes to forests after anthropogenic management [20,21].

This study aimed to determine: (1) the variations in the sequestered structural carbon of hybrid white spruce (*Picea glauca* x *engelmannii*) in natural and planted stands over time; (2) how variations in carbon relate to ring width (RW), earlywood width (EW), latewood width (LW), density and cell wall thickness measurements at the annual scale; and (3) how variations in both carbon and wood properties compare with changes in the temperature and precipitation in central interior BC.

2. Materials and Methods

Hybrid white spruce trees were selected from six natural (N1–N6) and six planted stands (P1–P6) from areas of central interior BC (Figure 1 and Table 1). One group of six stands (N1–N3; P1–P3) was selected from the John Prince Research Forest (FSJ), located north of the town of Fort St. James, where each stand sampled was within 5 km of another. The second group of six stands (N4–N6; P4–P6) were within 200 km from Prince George (PG) (Figure 1). The target sample sites were dominant stands of hybrid white spruce naturally grown or planted approximately 40 years prior to sampling. We described and classified the sites sampled according to BC's Biogeoclimatic Ecosystem Classification system [22]. The biogeoclimatic variant of each site, which describes the temperature and moisture variation of the area in comparison to similar sites, was determined from each site characteristics, including dominant vegetation, aspect and topography. PG stands were in the willow–wet–cool (wk1) and very-wet–cool (vk1) variants of the Sub Boreal (SBS) biogeoclimatic zone, characterized by high precipitation and cooler temperatures. FSJ stands were in the Stuart–dry–warm (dw3) variant of the SBS zone, characterized by lower snow packs and warmer temperatures [23].

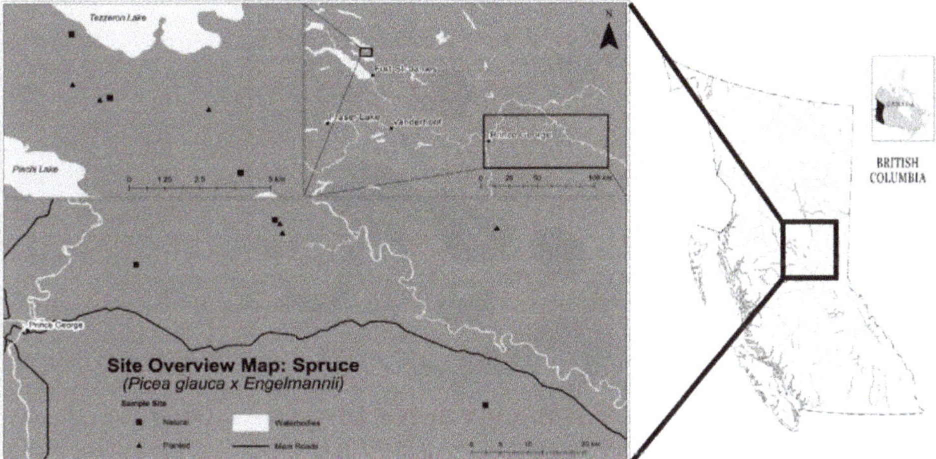

Figure 1. Overview map of the natural (squares) and planted (triangles) hybrid white spruce stands sampled for the study, near Prince George and Fort St. James, British Columbia, Canada.

Table 1. Site and stand characteristics of the natural (N1–N6) and planted (P1–P6) hybrid white spruce research samples surrounding the John Prince Research Forest (FSJ) and Prince George (PG), collected in 2016–2017. Note that not all the cores collected from the FSJ sites were suitable for analysis and one 12 mm core was collected in addition to the 5 mm cores. DBH = diameter at breast height.

	Site	Latitude	Longitude	Elevation (m)	Slope (%)	Mean Age (Years)	Number of 5 mm Tree Cores	Mean DBH (cm)
				Geographic Information			**Tree Characteristics**	
FSJ	N1	54 38′50.80	124 23′35.1	768	<2	101	40	27
	N2	54 39′46.60	124 24′36.6	833	<5	119	35	39
	N3	54 36′58.21	124 19′05.5	727	0	52	34	31
	P1	54 38′47.50	124 23′68.2	801	0	28	40	18
	P2	54 38′48.00	124 24′34.5	866	20	31	40	22
	P3	54 38′14.17	124 20′05.5	802	0	25	36	19
PG	N4	54 04′58.90	122 01′32.3	730	0	93	40	41
	N5	54 46′33.90	121 29′14.6	1113	0	145	40	48
	N6	54 01′01.00	122 24′54.5	707	0	154	40	32
	P4	54 04′05.90	121 26′48.8	843	<5	30	40	26
	P5	54 05′19.90	122 01′31.8	713	<10	30	40	28
	P6	54 04′01.10	122 01′09.7	689	0	33	40	23

Twenty dominant trees in each stand were selected for sampling. Sampled trees were at least 5 m apart to minimize spatial autocorrelation [24]. Trees with scars, fire or insect damage, split tops and abnormal growth patterns were avoided to minimize growth abnormalities in the tree series collected. Areas near roads or with open edges were also avoided to minimize non-climatic influences on tree growth [16]. Four 5 mm cores were collected from each tree at breast height and spaced 90 degrees around the stem (cores were collected at 30 cm aboveground for younger plantation trees). An additional 12 mm core from each tree was collected at the FSJ sites, from directly below one of the 5 mm cores. Surrounding vegetation, slope, elevation, flowing or standing water, diameter-at-breast-height and GPS site and tree location were recorded.

Of the 120 12 mm cores sampled in FSJ, 89 undamaged cores were selected for SilviScan analysis. Resins were removed from the cores selected via 12 h Soxhlet acetone extraction [25,26]. After extraction, the cores were conditioned at 40% relative humidity and 20 °C to obtain an 8% moisture content equilibrium. Once at 8% moisture content, the cores were cut into 2 mm × 7 mm (tangential × radial) pith-to-bark laths, using a twin-blade saw. Laths were analyzed using the SilviScan system at FPInnovations in Vancouver, BC, Canada. SilviScan analysis included (i) the image analysis of radial and tangential cell dimensions using optical microscopy, (ii) X-ray densitometry to provide measurements of wood density every 25 microns along the wood samples, and, (iii) X-ray diffractometry yielding measurements of microfibril angle at 5 mm increments [27].

The four 5 mm increment cores from FSJ and PG were dried, labelled and cross-dated using the Yamaguchi list method [28]. Two of the four 5 mm cores were sanded with progressively finer grit sand paper, and were scanned using an Epson 1640XL flatbed scanner at 1200 DPI (dots per inch) for the visual assessment of RW, EW and LW widths with WinDendro image analysis. Each core was reviewed to determine the accuracy of WinDendro RW, EW and LW auto-measurements, and corrections were performed manually. The other half of the 5 mm cores from each site were cut into 5 year increments (using only the last 30 years for the planted stands and 80 years for the natural stands) and processed for structural carbon content analysis. Although one-year increments were initially sought, annual increments did not provide enough wood mass for the percent carbon measurements. The 5 year sections of 20 cores were grouped together for each site and were analyzed as an aggregate sample. For example, all 20 cores' sections with rings dating from 2010–2015 were grouped together to make one sample for carbon analysis. The wet weights of the aggregated samples were measured, following which the resins were extracted from samples using a Soxhlet acetone extraction, at 110 °C for 1.5 h. Once dry, the sample dry weights were recorded, and the samples were ground into a powder with a Wiley mill grinder. Four 4–5 mg replicates from each aggregate sample were created; each was mixed with 10 mg of catalyst, valdium peroxide, and placed into a small container. Each replicate was analyzed with the PerkinElmer 2400 Series II CHNS/O Elemental Analyzer (2400 Series II, PerkinElmer, Waltham, MA, USA) yielding measurements of structural carbon content as a percentage of the total dry wood matter. We averaged the replicate measurements of percent carbon content to obtain a mean value for each 5 year segment. This process was repeated for each 5 year aggregate sample across the chronology, and for all sites.

The individual mean, minimum, and maximum density and cell wall thickness measurements were obtained from the SilviScan data of all the cores from natural and planted stands in FSJ only (N1–N3; P1–P3), due to the cost of analysis. Annual RW, EW, LW, density and cell wall thickness values from each core were averaged into 5 year increment values to correspond to 5 year carbon value increments (i.e., 1937–1941 to 2012–2016). The 20 cores representing a stand were then averaged for each 5 year interval to obtain the average stand-level chronologies of RW, EW, LW, density, and cell wall thickness over the time series. Stand-level and regional-level percent carbon and average RW, EW, LW and mean, minimum, and maximum cell wall thickness and density values were tested for normality using skewness, kurtosis and Shapiro–Wilk values prior to statistical tests. Shapiro–Wilk values for percent carbon were used to determine normality rather than skewness and kurtosis due to small sample sizes. Data failing one test were assessed using histograms to determine the severity of skew. Data failing all tests were transformed where possible or assumed non-normal. Data that were unable to be normalized were removed from further analysis.

Several one-way ANOVAs with Bonferroni post-hoc (alpha = 0.05) tests were conducted to determine the significant differences of mean percent carbon content between natural (N1 vs. N2 vs. N3) and planted stands (P1 vs. P2 vs. P3) (Table 2). Residuals of ANOVA tests were checked for normality. Regional data sets were created for the natural and planted stands of percent carbon and the mean, minimum and maximum density, and cell wall thickness, where no significant differences existed, by combining the data from the three sites together. An independent *t*-test analysis was conducted to determine if there was a significant difference between the regional data sets of mean percent carbon content in the natural and planted stands (Table 2). Correlation statistics were calculated for the regional data sets and natural and planted stand values, comparing the mean percent carbon values to the average RW, EW, LW and mean, maximum and minimum density and cell wall thickness values over time. We used a Pearson's or Spearman Rank correlation coefficient (R) (Table 3).

Historical climate information was obtained from the Adjusted Historical Canadian Climate Data for Fort St. James, nearest meteorological station to FSJ sites (Station #1092970, Latitude 54°45, Longitude-124°25, 686 m elevation), and Prince George, nearest meteorological station to the PG sites (Station #1096439, Latitude 53°88, Longitude-122°67, 680 m elevation). We investigated the following climate variables: monthly mean temperature, total monthly precipitation, and the winter (previous December, current January and February), spring (current March, April, May) and summer (current June, July, August), and previous autumn (previous September, November, December) seasonal averages. Random missing values within the climate data were calculated by averaging four surrounding points or filled with modelled climate data from Climate BC (http://cfcg.forestry.ubc.ca/projects/climate-data/climatebcwna/) for large gaps in data. Temperature and precipitation data were averaged into 5 year intervals for comparison with 5 year average percent carbon data.

Table 2. ANOVA and *t*-test results (mean (standard deviation)) between the percent carbon (% C) of natural (N1, N2, N3) and planted (P1, P2, P3) stands and the regionally averaged natural (N) and planted (P) % C from stands in the John Prince Research Forest with Bonferroni (*) post hoc. Different letters indicate significant differences among the groups ($p < 0.05$).

	Stand Level				**Regional Level**	
	% C (SD) *			**% C (SD) ***		**% C (SD)**
N1	45.34 a (0.87)	P1	42.21 a (0.45)	N	42.23 a (0.24)
N2	42.38 b (0.22)	P2	44.11 b (0.62)	P	41.89 a (0.73)
N3	42.06 b (0.59)	P3	41.57 a (0.96)		

Percent carbon measurements were correlated to the climate data (mean previous monthly May–December and mean current monthly January–September, and the previous autumn, winter, spring, and summer temperature and precipitation) values using Pearson's correlation coefficient (R) or Spearman's Rank coefficient for non-parametric data that could not be normalized (Table 3). In addition, correlation statistics were determined for the data from natural stands that were truncated to the last 30 years of growth (N(X)trunc). Truncated natural stand chronologies were compared with planted and entire natural stand chronologies. Partial correlation was used to determine the spurious correlations when relationships between percent carbon were found to both temperature and precipitation within the same months/seasons.

Table 3. Pearson's correlation coefficient between the current and previous (italics) average monthly and seasonal temperature (°C) (white) and precipitation (mm) (grey shading) and percent carbon for natural (N) and planted (P) stands surrounding the John Prince Research Forest (FSJ; N1–N3, P1–P3) and Prince George (PG; N4–N6, P4–P6) ** $p = 0.01$; * = 0.05 level. Blank cells indicate no significant relationships.

	FSJ							PG				
	P1	P2	P3	N1trunc	N2	N2trunc	N3	P4	P5	P6	N5	N6
Jan	-	-	-	-	-	-	−0.70 *	-	-	-	0.66 *	-
Mar	-	-	-	-	-	-	−0.64 *	-	-	-	-	-
Apr	-	-	-	-	-	-	−0.74 *	-	-	-	-	-
Jul	-	-	-	-	-	-	-	−0.82 *	-	−0.87 *	0.64 **	-
Jul	-	-	-	-	-	−0.82 *	−0.75 **	-	-	−0.82 *	0.76 **	-
Aug	-	-	-	-	-	-	−0.66 *	-	-	-	-	-
Aug	-	-	-	-	-	−0.78 *	-	-	-	-	0.64 **	-
Sept	-	-	-	-	-	-	-	−0.87 *	-	-	-	-
Sept	0.98 **	-	0.86 **	-	-	-	-	-	-	-	-	-
Nov	-	-	-	-	-	0.77 *	-	-	-	-	-	-
Dec	-	-	-	-	-	-	-	-	-	-	-	-
Spring	-	-	-	-	-	-	−0.8 *	-		-	-	-
Summer	-	-	-	-	-	-	−0.68 *	-	-	-	-	-
Winter	-	-	-	-	-	-	−0.75 **	-	-	-	-	-
Feb	-	-	0.84 *	-	-	-	-	-	-	-	-	-
Mar	-	-	-	−0.85 *	-	-	-	-	-	-	-	-
May	-	0.77 *	-	-	-	-	-	-	-	-	-	-
Jun	-	-	-	-	-	-	-	-	-	-	-	0.79 *
Sept	-	-	-	-	0.58 **	0.87 **	-	-	-	-	-	-
Nov	-	-	-	-	-	-	-	−0.78 *	-	-	-	-
Spring	-	-	-	-	-	-	-	-	−0.86 *	-	-	-
Winter	-	-	0.97 **	0.9 **	-	-	-	-	-	-	-	-

Regression analysis was completed where the significant Pearson's correlation coefficients were detected to further elucidate the relationships between the variables investigated. Only significant $R^2 > 0.40$ are reported [29]. Values for carbon, modelled based on the meteorological measurements were correlated back to the measured carbon values to verify the accuracy of the climate variable in predicting carbon values. Coordination between the measured and modelled carbon values were also visually assessed over time to determine the overall accuracy of the relationships modelled.

3. Results

3.1. Percent Carbon vs. Radial Growth Variables

Percent carbon was only determined for 30 years of growth for the planted stands and the last 80 years of growth for the natural ones, which resulted in greater proportions of juvenile wood in planted stand chronologies. Average carbon values for the PG natural stands ranged from 42.23% (+/−1.13) to 45.21% (+/−1.27), while we measured carbon values to range between 43.24% (+/−0.75) and 41.24% (+/−0.38) for the PG planted stands. Percent carbon values for FSJ stands a shared similar range to the PG stands (Table 2) and were further explored in comparison to the wood properties.

Percent carbon FSJ chronologies were normally distributed. Average percent carbon was statistically different among natural stands at a 5% confidence level ($F = 144.97$); post hoc comparisons indicated that the mean percent carbon of N1 was significantly different from the mean percent carbon of N2 and N3 ($p < 0.0001$) with no significant difference seen between N2 and N3 ($p = 0.589$) (Table 2).

Planted stands showed statistically different percentages of carbon at a 5% confidence level ($F = 20.98$); post hoc comparisons indicated that the mean percent carbon of P2 was significantly different from mean percent carbon of P1 and P3 ($p < 0.0001$) with no significant difference between the P1 and P3 ($p = 0.401$) (Table 2). Independent sample t-test results indicated no significant difference between regional-level mean percent carbon of natural stands (average of N2 and N3) and the planted stands (average of P1 and P3) ($t = 1.950$, p-value $= 0.070$, two-tailed) in FSJ (Table 2).

Percent carbon and wood properties were significantly correlated over time in stands N2, N3, and P1 (Figure 2). Relationships between the wood properties and percent carbon values by site are shown in Figure 2. Pearson[P] and Spearman Rank[S] correlations statistics determined significant correlations (** $p = 0.01$; * $p = 0.05$) between the percent carbon and wood properties over time in stands N2, N3, and P1 as follows: C vs. RW = 0.691 ** (N2); −0.855 ** (P1); C vs. EW = 0.624 ** (N2); C vs. LW = 0.602 ** (N2); C vs. MeanD = −0.592 * (N2); C vs. MinD = −0.674 ** (N2); C vs. MaxD = 0.712 ** (N2); C vs. MeanCWT = −0.688 ** (N2); C vs. MinCWT = −0.623 * (N2); −0.673 * (N3); C vs. MaxCWT = −0.712 ** (N2) (Figure 2).

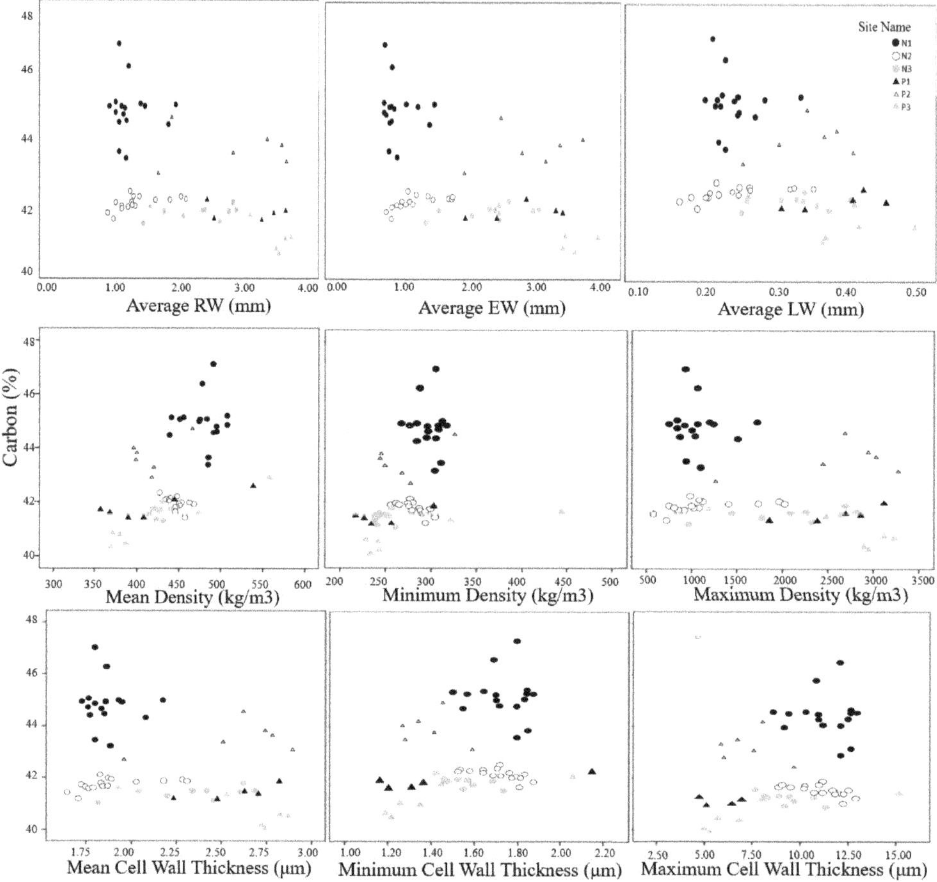

Figure 2. Comparison between the average percent (%) carbon (C) and ring width (RW), earlywood (EW), latewood (LW), and mean maximum minimum cell wall thickness (CWT) (μm) and density (D) (kg/m³) values of the natural (N1, N2, N3) and planted (P1, P2, P3) hybrid white spruce stands depicted by colour and shape. Each point represents a 5 year average within each series for the percent carbon and wood properties. Note that not all the axes are at the same scale.

3.2. Percent Carbon vs. Climate

Climate conditions recorded at the FSJ and PG weather stations have changed over the last 100 years. Historical climate records indicate that the mean annual precipitation has ranged from 282–770 mm in FSJ, and 368–934 mm in PG. Average annual precipitation and annual average temperatures have been recorded as 465 mm and 2.8 °C for FSJ, and 633 mm and 3.7 °C for PG. Mean average temperature has increased since 1920 by 1.2 °C and 0.4 °C in Fort St James and Prince George, respectively. Total annual precipitation since 1920 has increased by 31.2 mm in PG and decreased by 24.6 mm in FSJ.

Percent carbon chronologies in the planted stands of PG were significantly negatively correlated to the previous year's July and September, and the current year's July and spring temperatures; whereas the planted stands of FSJ were positively correlated to the current year's September temperature (Table 3). In the natural stands, the chronologies of the percent carbon from the FSJ were generally negatively correlated to the aforementioned climate variables, while the chronologies of percent carbon from PG were generally positively correlated to several previous and current monthly temperatures (Table 3). Only one of the truncated natural carbon chronologies was negatively correlated to the current July and August temperatures and positively correlated to the previous September temperatures (from FSJ), and no significant correlations were found between the truncated natural chronologies and temperature from PG.

Chronologies of percent carbon in the planted stands of FSJ were significantly positively correlated to the current February, May, and winter precipitation (Table 3). The chronologies of the percent carbon from natural stands of FSJ and PG were positively correlated to current June and September precipitation (Table 3); while the truncated natural chronologies of percent carbon from FSJ were significantly negatively correlated with the current year March precipitation and positively correlated with winter precipitation. We did not find any significant relationships between precipitation variables and the truncated carbon chronologies from the natural stands of PG (Table 3).

Although numerous significant relationships were found between the climate and chronologies of percent carbon from both the FSJ and PG using correlation and regression analyses, we showed only the strongest relationships found in natural stand percent carbon modelled from climate variables in Table 4 and Figure 3. Planted stand chronologies were not used for modelling carbon due to the unusually high R^2 values observed, resulting from a small sample size.

Table 4. R^2 values from the regression analysis of percent carbon on total monthly precipitation (mm) (Precip) and average monthly temperature (°C) (Temp) for the natural stands surrounding John Prince Research Forest (N2–N3) and Prince George (N4–N6) ** $p = 0.01$; * $= 0.05$ level. Correlation coefficients (Pearson's R) and p-values between measured (X_{mea}) and modelled (X_{mod}) percent carbon.

Month	Site	Carbon (R^2)	Pearson's R (X_{mea} v X_{mod})	p-Value
Jan temp	N3	0.485 **	0.696	0.017
Jan temp	N5	0.432 **	0.657	0.006
Mar precip	N1^{trunc}	0.718 **	0.848	0.016
Jul temp	N2^{trunc}	0.669 *	0.818	0.025
Jul temp	N3	0.555 **	0.745	0.009
Aug temp	N5	0.574 **	0.758	0.001
Previous Aug temp	N3	0.439 *	0.663	0.026
Spring temp	N3	0.639 **	0.800	0.003

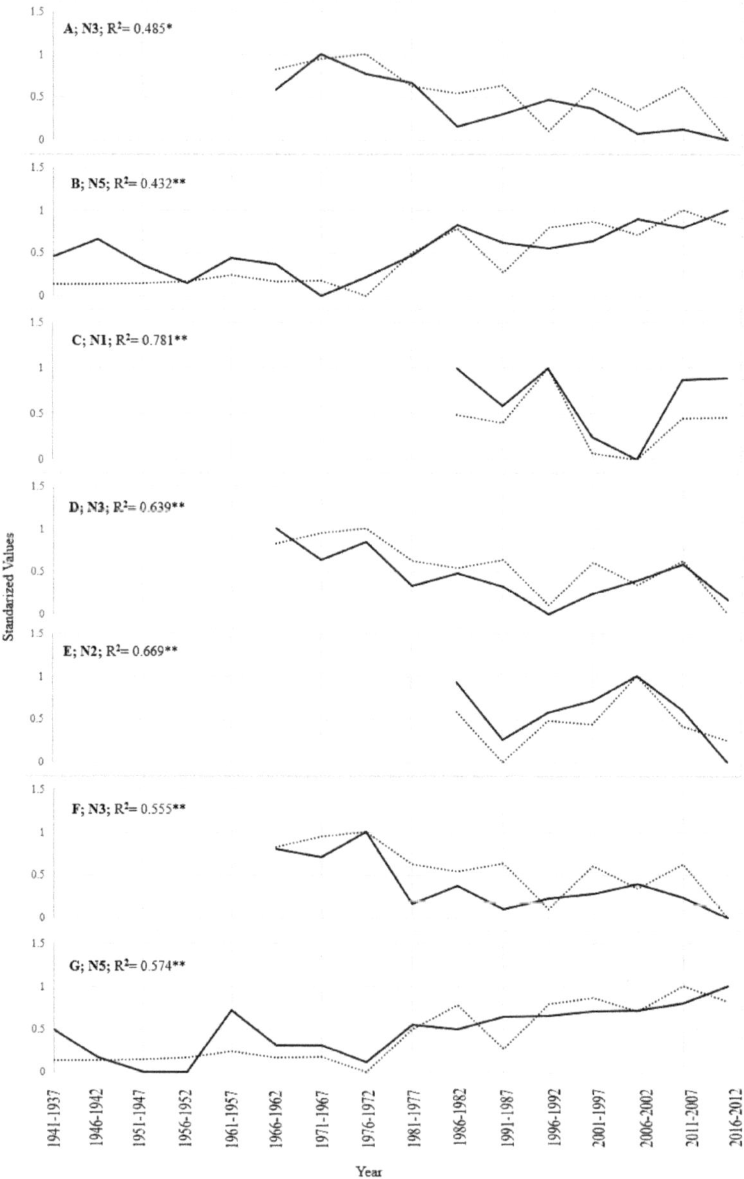

Figure 3. Measured (dotted line) vs. modelled (solid line) normalized percent carbon values from a natural stand of hybrid white spruce within the John Prince Research Forest (N1–N3) and near PG (N5). Data modelled from current year January temperature (**A,B**), current year March precipitation (**C**), average spring temperature (**D**), and current July temperature (**E–G**) from Fort St. James and Prince George climate stations. R^2 values are presented with ** $p = 0.01$ and * $p = 0.05$. Note that not all axes are at the same scale.

4. Discussion

This study sought to understand how the structural carbon of hybrid white spruce (*Picea glauca* × *engelmannii*) in natural and planted stands varied over time, and how these variations in carbon related to RW, EW, LW, density, and the cell wall thickness properties of hybrid white spruce wood. We were also interested in determining how variations in carbon allocation in hybrid white spruce corresponded with changes in the temperature and precipitation of interior British Columbia, Canada.

4.1. Percent Carbon vs. Radial Growth Variables

We provide empirical data on the structural carbon content of hybrid white spruce trees in the interior region of British Columbia; an area that is representative of a large boreal forest region extending across the Canadian landscape [30]. These carbon measurements, and relationships with climate and cell property variables, add significantly to what is known about the carbon content of forests based on existing models that use DBH and biomass [5,6]. This type of measured data can be used to improve the modelling of wood carbon content and our understanding of the variation that exists in carbon values within the same species across the landscape. We can also better appreciate the amount of carbon that remains sequestered long term in wood products, by specifically identifying the structural carbon element, as many volatile carbon components are lost in manufacturing processes [31]. We found that the structural component of carbon made up between 41–48% of the biomass of our wood samples; this carbon will remain as fixed products, or while standing in the forest, until the wood interacts with soil decomposers and is decayed.

The relationships we identified between the wood properties and carbon indicate that the RW, EW, and LW width values may be good indicators of percent carbon variation in natural and planted stands of BC hybrid white spruce [32]. Average RW, EW, and LW are easily measured using inexpensive standard dendrochronological techniques, and are therefore an advantageous way of capturing the estimates of carbon, where the relationships between these variables indicate this possibility. Although there is evidence that in some species tree diameter at breast height (DBH) is a good variable for carbon estimations, as an increasing tree diameter allows for increased biomass and thus increased carbon [2,33,34], this trend is not consistent across studies [33,34]. Therefore, perhaps in cases where DBH is not satisfactorily accurate, RW, or EW and LW could be used as a proxy. Furthermore, the relationships between carbon and density and cell wall thickness could provide even greater accuracy in cases where structural carbon estimates are simply determined by biomass as a function of tree height and DBH [1,35–38].

By comparing the carbon content of wood to easily measured wood properties, we can better understand their relatedness in specific environments, and apply this knowledge towards the use of these properties as proxies for carbon content. We can also better understand the natural genetically and environmentally controlled variation that exists between the trees and stands in terms of their carbon content. Significant correlations between the percent carbon and average RW, EW, and LW and mean, maximum, and minimum cell wall thickness and density values of the natural and planted stands were present in three out of the six stands measured (N2, N3 and P1 stands), with the majority of the relationships found with N2. Carbon could have been respired or used in other metabolic processes at different rates in a stand specific way, creating small differences in the amount of structural carbon measured in the stems. Alternatively, the trees at N2 may have allocated carbon differently within the tree, for example, to the cells of the stem, vs. needles, branches or roots, resulting in variation in the strength of the relationships between the carbon and wood properties measured in our stem cores. Inter-stand differences in the concentrations of cellulose, lignin and non-structural carbohydrates are also possible [1]. Parameters such as crown size, crown closure, and photosynthetic rates that could be used to determine these differences were not collected within this study; future research should consider the incorporation of these measurements in the experimental design. Statistical comparisons of carbon content in some stands were insignificant, likely due to small sample size (tree age and reduction from annual to 5-year increment measurements for carbon analysis mass requirements were limiting factors).

Relationships between the percent carbon and radial growth variable values in N2 suggest that higher amounts of cellulose and lignin (as represented by a thicker cell wall and denser wood) correspond to greater proportions of percent carbon; maximum values appeared to have stronger relationships than the mean and minimum measurement values (Figure 2). These results are similar to previous work in western Canada and Alaska, showing correlations between forest productivity and LW (max) and other studies relating carbon to biomass and density, with maximum values of cell wall thickness and density as the best predictors of percent carbon [1,2,6,32,38,39]. Mean and minimum cell wall thickness and density measurements did not share the same relationships with percent carbon as shown in the maximum values. In fact, the data from the N2 and N3 stands suggested an opposite trend—that even if the minimum measurement of cell wall material, or the mean amount within a given year, is increased, if the maximum amount of cell wall material is lower in that year, then the overall cell wall thickness will correspond to reduced proportions of structural carbon. Therefore, we conclude that maximum cell wall thickness and density values are the most valuable to capture for relating to structural carbon accumulation.

We investigated the potential differences that exist between carbon contents in plantation hybrid white spruce and naturally grown hybrid white spruce in central interior BC. Carbon contents in these natural and planted stands were difficult to compare due to the differences in age and a lack of data points. The oldest plantation hybrid white spruce stands in the interior BC were approaching 40 years, while the naturally grown hybrid white spruce trees were as old as the last natural disturbance (~80–150 years). Thus, most naturally-grown, dominant, canopy spruce trees are much older than 40 years and have proportionally more mature wood to juvenile wood in comparison to the 30–40 year-old plantation trees. We were unable to sample natural hybrid white spruce stands at 40 years of age, to match the age of the plantation stands, because naturally regenerated, 40-year old hybrid white spruce are usually found in the sub-canopy of more dominant tree types, and are limited in growth by competition rather than climate. Therefore, we observed the growth rates with the climate in last 40 years of growth of natural stands as a truncated chronology, and the last 40 years of growth in the plantation stands to keep the climate period consistent, even though the juvenile wood (JW): mature wood (MW) ratio was different between the stand types.

Based on the structural cell properties alone, fast-growing, thin-walled, and low-density cells typical of JW should have a lower structural percent carbon content than denser, thicker cells, found in MW [40]. Conversely, juvenile wood can contain higher percent carbon than mature wood where larger proportions of lignin and extractive concentrations exist [41,42]. Because our samples had chemical extractives, or non-structural carbon removed, we expected the older, natural stands to contain higher proportional amounts of carbon due to higher MW: JW. However, we found an insignificant difference between the average percent of carbon in the natural vs. planted stands at the regional level. There are a few possible explanations for these findings. The first is that the duration of production of juvenile wood was shorter in the plantation trees than in the naturally grown trees, which would lead to similar MW: JW between the two stand types. Naturally regenerated hybrid white spruce would have germinated from locally adapted seed sources, while plantation trees would have genetically originated from the region, but not the specific location where they were planted, potentially leading to a slower growing tree. Furthermore, genetically improved spruce tree stock that is commonly used today, and is fast growing in cut-block openings, was not available in the 1970s when the trees we measured were planted.

Another possible reason for the lack of significant differences between the carbon contents in the natural vs. planted trees is that not all the non-structural compounds were extracted. If some extractives remained in the wood, and there were more extractives in the JW to begin with, then the differences in structural carbon could have been muted by the variation in extractives. Samples analyzed for carbon content were extracted using a Soxhlet acetone apparatus, for 1.5 h. It is suggested that future research using this technique employs a longer extraction time, which will potentially yield more significant results in comparing the structural carbon content between stands. There is a lack of standardized sampling protocols to prepare samples for carbon measurements; variations include kiln-drying [43], freeze-drying, and oven-drying at varying temperatures, such as 105 °C [3] and 70 °C [44,45]. These variations in sample preparation make cross-study comparisons difficult to interpret [31,45]. These protocols require further investigation for the development of an optimal, standard method.

4.2. Percent Carbon vs. Climate

We found variation in the percent carbon between the individual stands in the FSJ region that may be attributed to geographical and environmental factors. Factors such as latitude, elevation, and topography, or site-specific differences such as soil water volume, crown cover, and nutrient availability have been shown to influence carbon contents [1,2,39]. Significantly higher percent carbon contents were observed in P2 and N1, likely due to the differences in these sites. The warmer south-facing slope of P2 received more sunlight, which likely expedited snow-melt and soil thaw in comparison with the flat topography of P1 and P3. The site conditions at P2 may have resulted in increased growing season length, directly enabling carbon accumulation [46]. Higher proportional carbon content of N1 may be attributed to the stream at this site that could stabilize or increase the soil moisture content. Increased moisture in N1 could counteract unfavourable conditions for growth such as rising temperatures coupled with reduced precipitation, as seen in the FSJ climate. These observations have been found in other studies with similar climatic conditions, such as trees grown in wet vs. dry conditions [47–49].

Percent carbon accumulation in the planted and natural stands in FSJ and PG was negatively influenced by rising temperatures with some site-specific differences in relationships with spring climate variation. We found that the percent carbon component of the wood sampled in the planted stands of the PG region, and the natural stands in the FSJ region, were statistically negatively related to multiple temperature variables, including the previous year's December, the current year's January, average spring, and previous year and current July temperatures. Increasing temperatures during winter months can reduce the length of snow cover and reduce the accumulated winter precipitation, or insulation, leading to deeper soil freezing [50]. These conditions can prevent or reduce the absorption of melting snow thus delaying bud burst and the percent carbon accumulation in spring months [51]. Additionally, increasing spring or summer temperatures beyond optimal growth thresholds have been shown to reduce or halt growth, and subsequent carbon accumulation, in previous studies of BC interior spruce [52,53]. Conversely, percent carbon measured from N5 was positively correlated to rising winter and summer temperatures in the PG region, likely due to the favourable site-specific growing conditions, including N5's higher elevation, relatively higher precipitation, and cooler average temperatures (SBS very–wet–cool). Thus, increasing percent carbon accumulations that coordinate with increasing temperatures are explainable in this case, and similar relationships have been found with increased forest productivity under favourable conditions [47,49].

Carbon accumulations in FSJ and PG were related to spring precipitation variation, albeit in different ways. Decreases in FSJ spring precipitation were statistically related to lower relative percent carbon values in the P2 chronology, and higher relative carbon values in the N1 chronology truncated to the last 30 years of growth (N1^{trunc}). Lower total precipitation in May could have negatively affected the percent carbon accumulation in P2 due to the south-facing slope, which could have caused increased rates of evaporation and transpiration and reduced soil moisture, and ultimately reduced carbon accumulation relative to the other planted stands. We found that decreases in March precipitation, typically falling as snow in this region, were correlated with greater carbon proportions in N1^{trunc}. Reduced March precipitation, or snow depth, could lead to earlier bud burst and an extension of growing season length and subsequent increased radial growth and carbon accumulation [54]. In PG, the results suggested that lower percent carbon values at P5 were related to increasing average spring precipitation. Increased and prolonged precipitation could have reduced cell production and subsequent carbon accumulation in P5, with decreased light availability for photosynthesis occurring with increased cloud cover [39].

Percent carbon measurements over the last 30 years, truncated from the natural stands in FSJ, were more strongly related to climate variation compared to full-length chronologies; these relationships were not seen in the natural stand data from the PG. This suggests that the temperature and precipitation variation in FSJ have become more influential in determining the percent carbon accumulation than earlier in the time series, an observation seen in radial growth and climate in other studies [55–57]. The historical climate of FSJ shows rising temperatures coupled with stark reductions in precipitation, which may explain the stronger relationships between climate variation and percent carbon in recent decades. Likely, these same relationships were not observed in the truncated percent carbon chronologies of PG natural stands because of the differences in climate regimes. Historical records of PG climate report roughly 200 mm higher average precipitation and 1 °C higher average temperature than FSJ climate. In recent decades, records in PG also show stable increases in precipitation that contrasts with stark decreases in FSJ. Higher and stable average precipitation coupled with increasing temperatures, as seen in the PG climate records, suggests that conditions are more favourable for growth than in the warmer, drier conditions of FSJ. This general difference in regional climate may explain the lack of significant relationships between the climate and natural stand chronologies around the PG. Results from this study indicate that trees in FSJ may be reaching growth–thresholds—warming coupled with a reduction in precipitation, trends that are not observed in PG [12,55,58].

We used current and one year lagged precipitation and temperature variables from the FSJ and PG climate stations to predict the percent carbon from natural hybrid white spruce stands in central interior BC. We found that these models were more difficult to apply to younger, planted stands, as small sample sizes lead to decreased reliability in the model statistics, increasing the likelihood of type II error [59]. Increased sampling error and outlier influences that question validity may also occur with small sample sizes [59]. However, the relationships presented between natural stands and carbon provide an example of a method that could form the basis for a novel approach to understanding climate effects on carbon allocation in forest stands based on a dendrochronological analysis. There are limited studies on variation in structural carbon content, and even fewer that use direct empirical measurements of structural carbon. To our knowledge, this is the only study that has used the measurements of structural carbon in dendrochronological applications to make stand-level estimates of wood carbon.

5. Conclusions

This study presents a novel approach for understanding the relationships between structural percent carbon and radial tree growth and climate. Results suggest that the maximum values of cell wall thickness and density may aid in improving the existing models of carbon estimations that are historically based on DBH and biomass. Investigation into the effect of temperature and precipitation on carbon allocation in natural and planted stands showed that rising temperatures were related to a reduction in carbon allocation; precipitation variation had site-specific differences.

Author Contributions: Conceptualization and methodology design, A.I. and L.J.W.; formal analysis, investigation, resources, data curation, writing—original draft preparation, A.I.; writing—review and editing, L.J.W. and K.L.; supervision, L.J.W. and K.W; funding acquisition, A.I. and L.J.W. All authors have read and agreed to published version of the manuscript.

Funding: This research was funded by Natural Resources Canada, Mitacs, Canfor Pulp Ltd., and UNBC.

Acknowledgments: Special thanks to Aita Bezzola for creating the site map for this manuscript.

Conflicts of Interest: The authors declare no conflict of interest.

References

1. Weber, J.C.; Sotelo Montes, C.; Abasse, T.; Sanquetta, C.R.; Silva, D.A.; Mayer, S.; Muniz, G.I.B.; Garcia, R.A. Variation in growth, wood density and carbon concentration in five tree and shrub species in Niger. *New For.* **2018**, *49*, 35–51.
2. Elias, M.; Potvin, C. Assessing inter- and intra-specific variation in trunk carbon concentration for 32 neotropical tree species. *Can. J. Res.* **2003**, *33*, 1039–1045.
3. Thomas, S.C.; Malczewski, G. Wood carbon content of tree species in eastern China: Interspecific variability and the importance of the volatile fraction. *J. Environ. Manag.* **2007**, *85*, 659–662.
4. Lachenbruch, B.; Mcculloh, K.A. Tansley review Traits, properties, and performance: How mechanical functions in a cell, tissue, or whole. *New Phytol.* **2014**, *204*, 747–764.
5. Zabek, L.M.; Prescott, C.E. Biomass equations and carbon content of aboveground leafless biomass of hybrid poplar in coastal British Columbia. *For. Ecol. Manag.* **2006**, *223*, 291–302.
6. Castaño-Santamaría, J.; Bravo, F. Variation in carbon concentration and basic density along stems of sessile oak (*Quercus petraea* (Matt.) Liebl.) and Pyrenean oak (*Quercus pyrenaica* Willd.) in the Cantabrian Range (NW Spain). *Ann. For. Sci.* **2012**, *69*, 663–672.
7. Allen, C.D.; Macalady, A.K.; Chenchouni, H.; Bachelet, D.; McDowell, N.; Vennetier, M.; Kitzberger, T.; Bigling, A.; Breshears, D.D.; Hogg, E.H.; et al. A global overview of drought and heat-induced tree mortality reveals emerging climate change risks for forests. *For. Ecol. Manag.* **2010**, *259*, 660–684.
8. Schwalm, C.R.; Williams, C.A.; Schaefer, K.; Arneth, A.; Bonal, D.; Buchmann, N.; Chen, J.; Law, B.L.; Lindroth, A.; Luyssaert, S.; et al. Assimilation exceeds respiration sensitivity to drought: A FLUXNET synthesis. *Glob. Chang. Biol.* **2010**, *16*, 657–670.
9. Haughian, S.R.; Burton, P.J.; Taylor, S.W.; Curry, C. Expected effects of climate change on forest disturbance regimes in British Columbia. *J. Ecosyst. Manag.* **2012**, *13*, 1–24.
10. Babst, F.; Alexander, M.R.; Szejner, P.; Bouriaud, O.; Klesse, S.; Roden, J.; Ciais, P.; Poulter, B.; Frank, D.; Moore, D.J.P.; et al. A tree-ring perspective on the terrestrial carbon cycle. *Oecologia* **2014**, *176*, 307–322.
11. McDowell, N.G.; Allen, C.D. Darcy's law predicts widespread forest mortality under climate warming. *Nat. Clim. Chang.* **2015**, *5*, 669–672.
12. Lo, Y.H.; Blanco, J.A.; Seely, B.; Welham, C.; Kimmins, J.P. Relationships between climate and tree radial growth in interior British Columbia, Canada. *For. Ecol. Manag.* **2010**, *259*, 932–942.
13. Fleming, S.W.; Whitfield, P.H. Spatiotemporal mapping of ENSO and PDO surface meteorological signals in British Columbia, Yukon, and southeast Alaska Spatiotemporal Mapping of ENSO and PDO Surface Meteorological Signals in British Columbia, Yukon, and Southeast Alaska. *Can. Meteorol. Oceanogr. Soc.* **2010**, *48*, 122–131.
14. Jiang, X.; Huang, J.; Stadt, K.J.; Comeau, P.G.; Chen, H.Y.H. Spatial climate-dependent growth response of boreal mixedwood forest in western Canada. *Glob. Planet. Chang.* **2016**, *139*, 141–150.
15. Millar, C.I.; Stephenson, N.L. Temperate forest health in an era of emerging megadisturbance. *Science* **2015**, *349*, 823–826.
16. Fritts, H.C. *Tree Rings and Climate*; Academic Press Inc.: New York, NY, USA, 1976.
17. Bouriaud, O.; Bréda, N.; Dupouey, J.-L.; Granier, A. Is ring width a reliable proxy for stem-biomass increment? A case study in European beech. *Can. J. For. Res.* **2005**, *35*, 2920–2933.
18. Grünwald, T.; Bernhofer, C. A decade of carbon, water and energy flux measurements of an old spruce forest at the Anchor Station Tharandt. *Chem. Phys. Meteorol.* **2007**, *59*, 387–396.
19. Liepiņš, J.; Lazdiņš, A.; Liepiņš, K. Equations for estimating above- and belowground biomass of Norway spruce, Scots pine, birch spp. and European aspen in Latvia. *Scand. J. For. Res.* **2018**, *33*, 58–70.
20. Davis, S.C.; Hessl, A.E.; Scott, C.J.; Adams, M.B.; Thomas, R.B. Forest carbon sequestration changes in response to timber harvest. *For. Ecol. Manag.* **2009**, *258*, 2101–2109.
21. Ter-Mikaelian, M.T.; Colombo, S.J.; Chen, J. Effect of age and disturbance on decadal changes in carbon stocks in managed forest landscapes in central Canada. *Mitig. Adapt. Strateg. Glob. Chang.* **2014**, *19*, 1063–1075.
22. Meidinger, D.; Pojar, J. *Ecosystems of British Columbia*; BC Ministry of Forests: Victoria, BC, Canada, 1991.
23. Delong, C.; Tanner, D.; Jull, M.J. *A Field Guide for Site Identification and Interpretation for the Southwest Portion of the Prince George Forest Region*; Province of British Columbia Minsitry of Forests: Prince George, BC, Canada, 1992.

24. Dale, M.R.T.; Fortin, M.J. *Spatial Analysis: A Guide for Ecologists*; Cambridge University Press: Cambridge, UK, 2014.

25. Jensen, W.B. The Origin of the Soxhlet Extractor. *J. Chem. Educ.* **2007**, *84*, 1913–1914.

26. Grabner, M.; Wimmer, R.; Gierlinger, N.; Evans, R.; Downes, G. Heartwood extractives in larch and effects on x-ray densitometry. *Can. J. For. Res.* **2005**, *35*, 2781–2786.

27. Evans, R.; Ilic, J. Rapid prediction of wood stiffness from microfibril angle and density. *For. Prod. J.* **2001**, *51*, 53.

28. Yamaguchi, D.K. A simple method for cross-dating cores from living trees. *Can. J. Res* **1991**, *21*, 414–416.

29. Wood, L.J.; Smith, D.J. Climate and glacier mass balance trends from AD 1780 to present in the Columbia mountians, British Columbia, Canada. *Holocene* **2012**, *23*, 739–748.

30. Coates, D.K.; Haeussler, S.; Lindeburgh, S.; Pojar, R.; Stock, A.J. *Ecology and Silviculture of Interior Spruce in British Columbia*; Government of Canada: British Columbia, BC, Canada, 1994; pp. 1–182.

31. Jones, D.A.; Hara, K.L.O. The influence of preparation method on measured carbon fractions in tree tissues. *Tree Physiol.* **2017**, *36*, 1177–1189.

32. Beck, P.S.A.; Andreu-Hayles, L.; D'Arrigo, R.; Anchukaitis, K.J.; Tucker, C.J.; Pinzón, J.E.; Goetz, S.J. A large-scale coherent signal of canopy status in maximum latewood density of tree rings at arctic treeline in North America. *Glob. Planet. Chang.* **2013**, *100*, 109–118.

33. Navarro, M.; Moya, R.; Chazdon, R.; Ortiz, E.; Vilchez, B. Successional variation in carbon content and wood specific gravity of four tropical tree species. *Bosque* **2013**, *34*, 9–10.

34. Clark, D.A.; Piper, S.C.; Keeling, C.D.; Clark, D.B. Tropical rain forest tree growth and atmospheric carbon dynamics linked to interannual temperature variation during 1984–2000. *Proc. Natl. Acad. Sci. USA* **2003**, *100*, 5852–5857.

35. Somogyi, Z.; Cienciala, E.; Mäkipää, R.; Muukkonen, P.; Lehtonen, A.; Weiss, P. Indirect methods of large-scale forest biomass estimation. *Eur. J. For. Res.* **2007**, *126*, 197–207.

36. Kearsley, E.; De Haulleville, T.; Hufkens, K.; Kidimbu, A.; Toirambe, B.; Baert, G.; Huygens, D.; Kebede, Y.; Defourny, P.; Bogaert, J.; et al. Conventional tree height-diameter relationships significantly overestimate aboveground carbon stocks in the Central Congo Basin. *Nat. Commun.* **2013**, *4*, 1–8.

37. Ali, A.; Yan, E.R.; Chen, H.Y.H.; Chang, S.X.; Zhao, Y.T.; Yang, X.D.; Xu, M.S. Stand structural diversity rather than species diversity enhances aboveground carbon storage in secondary subtropical forests in Eastern China. *Biogeosciences* **2016**, *13*, 4627–4635.

38. Zhang, Q.; Wang, C.; Wang, X.; Quan, X. Carbon concentration variability of 10 Chinese temperate tree species. *For. Ecol. Manag.* **2009**, *258*, 722–727.

39. Martin, A.R.; Thomas, S.C. A reassessment of carbon content in tropical trees. *PLoS ONE* **2011**, *6*, e23533.

40. Bao, F.C.; Jiang, Z.H.; Jiang, X.M.; Lu, X.; Luo, X.Q.; Zhang, S.Y. Differences in wood properties between juvenile wood and mature wood in 10 species grown in China. *Wood Sci. Technol.* **2001**, *35*, 363–375.

41. Zobel, B.J.; van Buijtenen, J.P. *Wood Variation: Its Causes and Controls*; Springer: Berlin/Heidelberg, Germany, 1989.

42. Bert, D.; Danjon, F. Carbon concentration variations in the roots, stem and crown of mature Pinus pinaster (Ait.). *For. Ecol. Manag.* **2006**, *222*, 279–295.

43. Lamlom, S.H.; Savidge, R.A. A reassessment of carbon content in wood: Variation within and between 41 North American species. *Biomass Bioenergy* **2003**, *25*, 381–388.

44. Gower, S.T.; Krankina, I.O.; Olson, R.J.; Apps, M.; Linder, S.; Wangi, C. Net primary production and carbon allocation patterns of boreal forest ecosystems. *Ecol. Appl.* **2001**, *11*, 1395–1411.

45. Wang, C.; Bond-Lamberty, B.; Gower, S.T. Carbon distribution of a well- and poorly-drained black spruce fire chronosequence. *Glob. Chang. Biol.* **2003**, *9*, 1066–1079.

46. Rossi, S.; Deslauriers, A.; Anfodillo, T.; Morin, H.; Saracino, A.; Motta, R.; Borghetti, M. Conifers in cold environments synchronize maximum growth rate of tree-ring formation with day length. *New Phytol.* **2006**, *170*, 301–310.

47. Guehl, J.M. Étude comparée des potentialités hivernales d'assimilation carbonée de trois conifères de la zone tempérée (*Pseudotsuga menziesii* Mirb., *Abies alba* Mill. et *Picea excelsa* Link.). *Ann. Sci.* **1985**, *42*, 23–38.

48. McMahon, M.S.; Parker, G.G.; Miller, D.R. Evidence for a recent increase in forest growth is questionable. *Proc. Natl. Acad. Sci. USA* **2010**, *107*, 3611–3615.

49. Hember, R.A.; Kurz, W.A.; Metsaranta, J.M.; Black, T.; Guy, R.D.; Coops, N.C. Accelerating regrowth of temperate-maritime forests due to environmental change. *Glob. Chang. Biol.* **2012**, *18*, 2026–2040.

50. Jarvis, P.; Linder, S. Constraints to growth of boreal forests. *Nature* **2000**, *405*, 904–905.

51. Knudson, D.V.; Lindsey, C. Type I and Type II errors in correlations of various sample sizes. *Compr. Psychol.* **2014**, *3*, 2165–2228.

52. Zhang, Q.; Alfaro, R.I.; Hebda, R.J. Dendroecological studies of tree growth, climate and spruce beetle outbreaks in central British Columbia, Canada. *For. Ecol. Manag.* **1999**, *121*, 215–225.

53. Oberhuber, W.; Gruber, A.; Kofler, W.; Swidrak, I. Radial stem growth in response to microclimate and soil moisture in a drought-prone mixed coniferous forest at an inner Alpine site. *Eur. J. For. Res.* **2014**, *133*, 467–479.

54. Peterson, D.W.; Peterson, D.L.; Ettl, G.L. Growth responses of subalpine fir to climate variability in the pacific northwest. *Can. J. For. Res.* **2002**, *32*, 1503–1517.

55. Camarero, J.J.; Gazol, A.; Sangüesa-Barreda, G.; Oliva, J.; Vicente-Serrano, S.M. To die or not to die: Early warnings of tree dieback in response to a severe drought. *J. Ecol.* **2015**, *103*, 44–57.

56. Martin-Benito, D.; Beeckman, H.; Cañellas, I. Influence of drought on tree rings and tracheid features of Pinus nigra and Pinus sylvestris in a mesic Mediterranean forest. *Eur. J. For. Res.* **2013**, *132*, 33–45.

57. Peñuelas, J.; Ogaya, R.; Boada, M.; Jump, A.S. Migration, invasion and decline: Changes in recruitment and forest structure in a warming-linked shift of European beech forest in Catalonia (NE Spain). *Ecography* **2007**, *30*, 829–837.

58. Lazarus, B.E.; Castanha, C.; Germino, M.J.; Kueppers, L.M.; Moyes, A.B. Growth strategies and threshold responses to water deficit modulate effects of warming on tree seedlings from forest to alpine. *J. Ecol.* **2018**, *106*, 571–585.

59. Li, T.; Ren, B.; Wang, D.; Liu, G. Spatial variation in the storages and age-related dynamics of forest carbon sequestration in different climate zones-evidence from black locust plantations on the loess plateau of China. *PLoS ONE* **2015**, *10*, e0121862.

Article

Growth, Carbon Storage, and Optimal Rotation in Poplar Plantations: A Case Study on Clone and Planting Spacing Effects

Yanhua Zhang [1], Ye Tian [1,2], Sihui Ding [1], Yi Lv [1], Wagle Samjhana [1] and Shengzuo Fang [1,2,*]

1   College of Forestry, Nanjing Forestry University, Nanjing 210037, China; yhzhang99@126.com (Y.Z.);
     tianyes@hotmail.com (Y.T.); dingsh94@163.com (S.D.); lvyi_njfu95@163.com (Y.L.);
     me.samjhana2012@gmail.com (W.S.)
2   Co-Innovation Center for Sustainable Forestry in Southern China, Nanjing Forestry University,
     Nanjing 210037, China
*   Correspondence: fangsz@njfu.edu.cn; Tel.: +86-25-85428603

Received: 6 July 2020; Accepted: 31 July 2020; Published: 3 August 2020

Abstract: Poplar, as the most widely cultivated fast-growing tree species in the middle latitude plain, provides important wood resources and plays an important role in mitigating climate change. In order to understand the response of growth, biomass production, carbon storage to poplar clones, planting spacings, and their interaction, a field trial was established in 2007. In 2018, we destructively harvested 24 sample trees for biomass measurements and stem analyses. Biomass production and carbon storage for the single tree of three clones enhanced as planting spacing increasing at the age of 13, but both the biomass production and carbon storage of clones NL-895 and NL-95 were higher than the clone NL-797 at the spacings of 6 × 6 m and 5 × 5 m. The average carbon concentration of the tested clones was in the order of stem > branches > leaves, and showed significant variation between different components ($p < 0.05$). Large spacing stimulated more biomass to be partitioned to the canopy. Based on the prediction values of tree volume growth by established Chapman–Richards models, the quantitative maturity ages of stand volume varied among the investigating plantations, ranging from 14 to 17 years old. Our results suggest that the selecting clones NL-895 and NL-95 with 6 × 6 m spacing would be recommended at similar sites for future poplar silviculture of larger diameter timber production, as well as for carbon sequestration.

Keywords: poplar clone; planting density; biomass production; carbon storage; Chapman–Richards model; quantitative mature

---

1. Introduction

As reported, planted forest area accounts for about 7% of the total forest area, but the industrial roundwood production from plantations represented 33.4% of global production from all types of forest [1]. Therefore, planted forests play an important role in the global and regional economies to secure industrial roundwood and wood fuel, and to mitigate climate change [1,2]. China owns the largest plantation area in the world, accounting for 24.82% of the global plantation areas [2–5]. The continuous increase of plantation area contributes significantly to the forest coverage of China, but the yield and quality in the plantations are not high in general, and the forest stock per hectare is only 69% of the world average level of 131 m³ [3]. Poplar plantations are no exception, although they play an important role in sustaining the commercial supply of forest products and fixing atmospheric CO_2 [6]. With the development of the economy, timber consumption level has increased dramatically in China. In 2018, China imported about 127.61 M m³ of major forest products, including logs, sawn timber, and wood pulp, totaling 47.01 billion dollars [4].

Regarding the observed increase in the atmospheric concentration of CO_2 and the global climate problems, one main option is to increase forest biomass for reducing the volume of atmospheric CO_2, which can be achieved by planting currently unforested land, or by improving the productivity of existing forests. Poplar, as one of the most widely cultivated timber and ecological commonweal tree species in the middle latitude plain of the world, has the characteristics of being fast-growing and having large biomass, diverse uses, and easy adaptability [7]. By 2015, China's poplar plantation area reached 8.54 M ha, accounting for 12.32% of the national plantation area [3]. Considering the huge carbon storage and the interest of the forest industry in poplar plantations, any fluctuations in their productivity could have major ecological and economic consequences. Also, in fast-growing, short rotation plantations, nutrient accumulation and export from the site has been considered for a long time, as nutrients are removed through frequent harvests [8,9]. Present studies have shown that poplar growth does not only depend on the growing season length [10], but also on the genotype [11,12], planting density [13,14], site conditions [15–17], and management strategies applied [18–20]. In practice, genotype and planting spacing as the most easily controlled factors are very important in the directional cultivation of plantations [21]. Choice of the best spacing system is closely related to the maximum production of biomass production, biomass partitioning, nutrient accumulation, and determination of duration of production cycle. Generally, the diameter and aboveground biomass of a single tree increases with increasing planting spacing [13,22], while the greatest biomass productivity has been achieved in higher density stands [14,23]. Meanwhile, the selection of a poplar genotype is critical to improving yield and productivity levels [9,11,12]; the crown ratio, branches traits, biomass allocated of the stem were also affected by cloning and spacing [13,23]. Fang et al. [23] evaluated growth dynamics, biomass production, and carbon storage in short-rotation poplar plantations, with four planting spacings (3×3 m, 3×4 m, 4×4 m. and 4×5 m) and three poplar clones (NL-80351, I-69, and I-72), but optimal rotation was not reported for the investigated plantations. Therefore, attempts were made in this study to understand the influence of new genotypes and planting spacings on the growth, aboveground biomass production and partitioning, carbon storage, and rotation length of poplar plantations, which can provide some references for optimizing the cultivation patterns at similar sites.

2. Material and Methods

2.1. Study Area and Experimental Design

The site is located in the Sihong forest farm, Jiangsu Province, China (33°16′ N, 118°18′ E). In this region, there is a mid-latitude climate, with a mean annual air temperature of 14.84 °C, while the mean temperatures is 27.4 °C in July and −7 °C in January, respectively [24]. The mean annual sunshine hour is between 2250 and 2350 h, and the mean annual precipitation is about 1000 mm, occurring mostly from June to August each year. Soils at this site were formed on fine sediments of Hongze Lake. The basic physical and chemical properties at the trial site are relatively consistent, with clay loam texture and middle soil fertility [24].

The experimental design consisted of a randomized, complete block with three replications, and the plantations were established in March 2007 with one-year-old rooted cutting of three poplar clone seedlings, including Nanlin-895 (NL-895), Nanlin-95 (NL-95), and Nanlin-797 (NL-797), a hybrid of clone I-69 (*Populus deltoides* Bartr. cv. "Lux") × clone I-45 (*P.* × *euramericana* (Dode) Guinier cv. I-45/51′). Planting spacing designs included two squares (6×6 m and 5×5 m) and two rectangles (4.5×8 m and 3.5×8 m). In total, 36 plots were randomly arranged and established, with each plot about 1200–1800 m^2 (50 trees per plot). However, in this study, we only choose two spacings (6×6 m and 5×5 m) of clones NL-95 and NL-797 and four spacings of clone NL-895 for detailed biomass investigation and stem analysis, so in total 24 plots were selected.

2.2. Tree Destructive Sampling and Biomass Investigation

In early October 2018, twelve years following the plantation's establishment, the diameter at breast height (1.3 m height above ground, *DBH* in cm) and tree height (*H* in m) of all trees were measured at each plot. For the biomass measurement, one tree (selected on the basis of mean diameter) was identified and harvested for destructive sampling in the selected plots, and in total 24 sample trees were harvested. Trees were cut at ground level and felled onto a large canvas. Each sample tree was divided into three components—stem, branches, and leaves—and fresh weights of all components were determined in the field; sub-samples were collected for moisture and carbon (C) analysis. The fresh biomass of branches and leaves were sampled and recorded in a stratified manner, where the lower, middle, and upper layers of the tree canopy were divided by equal canopy height. Both woody and non-woody biomass were dried to a constant weight at 60 °C.

Stems were cut at 1.3 m and 3.6 m, and then cut into 2 m long segments in this experiment, and stem samples (stemwood + stembark) were obtained from disks cut at 1.3 m (breast height), 5.6 m, 11.6 m, and 17.6 m (nearly 25%, 50%, and 75% of the total height, respectively) for each sample tree. Total stem biomass was obtained by summing the values for all bolts of each sample tree, while the total aboveground biomass of the sample tree was estimated by summing the values for all components of each sample tree.

2.3. Stem Analysis

Stem analysis was conducted to estimate tree growth dynamics in different treatments, and a total of 24 sample trees were harvested for stem analysis. Tree height was measured again after harvest by the clinometer. A stem disc was taken from each section (0 m, 1.3 m, 3.6 m, 5.6 m, etc.) for analysis of the number and width of tree rings. The wood discs were polished to increase visualization of growth rings [25,26]. North–south and east–west transects were selected from the center of the disc to the edges of the stem and marked with permanent marker, and the number of rings were counted and then averaged for each stem. The number of tree rings of the base discs of the stem analysis trees were counted and compared with the documented age of plantations, in order to validate estimations of tree age based on annual tree rings. Comparison of the number of tree rings between stem discs at different tree heights was used to estimate tree height growth over tree ages.

The tree volume was estimated according to following formula (1) [27]:

$$v = 2.6 \cdot \pi \cdot \frac{d_1^2}{4} + 2 \cdot \pi \cdot \frac{d_2^2 + d_3^2 + \cdots + d_{n-1}^2}{4} + h_{top} \cdot \frac{\pi}{3} \cdot \frac{d_n^2}{4} \tag{1}$$

in which 2.6 and 2 are height of the bolts; d_1 is the diameter at 1.3 m, marked number 1; $d_2, \ldots, d_{n-1}$ are the cross-sectional diameters at every tree height where discs were cut, except for number 1 and the top disc; d_n is diameter of the top disc cut from the stem analysis tree; and h_{top} is height of the top section (the cone section from the last disc to the top of the tree) of the stem analysis tree.

2.4. Modeling of Tree Growth Dynamics

In order to select the most suitable, nonlinear model for fitting and accurately evaluating the growth dynamics of *DBH*, *H*, and tree volume, seven theoretical functions (Table 1) that are widely used in forestry modeling were evaluated based on the data obtained from stem analysis.

Matlab2014a (Math Works Inc., Natick, MA, USA) was used to calculate nonlinear regression, and the coefficient of determination (R^2), residual sum of squares (RSOS), root mean square error (RMSE), and Akaike's information criterion (AICc) were calculated simultaneously (Table S1) [28]. Since the models are nonlinear, it is helpful to use RMSE and RSOS as a measure of accuracy, together with R^2. Since the models we test vary from one to four parameter models, the AICc is an important measurement for comparison and model evaluation. In addition, growth dynamics of the poplar plantations generally followed the three growth stages of "slow–fast–slow". Then, seven models were respectively used to take the second derivative of poplar volume to get X_1 and X_3, and the third derivative to get X_2 (the year with the highest periodic annual increment (PAI)); these represent the three turning points of poplar growth. Meanwhile, we calculated the second and third derivatives of the smoothed curve of the stem analysis measurement value, and also got three growth turning points. Finally, we compared these three growth turning points of seven models with the measurement curve (Table S2), and the best growth model was selected to evaluate the tree growth dynamics, based on the lowest error and best biologically meaning.

Table 1. Seven theoretical functions used to fit the diameter at breast height (*DBH*), tree height (*H*), and tree volume growth in the study.

NO.	Function Name	K *	Function *	Reference
1	Mitscherlich	2	$Y(t) = a[1 - \exp(-at)]$	Li et al. [29]
2	Logistic	3	$Y(t) = a/[1 + b\exp(-ct)]$	Yoshimoto [30]
3	Gompertz	3	$Y(t) = a\exp[-b\exp(-ct)]$	Yoshimoto [30]
4	Johnson Schumacher	3	$Y(t) = a\exp[-b/(t+c)]$	Fang et al. [31]
5	Korf	3	$Y(t) = a\exp[-bt^{-c}]$	Li et al. [29]
6	Chapman-Richards	3	$Y(t) = a[1 - \exp(-bt)^c]$	Liu and Li [32]
7	Weibull	4	$Y(t) = a[1 - b\exp(-ct^d)]$	Deng et al. [33]

* K is the number of fixed model parameters; Y is the total tree *DBH* in cm, *H* in m, and volume in m^3; t is the tree age in years, and the fixed model parameters are a, b, and c.

2.5. Carbon Storage

The quantification of carbon storage aboveground of poplar trees was assessed using the samples collected in the destructive assessment. We collected seven samples for each sampling tree, including three leaf samples, three branch samples, and one stem sample, all collected at the tree height of 11.6 m, and a total 168 samples were collected for carbon concentration measurement. After measuring the water content, all samples were macerated in a pulverizer with a 270-mesh sieve, and then placed into sealed sacks. The carbon concentration was measured by using a universal Elementar analyzer (Vario micro cube, Germany). Samples with 0.5 mg were allocated into tin capsules and completely incinerated at 1200 °C [34]. The mean carbon storage of the sample tree for each treatment was calculated using three replications, and each sample tree was obtained from the biomass of each component (leaf, branch, and stem) and the average carbon concentrations in that component. The carbon storage was calculated using the mean carbon concentration of the sample trees for each treatment, multiplying it with their standing biomass per hectare.

2.6. Optimal Rotation Age

Determining the optimal rotation age of the plantations is a key technique in plantation management practices. During this practice, the quantitative maturity age is mainly considered, where the periodic annual increment (PAI) is equal to mean annual increment (MAI) of the sample trees' volume growth.

2.7. Statistical Analysis

Data are reported as the mean ± standard error (SE), and all tests were statistically analyzed with the software Matlab2014a and SPSS19.0 (SPSS Inc., Chicago, IL, USA). One-way analysis of variance (ANOVA) was used to examine the significant differences in aboveground biomass, carbon storage of different clones, and planting spacings. We used Tukey tests to examine differences between means with significant results. Two-way ANOVA was performed to determine the significant dependence of *DBH*, *H*, and tree volume on clones, planting spacings, and their interactions. All statistical tests were considered significant at $p < 0.05$.

3. Results

3.1. Tree Growth and Stand Volume

Significant ($p < 0.05$) variations in *DBH*, *H*, tree volume, and stand volume were observed in the plantations at 13 years of age (Figure 1 and Table 2). *DBH* and *H* of clone NL-895 and the tree volume of clones NL-895 and NL-95 were shown to have significant differences between different planting spacings, while stand volume did not show any significant differences between planting spacings of the three clones. *DBH*, *H*, and the tree volume of clones were consistently higher in the large spacings (6 × 6 m and 4.5 × 8 m) than the narrow spacings (5 × 5 m and 3 × 8 m).

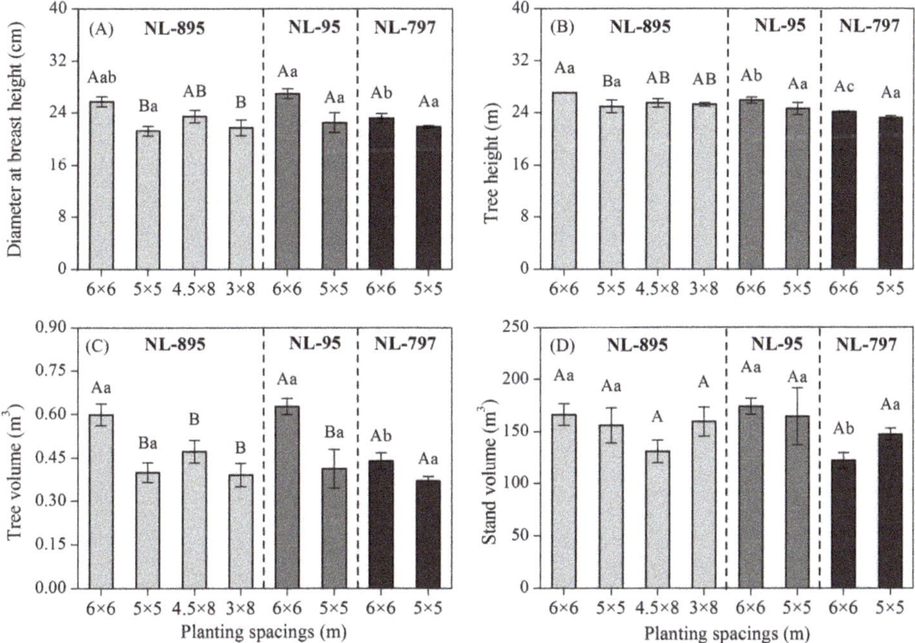

Figure 1. Effects of poplar clones and planting spacings on diameter at breast height (*DBH*, **A**), tree height (**B**), tree volume (**C**), and stand volume (**D**) in 13-year old poplar plantations. Different capital letters indicate significant differences ($p < 0.05$) by Tukey test between planting spacings in each clone, while different small letters indicate significant differences ($p < 0.05$) by Tukey test between clones within each planting spacing. The error bars indicate the standard error.

There was no significant interactive effect of clone and planting spacing over 13 years of poplar growth ($p > 0.05$; Table 2). Stem analysis data showed that initially, clone and planting spacing did not show any significant effects on *DBH*, *H*, and tree volume; however, these parameters varied significantly with increasing age. Planting spacing showed a significant effect on *DBH* and tree volume growth after six and five growing seasons respectively, whereas no significant effects on tree height growth were detected in most cases (Table 2). However, both poplar clone and planting spacing showed a significant effect on tree volume after the six growing seasons.

Table 2. *F*-values of two-way ANOVA for effects of different clone and planting spacing on poplar growth indicators at the stand age of 13 years.

Index	Treatment	df	Stand Age (years)												
			1	2	3	4	5	6	7	8	9	10	11	12	13
Diameter	Clone (C)	2	0.35	0.11	0.34	0.58	0.72	1.37	2.40	2.72	3.25	3.55	**3.95 ***	**4.12 ***	**3.79 ***
At breast height	Spacing (S)	3	0.86	0.76	0.74	0.54	1.25	2.75	**3.82 ***	**4.73 ****	**5.89 ****	**7.15 ****	**8.65 ****	**9.17 ****	**8.23 ****
	C × S	2	0.76	1.32	0.56	0.19	0.79	1.09	1.01	1.28	1.34	1.79	2.30	2.54	2.56
Tree height	C	2	0.27	0.82	0.79	0.48	1.27	0.79	0.70	1.32	1.60	**5.85 ***	**13.41 ***	**9.93 ***	**8.89 ***
	S	3	0.57	0.15	0.41	0.64	1.75	3.14	1.33	0.64	0.74	3.18	3.16	3.09	**3.66 ***
	C × S	2	0.27	0.85	0.42	1.30	0.84	0.97	0.77	0.67	0.21	0.67	2.10	0.96	0.55
Tree volume	C	2	2.04	0.09	0.26	0.49	1.19	2.36	**4.08 ***	**4.92 ***	**5.34 ***	**6.26 ***	**7.21 ***	**7.10 ***	**6.33 ***
	S	3	2.77	0.51	0.44	0.47	1.28	**3.25 ***	**4.87 ***	**6.81 ***	**8.74 ***	**10.72 ***	**12.83 ***	**12.77 ***	**11.229 ***
	C × S	2	2.05	1.18	0.90	0.46	0.55	0.82	0.92	1.30	1.49	2.04	2.68	2.80	2.73

* $0.01 < p < 0.05$; ** $p < 0.01$; significant results ($p < 0.05$) are shown in bold.

3.2. Biomass Production and Partitioning

There existed a significant difference in aboveground tree biomass between different spacings of clones NL-895 and NL-95 ($p < 0.05$; Figure 2A). The higher tree biomass accumulated in the large spacings (6 × 6 m and 4.5 × 8 m) for each poplar clone; this biomass was 45.42~65.51% higher than that in the narrow spacings (5 × 5 m and 3 × 8 m). The aboveground biomass of clones NL-895, NL-95 and NL-797 were 344.18 kg tree^{-1}, 337.71 kg tree^{-1}, and 249.04 kg tree^{-1} in the 6 × 6 m spacing stand, which was 63.75%, 56.30%, and 24.85% higher than in the 5 × 5 m spacing, respectively. The aboveground biomass of NL-895 and NL-95 was 37.85% and 35.60% higher than NL-797 in the 6 × 6 m spacing stand ($p < 0.05$), respectively, but no significant differences were observed between the three clones in the 5 × 5 m spacing stand, suggesting that the difference of aboveground biomass between three clones tend to be reduced as the planting spacing decreases.

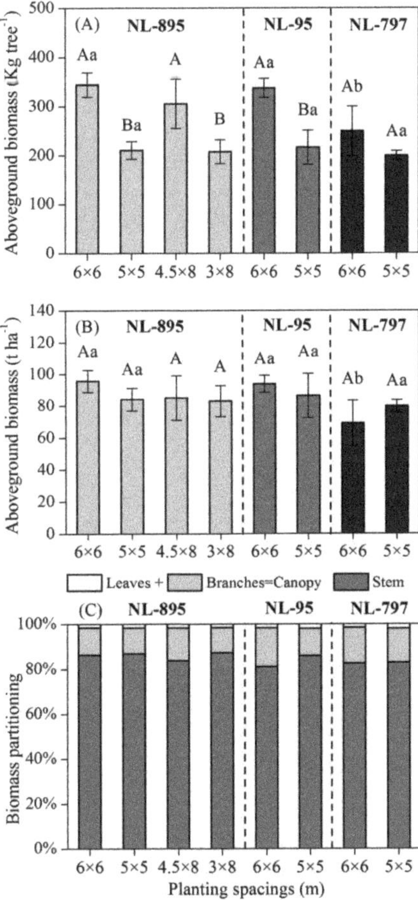

Figure 2. Effects of poplar clones and planting spacings on aboveground biomass of per sample tree (**A**) and per hectare (**B**) in 13-year-old poplar plantations. Different capital letters indicate significant differences ($p < 0.05$) by Tukey test between planting spacings for each clone, and different small letters indicate significant differences ($p < 0.05$) by Tukey test between clones within each planting spacing. Last is the biomass partitioning of poplar in eight treatments (**C**). The error bars indicate the standard error.

The total aboveground biomass of clones NL-895, NL-95, and NL-797 were not significantly ($p > 0.05$) influenced by planting spacing (Figure 2B). A significant variation in total aboveground biomass was observed among the three clones in the 6×6 m spacing stand ($p < 0.05$), but not in the 5×5 m spacing stand ($p > 0.05$; Figure 2B). The total aboveground biomass production in the 6×6 m spacing stand reached 95.68 t ha^{-1} for NL-895, 93.88 t ha^{-1} for NL-95, and 69.23 t ha^{-1} for NL-797, respectively. Contrary to the clones NL-895 and NL-95, the total aboveground biomass of clone NL-797 in the 5×5 m spacing stand was higher than 6×6 m, indicating that the stem number of per hectare was the main driver of total aboveground biomass production in clone NL-797.

The partitioning of biomass in leaves, branches, and stems was also influenced by clone and planting spacing (Figure 2C). Most of the aboveground biomass was allocated to stem production, but more biomass was partitioned to canopy biomass (leaves and branches) in the 6×6 m spacing when compared to the 5×5 m. The percentage of canopy biomass averaged across three poplar clones was 12.88% greater in the 6×6 m spacing stand than the 5×5 m one, while the percentage of canopy biomass averaged between the 6×6 m and 5×5 m spacings for clones NL-895, NL-95, and NL-797 was 13.25%, 16.44%, and 17.25%, respectively.

As shown in Figure 3, poplar clones and planting spacings significantly affected the vertical biomass distribution in the canopy ($p < 0.05$). The biomass of all tree components in 6×6 m and 4.5×8 m spacings was higher than that in the spacings of 5×5 m and 3×8 m in the same segment or canopy layer (Figure 3). For three clones, the greatest biomass of the branches was observed in the middle layer at any planting spacings, followed by the lower and upper layers (Figure 3E–H), while the biomass of the leaves was concentrated in the middle and upper layers (Figure 3I–L).

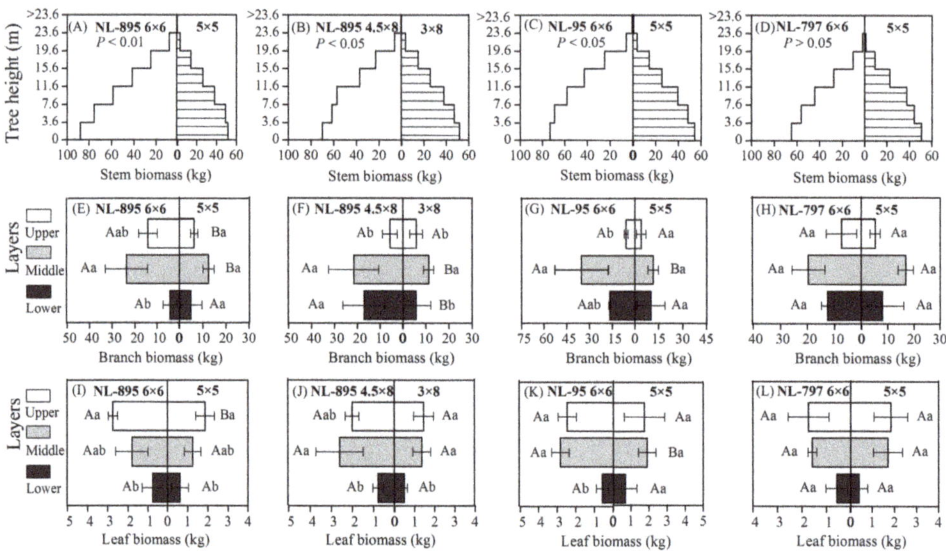

Figure 3. Effects of poplar clones and planting spacings on the vertical biomass distribution in stem (**A–D**), branches (**E–H**), and leaves (**I–L**) for the 13-year-old trees. Different capital letters indicate significant differences ($p < 0.05$) by Tukey test between the two spacing treatments at the same layer. Different small letters indicate significant differences ($p < 0.05$) by Tukey test between the lower, middle, and upper layers of crown.

3.3. Carbon Concentration and Storage

The average carbon concentration of the different components was in the order of stem (46.66%) > branch (46.34%) > leaf (40.96%) (Table 3). ANOVA revealed significant differences between the values of carbon concentration among these three components ($p < 0.05$). However, the poplar clones and planting spacings did not significantly affect carbon concentrations in the branches, whereas the leaf carbon concentration in clone NL-797 was significantly higher than that in clone NL-895 (Table 3).

Table 3. The average carbon concentrations of different biomass components from the poplar plantations with various treatments (%).

Poplar Clone	Planting Spacing (m)	Biomass Components		
		Leaf	Branch	Stem
NL-895	6 × 6	39.79 ± 0.14 Cb	46.11 ± 0.28 Ba	47.76 ± 0.25 Aa
	5 × 5	39.85 ± 0.23 Bb	46.08 ± 0.29 Aa	46.69 ± 0.26 Aa
	4.5 × 8	40.11 ± 0.41 B	46.01 ± 0.25 A	46.88 ± 0.09 A
	3 × 8	40.92 ± 0.58 B	46.41 ± 0.39 A	46.66 ± 0.27 A
NL-95	6 × 6	41.23 ± 0.91 Bab	46.15 ± 0.4 Aa	46.01 ± 0.21 Ab
	5 × 5	41.11 ± 0.64 Bab	46.4 ± 0.19 Aa	46.09 ± 0.08 Aa
NL-797	6 × 6	42.35 ± 0.4 Ba	46.88 ± 0.18 Aa	46.66 ± 0.09 Ab
	5 × 5	42.29 ± 0.31 Ba	46.67 ± 0.25 Aa	46.56 + 0.21 Aa

Note: different capital letters indicate significant differences ($p < 0.05$) between biomass components in each treatment, while different small letters indicate significant differences ($p < 0.05$) between clones for each same planting spacing.

The carbon storage in the stems, branches, and leaves was 84.60%, 13.84%, and 1.56% at the stand age of 13 respectively, where the values were averaged in all treatments (Figure 4). However, there was a significant difference in aboveground carbon storage among the plantations of four planting spacings for NL-895 ($p < 0.05$), and the highest carbon storage was obtained in the NL-895 plantation of 6 × 6 m spacing, reaching 45.39 Mg C ha^{-1}, which was 23.81% higher than that in the 3 × 8 m spacing stand. Also, a significant variation in carbon storage was also observed among the three clones in the 6 × 6 m spacing ($p < 0.05$), and the carbon storage of clone NL-895 was 40.50% higher than that of clone NL-797 in the 6 × 6 m spacing. It is worth noting that carbon storage in the NL-895 plantations in square configurations (5 × 5 m and 6 × 6 m) was higher than in those with rectangular configurations (3 × 8 m, 4.5 × 8 m) at equal or similar planting density (Figure 4).

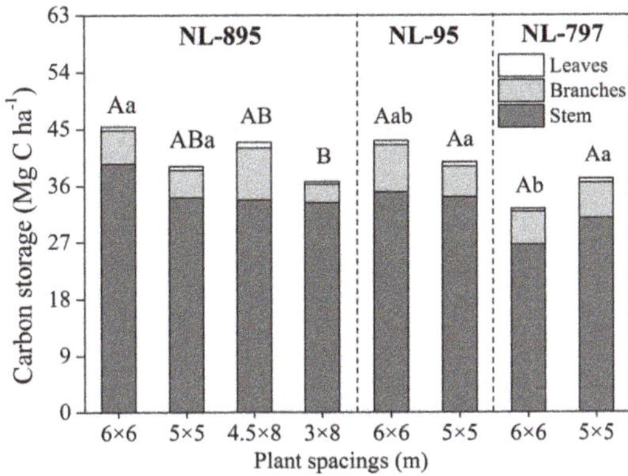

Figure 4. Effects of poplar clones and planting spacings on aboveground carbon storage in 13-year-old poplar plantations. Different capital letters indicate significant differences ($p < 0.05$) by Tukey test between spacings for each clone, while different small letters indicate significant differences ($p < 0.05$) by Tukey test between clones within each planting spacing.

3.4. Quantitative Maturity Age

Based on the values of R^2, RSOS, RMSE, and AICc for the seven growth models (Table S1) and comparison of the prediction results from the models and measured data (Table S2), the Chapman–Richards model was selected as the best function to predict poplar growth dynamics for each poplar clone at each planting spacing in the present study. According to the established Chapman–Richards model, the curves of tree volume and stand volume growth are fitted to predict the tree growth of different treatments (Figures 5 and 6). After 13 growing seasons, the highest tree volume growth was predicted in the NL-95 plantation with 6 × 6 m spacing, followed by the NL-895 plantation of 6 × 6 m spacing; the lowest growth appeared in the NL-797 plantation of 5 × 5 m spacing (Figure 5A). However, the highest stand volume was estimated in the NL-95 plantation of 6 × 6 m spacing, followed by the NL-95 plantation of 5 × 5 m spacing, while the lowest growth was detected in NL-797 plantation of 6 × 6 m spacing (Figure 5B).

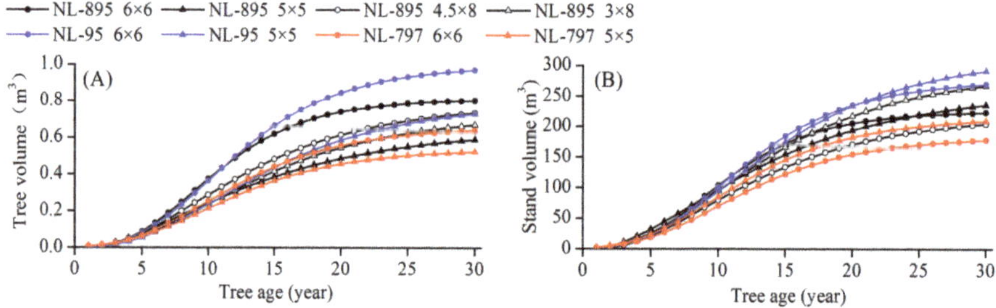

Figure 5. Predictions of tree volume (**A**) and stand volume (**B**) in different treatments of poplar clones and planting spacings by the fitted Chapman–Richards model (up to 30 years).

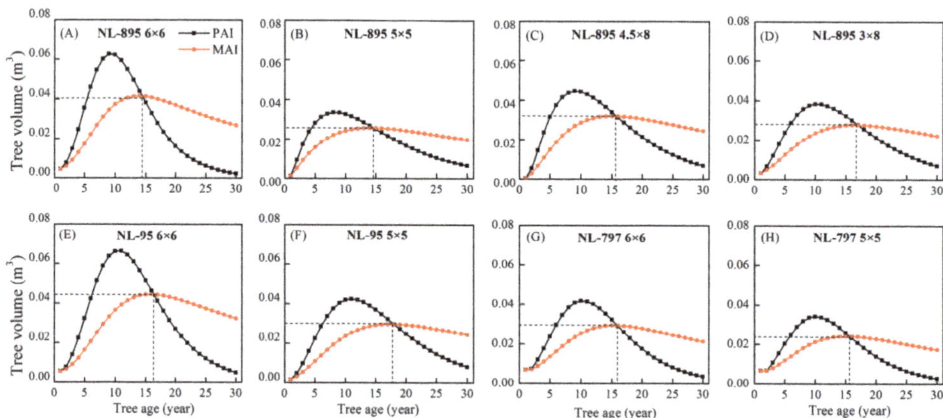

Figure 6. The mean annual increment (MAI) and periodic annual increment (PAI) of sample trees of four planting spacings of NL-895 (**A–D**) and two planting spacings of NL-95 (**E,F**) and NL-797 (**G,H**), which were calculated using Chapman–Richards model prediction data (up to 30 years).

In term of mean annual increment (MAI) and periodic annual increment (PAI) predicted from sample trees, the quantitative maturity ages in the plantations of different treatments ranged from 14 to 17 years (Figure 6), indicating that poplar clones and planting spacings had effects on the quantitative maturity ages. For example, the quantitative maturity age in the NL-895 plantation with 6 × 6 m spacing was 14 years old, while the maturity age was 16 years old in the NL-95 plantation with 6 × 6 m spacing.

4. Discussion

Poplar plantations are gaining increased attention as a source of timber production for producing wood products, such as plywood, pulp, and paper, in the world [12,35]. However, plywood, veneer, and fiberboard are the most important outlet for poplar plantations. Previous studies have indicated differences in survival, growth, and productivity between different poplar genotypes (clones) and planting densities under various growing conditions [11,36,37]. It has also been demonstrated that the optimal rotation length of poplar plantations can vary depending on the requirements of the particular target wood product [12,38,39]. Some studies have reported that wood density, microfibril angle, cellulose content, fiber diameter, and the ratio of fiber length to diameter were not only significantly affected by poplar clones and planting densities, but also by the stand ages [39,40]. As a general tendency, basic wood density, fiber length, fiber diameter, and cellulose content showed an increasing tendency along the direction from pith to bark among the growth rings, and a general increasing trend with the increasing height of trees [39–42]. However, after eight growing seasons, these wood properties reached relative stability [39,40], which means that the technical maturity of poplar plantations is about eight years for plywood and fiberboard. Furthermore, many studies have also shown that the static modulus of elasticity (MOE) and modulus of rupture (MOR) were lower for laminated veneer lumber (LVL) made from juvenile veneers than for LVL made from mature veneers [43,44]. Therefore, adopting a silvicultural regime of longer rotation would improve the mechanical properties of LVL. The possible reasons are these LVL properties are affected by wood density and fiber length [42,45].

The optimal rotation length in poplar plantations should be determined based on technical maturity and quantitative maturity. For example, our results indicate that the quantitative maturity age of clones NL-895 and NL-95 in the 6 × 6 m spacing was 14 years and 16 years, with stand volumes reaching 162.95 m^3 ha^{-1} and 200.19 m^3 ha^{-1}, respectively. However, according to the requirement of *DBH* for plywood timber production (*DBH* ≥ 26.0 cm) [46], only the clone NL-95 in the 6 × 6 m spacing was observed approximately to achieve the standard of technical maturity for the plywood target. Considering the quantitative maturity age and technical maturity age, a longer rotation age of poplar plantations was observed at the study site when compared to the results from other studies. Huang et al. [38] observed that the rotation age was 11 years old in 5 × 6 m planting spacing, and Tang [47] reported that the quantitative mature age was 13 years old in planting spacing of 4 × 6 m. The reasons probably are due to the interplay between genotypes and site conditions, which largely affects the growth rate of poplar plantations [9,15]. In addition, Tun et al. [22] suggested that better stem roundness was found in the square configurations. Overall, we suggest that the clone NL-95 with 6 × 6 m spacing would be selected for high-value, larger diameter timber production, and its optimal rotation length for plywood timber should be about 20 years at the research site. Notably, the economic maturity age, as an important factor to decide the optimal rotation age, should be carefully considered in the future research.

Forest growth plays a key role in the carbon balance of terrestrial ecosystems [48,49]. However, carbon concentration is generally assumed to be constant in carbon storage assessments, and the variation in carbon concentration between the tree components was neglected in a large-scale study [50]. Our results confirmed that the carbon concentration of tree components was significant different, in the order of stem > branch > leaf. However, the carbon concentration showed less correlation with poplar clones and planting spacings (Table 3), which is in agreement with the results from [51,52]. However, the carbon concentrations from the present study are lower than those from other poplar plantations [51,53], where the leaf carbon concentration of the poplar was over 42.9%. The possible reason is that poplar genotype, site condition, geographical area, stand age, and sampling time would create this difference in carbon content. However, some studies have indicated that plant carbon concentration was in the range of 24.95–55.44%, with an average of 43.63%, while the average carbon concentration for different components was flower (48.52%) > fruit (47.19%) > branch (45.42) > stem (44.48%) > leaf (43.36%) > root (42.88%) [54,55]. Furthermore, the average carbon concentration was in the order of high latitude area (50.30%) > low latitude area (45.30%) > middle latitude area

(39.68%) [54]. Our research site was located in the middle latitude area, and carbon concentrations were within the range. Thereby, the biomass partitioning of poplar trees can alter carbon storage in the poplar plantations to some extent, and play an important role in forest ecosystem carbon cycling [49]. Our results (Figure 2) agree with the well-known fact that growth performance and biomass partitioning of poplar plantations are affected by clones and planting spacings [9,15,22]. For example, poplar plantations of three clones at low density showed higher leaves and branches biomass, probably due to the light availability, which likely stimulated growth and production of branches and leaves [56]. However, the branch biomass for all tested clones was the highest in the middle layer, followed by lower and upper layers, while the leaf biomass was concentrated in the middle and upper layers (Figure 3), which is beneficial for improving solar radiation interception efficiency under the inter-tree competition [57].

Aboveground carbon storage of poplar plantations was mainly determined by productivity and biomass partitioning. The present study shows that the *DBH*, *H*, tree volume, and aboveground biomass of the three clones for a single tree was enhanced as planting spacing increased, and the biomass in the NL-895 plantations of square configurations (5 × 5 m and 6 × 6 m) was higher than those with rectangular configurations (3.0 × 8.0 m and 4.5 × 8.0 m) at equal or similar planting densities. However, both stand volume and total aboveground biomass per area for clone NL-797 were higher in the 5 × 5 m spacing stand than that in the 6 × 6 m spacing, contrary to that of clones NL-895 and NL-95. This presumably is due to a lower growth rate in clone NL-797 compared to clones NL-895 and NL-95, as well as the stem number per hectare being the main driver of aboveground biomass accumulated at the measured period.

Generally, the variation of carbon storage in clones NL-895, NL-95, and NL-797 was consistent with the total aboveground biomass (Figures 2 and 4). In the current research, stems and branches accounted for 84.6% and 13.84% of the total aboveground biomass, respectively, which are congruent with previous reports that most biomass accumulates in stem and branches. Additionally, in different planting spacings, the proportion of leaf and branch biomass decreases with forest age, competition, and resource availability [34,51]. This means, thus, that rotation length affects the carbon storage in the plantations, and that the stems are the main component of aboveground carbon storage with regard to prolonging the rotation. Based on the results of PAI in the present study, we speculate that the carbon storage of poplar plantations would still increase after reaching the quantitative maturity age (Figure 6), and suggest that the optimal rotation length for carbon sequestration should be longer than that of the quantitative maturity age.

5. Conclusions

In conclusion, poplar clones and planting spacings significantly affected the growth, biomass accumulation, biomass partitioning, and aboveground carbon storage in the poplar plantations, while the NL-895 plantations with the square configurations produced higher biomass than the plantations with the rectangular configurations at equal or similar planting densities. Poplar clones and planting spacings also significantly affected the vertical biomass distribution in the canopy. More biomass was partitioned to canopy biomass (leaves and branches) in the plantations with wide spacing when compared to the narrow ones. However, the greatest biomass of the branches was observed in the middle layer, whereas the biomass of the leaves was concentrated in the middle and upper layers. Based on the growth prediction of the established Chapman–Richards model, the quantitative maturity ages of stand volume varied among the investigating plantations, ranging from 14 to 17 years old. Our results suggest that selecting clones NL-895 and NL-95 with 6 × 6 m spacing would be recommended for future poplar silviculture of larger diameter timber production, and the optimal rotation length for plywood timber and carbon sequestration should be about 20 years at similar sites.

Supplementary Materials: The following are available online at http://www.mdpi.com/1999-4907/11/8/842/s1, Table S1: Parameters of poplar *DBH*, *H*, and tree volume models with different treatments; Table S2: Comparison of models and measured data of poplar tree volume for smoothed data set, X_1, X_2, and X_3.

Author Contributions: S.F. and Y.T. conceived and designed this experiment; Y.Z., S.D., Y.L., and W.S. conducted field investigations and laboratory work; Y.Z. conducted data analysis and wrote the initial draft of the manuscript, and S.F. revised the draft. All authors contributed to the revising of the final version of the manuscript. The authors approved the final version for publication, and agree to be held accountable for the content therein. All authors have read and agreed to the published version of the manuscript.

Acknowledgments: The study described in this paper was supported financially by the National Key Research and Development Project (2016YFD0600402). We thank Linlin Wang and Jian Qin for assistance with the carbon–nitrogen analyzer, and Guangzhen Qin and Xiliang Yue for their assistance with the laboratory work.

Conflicts of Interest: The authors declare no conflict of interest.

References

1. Jürgensen, C.; Kollert, W.; Lebedys, A. *Assessment of Industrial Roundwood Production from Planted Forests*; Planted Forests and Trees Working Papers (FAO) eng no. FP/48/E; FAO: Rome, Italy, 2014.

2. Brown, S. Measuring carbon in forests: Current status and future challenges. *Environ. Pollut.* **2002**, *116*, 363–372. [CrossRef]

3. Forest Resources Management Division. *National Forest Resources Statistics-Main Results of the Eighth National Forest Resources Inventory in China*; State Forestry Bureau: Beijing, China, 2015.

4. Forest Resources Management Division. *National Forestry and Grassland Development Statistical Bullet in 2018*; State Forestry Bureau: Beijing, China, 2019.

5. Keenan, R.J.; Reams, G.A.; Achard, F.; de Freitas, J.V.; Grainger, A.; Lindquist, E. Dynamics of global forest area: Results from the FAO global forest resources assessment 2015. *Ecol. Manag.* **2015**, *352*, 9–20. [CrossRef]

6. Malhi, Y.; Meir, P.; Brown, S. Forests, carbon and global climate. *Philos. Trans. A Math. Phys. Eng. Sci.* **2002**, *360*, 1567–1591. [CrossRef]

7. Nassi, O.D.N.; Guidi, W.; Ragaglini, G.; Tozzini, C.; Bonari, E. Biomass production and energy balance of a 12-year-old short-rotation coppice poplar stand under different cutting cycles. *GCB Bioenergy* **2010**, *2*, 89–97. [CrossRef]

8. Hopmans, P.; Stewart, H.T.L.; Flinn, D.W. Impacts of harvesting on nutrients in a eucalypt ecosystem in southeastern Australia. *Ecol. Manag.* **1993**, *59*, 29–51. [CrossRef]

9. Swamy, S.L.; Mishra, A.; Puri, S. Comparison of growth, biomass and nutrient distribution in five promising clones of *Populus deltoides* under an agrisilviculture system. *Bioresour. Technol.* **2006**, *97*, 57–68. [CrossRef] [PubMed]

10. Piao, S.; Friedlingstein, P.; Ciais, P.; Viovy, N.; Demarty, J. Growing season extension and its impact on terrestrial carbon cycle in the Northern Hemisphere over the past 2 decades. *Glob. Biogeochem. Cycle* **2007**, *21*, GB3018. [CrossRef]

11. Guo, X.Y.; Zhang, X.S. Performance of 14 hybrid poplar clones grown in Beijing, China. *Biomass Bioenergy* **2010**, *34*, 906–911. [CrossRef]

12. Fortier, J.; Gagnon, D.; Truax, B.; Lambert, F. Biomass and volume yield after 6 years in multiclonal hybrid poplar riparian buffer strips. *Biomass Bioenergy* **2010**, *34*, 1028–1040. [CrossRef]

13. Benomar, L.; DesRochers, A.; Larocque, G.R. The effects of spacing on growth, morphology and biomass production and allocation in two hybrid poplar clones growing in the boreal region of Canada. *Trees* **2012**, *26*, 939–949. [CrossRef]

14. Khan, G.S.; Chaudhry, A.K. Effect of spacing and plant density on the growth of poplar (*Populus deltoides*) trees under agro-forestry system. *Pak. J. Agric. Sci.* **2007**, *44*, 321–327.

15. Nelson, N.D.; Berguson, W.E.; McMahon, B.G.; Cai, M.; Buchman, D.J. Growth performance and stability of hybrid poplar clones in simultaneous tests on six sites. *Biomass Bioenergy* **2018**, *118*, 115–125. [CrossRef]

16. Bogeat-Triboulot, M.; Brosché, M.; Renaut, J.; Jouve, L.; Le Thiec, D.; Fayyaz, P.; Vinocur, B.; Witters, E.; Laukens, K.; Teichmann, T. Gradual soil water depletion results in reversible changes of gene expression, protein profiles, ecophysiology, and growth performance in *Populus euphratica*, a poplar growing in arid regions. *Plant Physiol.* **2007**, *143*, 876–892. [CrossRef] [PubMed]

17. Pinno, B.D.; Thomas, B.R.; Bélanger, N. Predicting the productivity of a young hybrid poplar clone under intensive plantation management in northern Alberta, Canada using soil and site characteristics. *New For.* **2010**, *39*, 89–103. [CrossRef]

18. Li, Y.C.; Li, C.; Li, M.Y.; Liu, Z.Z. Influence of variable selection and forest type on forest aboveground biomass estimation using machine learning algorithms. *Forests* **2019**, *10*, 1073. [CrossRef]

19. Ceulemans, R.; Deraedt, W. Production physiology and growth potential of poplars under short-rotation forestry culture. *Ecol. Manag.* **1999**, *121*, 9–23. [CrossRef]

20. Czapowskyj, M.M.; Safford, L.O. Site preparation, fertilization, and 10-year yields of hybrid poplar on a clearcut forest site in eastern Maine, USA. *New For.* **1993**, *7*, 331–344.

21. Dickmann, D.I.; Isebrands, J.G.; Eckenwalder, J.E.; Richardson, J. *Poplar Culture in North America*, 1st ed.; NRC Research Press: Ottawa, ON, Canada, 2002; ISBN 0660189887.

22. Tun, T.N.; Guo, J.; Fang, S.Z.; Tian, Y. Planting spacing affects canopy structure, biomass production and stem roundness in poplar plantations. *Scand. J. Res.* **2018**, *33*, 464–474. [CrossRef]

23. Fang, S.Z.; Xu, X.Z.; Lu, S.X.; Tang, L.Z. Growth dynamics and biomass production in short-rotation poplar plantations: 6-year results for three clones at four spacings. *Biomass Bioenergy* **1999**, *17*, 415–425. [CrossRef]

24. Yan, Y.F.; Fang, S.Z.; Tian, Y.; Deng, S.P.; Tang, L.Z.; Chuong, D.N. Influence of tree spacing on soil nitrogen mineralization and availability in hybrid poplar plantations. *Forests* **2015**, *6*, 636–649. [CrossRef]

25. Menezes, M.; Berger, U.; Worbes, M. Annual growth rings and long-term growth patterns of mangrove trees from the Bragança peninsula, North Brazil. *Wetl. Ecol. Manag.* **2003**, *11*, 233–242. [CrossRef]

26. Weiskittel, A.R.; Hann, D.W.; Kershaw, J.A.; Vanclay, J.K. *Forest Growth and Yield Modeling*, 1st ed.; John Wiley and Sons: Manhattan, NY, USA, 2011; ISBN 978-0-470-66500-8.

27. Phan, S.M.; Nguyen, H.T.T.; Nguyen, T.K.; Lovelock, C. Modelling above ground biomass accumulation of mangrove plantations in Vietnam. *Ecol. Manag.* **2019**, *432*, 376–386. [CrossRef]

28. Burnham, K.P.; Anderson, D.R. Multimodel inference: Understanding AIC and BIC in model selection. *Sociol. Method Res.* **2004**, *33*, 261–304. [CrossRef]

29. Li, F.R.; Zhao, B.D.; Su, G.L. A derivation of the generalized Korf growth equation and its application. *J. For. Res.* **2000**, *11*, 81–88.

30. Yoshimoto, A. Application of the logistic, gompertz, and richards growth functions to gentan probability analysis. *J. For. Res.* **2001**, *6*, 265–272. [CrossRef]

31. Fang, L.; Liu, S.Q.; Huang, Z.Y. Uncertain Johnson–Schumacher growth model with imprecise observations and k-fold cross-validation test. *Soft Comput.* **2020**, *24*, 2715–2720. [CrossRef]

32. Liu, Z.G.; Li, F.R. The generalized Chapman-Richards function and applications to tree and stand growth. *J. For. Res.* **2003**, *14*, 19–26.

33. Deng, C.; LV, Y.; Lei, Y.C.; Zhang, J. Study on individual tree diameter growth models with the relative diameter as competition indicator. *For. Resour. Manag.* **2011**, 40–43. [CrossRef]

34. Schwerz, F.; Eloy, E.; Elli, E.F.; Caron, B.O. Reduced planting spacing increase radiation use efficiency and biomass for energy in black wattle plantations: Towards sustainable production systems. *Biomass Bioenergy* **2019**, *120*, 229–239. [CrossRef]

35. Balatinecz, J.J.; Kretschmann, D.E.; Leclercq, A. Achievements in the utilization of poplar wood–guideposts for the future. *For. Chron.* **2001**, *77*, 265–269. [CrossRef]

36. Ghezehei, S.B.; Nichols, E.G.; Hazel, D.W. Early clonal survival and growth of poplars grown on North Carolina Piedmont and Mountain Marginal Lands. *Bioenergy Res.* **2016**, *9*, 548–558. [CrossRef]

37. Pliura, A.; Zhang, S.Y.; Mackay, J.; Bousquet, J. Genotypic variation in wood density and growth traits of poplar hybrids at four clonal trials. *Ecol. Manag.* **2007**, *238*, 92–106. [CrossRef]

38. Huang, Q.F.; Sun, Q.X.; Wu, Z.M.; Xiang, Y. A study on principal felling age of poplar plantation on Changjiang River beach. *Sci. Silvae Sin.* **2002**, *39*, 154–158.

39. Fang, S.; Yang, W.Z. Interclonal and within-tree variation in wood properties of poplar clones. *J. Res.* **2003**, *14*, 263–268.

40. Fang, S.Z.; Yang, W.Z.; Fu, X.X. Variation of microfibril angle and its correlation to wood properties in poplars. *J. Res.* **2004**, *15*, 261–267.

41. Debell, D.S.; Singleton, R.; Harrington, C.A.; Gartner, B.L. Wood density and fiber length in young *populus* stems: Relation to clone, age, growth rate, and pruning. *Wood Fiber Sci.* **2002**, *34*, 529–539.

42. De Boever, L.; Vansteenkiste, D.; Van Acker, J.; Stevens, M. End-use related physical and mechanical properties of selected fast-growing poplar hybrids (*Populus trichocarpa* × *P. deltoides*). *Ann. Sci.* **2007**, *64*, 621–630. [CrossRef]

43. Rahayu, I.; Denaud, L.; Marchal, R.; Darmawan, W. Ten new poplar cultivars provide laminated veneer lumber for structural application. *Ann. Sci.* **2015**, *72*, 705–715. [CrossRef]

44. Nazerian, M.; Ghalehno, M.D.; Kashkooli, A.B. Effect of wood species, amount of juvenile wood and heat treatment on mechanical and physical properties of laminated veneer lumber. *J. Appl. Sci.* **2011**, *11*, 980–987. [CrossRef]

45. Haouzali, H.E.; Marchal, R.; Bléron, L.; Kifani-Sahban, F.; Butaud, J. Mechanical properties of laminated veneer lumber produced from ten cultivars of poplar. *Eur. J. Wood Prod.* **2020**. [CrossRef]

46. Jiang, B.; Yuan, W.G.; Qi, L.Z.; Zhu, J.R. Studies on key techniques for culturing large timber of poplar in flood land. *Sci. Silvae Sin.* **2001**, *38*, 68–75.

47. Tang, W.P. Density Control Techniques and Successive Planting Effect of Poplar Plantation in Jianghan Plain. Ph.D. Thesis, Beijing Forestry University, Beijing, China, 2009.

48. Babst, F.; Bouriaud, O.; Papale, D.; Gielen, B.; Janssens, I.A.; Nikinmaa, E.; Ibrom, A.; Wu, J.; Bernhofer, C.; Kostner, B. Above-ground woody carbon sequestration measured from tree rings is coherent with net ecosystem productivity at five eddy-covariance sites. *New Phytol.* **2014**, *201*, 1289–1303. [CrossRef] [PubMed]

49. Litton, C.M.; Raich, J.W.; Ryan, M.G. Carbon allocation in forest ecosystems. *Glob. Chang. Biol.* **2007**, *13*, 2089–2109. [CrossRef]

50. Fang, J.Y.; Chen, A.P.; Peng, C.H.; Zhao, S.Q.; Ci, L.J. Changes in forest biomass carbon storage in China between 1949 and 1998. *Science* **2001**, *292*, 2320–2322. [CrossRef] [PubMed]

51. Fang, S.Z.; Xue, J.N.; Tang, L. Biomass production and carbon sequestration potential in poplar plantations with different management patterns. *J. Environ. Manag.* **2007**, *85*, 672–679. [CrossRef] [PubMed]

52. Hegazy, S.S.; Aref, I.M.; Al-Mefarrej, H.; El-Juhany, L.I. Effect of spacing on the biomass production and allocation in *Conocarpus erectus L.* trees grown in Riyadh, Saudi Arabia. *Saudi J. Biol. Sci.* **2008**, *15*, 315–322.

53. Fang, S.Z.; Li, H.L.; Sun, Q.X.; Chen, L.B. Biomass production and carbon stocks in poplar-crop intercropping systems: A case study in northwestern Jiangsu, China. *Agrofor. Syst.* **2010**, *79*, 213–222. [CrossRef]

54. Zheng, W.J.; Bao, W.K.; Gu, B.; He, X.; LI, L. Carbon concentration and its characteristics in terrestrial higher plants. *Chin. J. Ecol.* **2007**, *26*, 307–313.

55. Demirbas, A. Relationships between lignin contents and fixed carbon contents of biomass samples. *Energy Convers. Manag.* **2003**, *44*, 1481–1486. [CrossRef]

56. Benomar, L.; Desrochers, A.; Larocque, G.R. Changes in specific leaf area and photosynthetic nitrogen-use efficiency associated with physiological acclimation of two hybrid poplar clones to intraclonal competition. *Can. J. Res.* **2011**, *41*, 1465–1476. [CrossRef]

57. Mitchell, C.P. Ecophysiology of short rotation forest crops. *Biomass Bioenergy* **1992**, *2*, 25–37. [CrossRef]

 *forests*

Article

Estimation of Aboveground Oil Palm Biomass in a Mature Plantation in the Congo Basin

Pierre Migolet [1,*], Kalifa Goïta [1], Alfred Ngomanda [2] and Andréana Paola Mekui Biyogo [3]

[1] Centre D'Applications et de Recherches en Télédétection (CARTEL), Département de Géomatique Appliquée, Université de Sherbrooke, 2500 Boulevard de l'Université, Sherbrooke, QC J1K 2R1, Canada; Kalifa.Goita@usherbrooke.ca

[2] Commissariat Général du Centre National de la Recherche Scientifique et Technologique (CENAREST), Libreville, B.P 842 Libreville, Gabon; ngomanda@yahoo.fr

[3] Independent Geomatics Consultant, Former Employee of Olam Palm Gabon, 79 rue du Radoux, 1430 Rebecq, Belgique; pamekui@yahoo.fr or andreana.mekui@olamnet.com

* Correspondence: Pierre.Migolet@usherbrooke.ca

Received: 30 March 2020; Accepted: 11 May 2020; Published: 12 May 2020

Abstract: Agro-industrial oil palm plantations are becoming increasingly established in the Congo Basin (West Equatorial Africa) for mainly economic reasons. Knowledge of oil palm capacity to sequester carbon requires biomass estimates. This study implemented local and regional methods for estimating palm biomass in a mature plantation, using destructive sampling. Eighteen 35-year-old oil palms with breast height diameters (DBH) between 48 and 58 cm were felled and sectioned in a plantation located in Makouké, central Gabon. Field and laboratory measurements determined the biomasses of different tree compartments (fruits, leaflets, petioles, rachises, stems). Fruits and leaflets contributed an average of 6% to total aboveground palm biomass, which petioles accounted for 8%, rachises for 13% and the stem, 73%. The best allometric equation for estimating stem biomass was obtained with a composite variable, formulated as $DBH^2 \times$ stem height, weighted by tissue infra-density. For leaf biomass (fruits + leaflets + petioles + rachises), the equation was of a similar form, but included the leaf number instead of infra-density. The allometric model combining the stem and leaf biomass yielded the best estimates of the total aboveground oil palm biomass (coefficient of determination (r^2) = 0.972, $p < 0.0001$, relative root mean square error (RMSE) = 5%). Yet, the model was difficult to implement in practice, given the limited availability of variables such as the leaf number. The total aboveground biomass could be estimated with comparable results using $DBH^2 \times$ stem height, weighted by the infra-density ($r^2 = 0.961$, $p < 0.0001$, relative RMSE (%RMSE) = 5.7%). A simpler model excluding infra-density did not severely compromise results ($R^2 = 0.939$, $p < 0.0003$, %RMSE = 8.2%). We also examined existing allometric models, established elsewhere in the world, for estimating aboveground oil palm biomass in our study area. These models exhibited performances inferior to the best local allometric equations that were developed.

Keywords: agro-industrial plantations; oil palms; aboveground biomass; allometric equations; Congo Basin; Gabon

---

1. Introduction

The world is becoming increasingly concerned with changes that are occurring in ecosystems and in the climate. These changes are manifested in and exacerbated by forest conversion into agricultural land. This process is a major driver of global deforestation. Results of the Global Forest Resources Assessment [1] have indicated that the total forest area has declined by about 3%, from a worldwide estimate of 4128 Mha in 1990 to 3999 Mha in 2015 [2]. These losses are responsible for increasing greenhouse gas (GHG) emissions and incurring substantial changes in the amounts of carbon that are stored in forested ecosystems [3]. Indeed, a global loss of forest biomass carbon in the order of 11.1 gigatons has been reported over 25 years (from 1990 to 2015); this represents a 3.8% decrease in storage from an initial 296 Gt estimate of the total forest C [1]. The intensification of forest- and agriculture-related economic activities largely explains this change. For example, tropical forested areas are being converted into oil palm plantations to meet economic (biofuel, palm oil), social (food, household products, cosmetics, soap), cultural (traditional medicines and other care products) or scientific needs [4–6]. An increase in these needs has led to the development and expansion of agro-industrial oil palm plantations in tropical Asia, Oceania, Africa and Amazonia [7–9].

Several species of oil palms have been naturalized throughout the world. Currently, the Afrotropical species *Elaeis guineensis* Jacq. (African oil palm) is the palm most extensively cultivated for oil production. The Neotropical *Elaeis oleifera* (Kunth) Cortés (American oil palms), which also produce oil, have been rarely exploited commercially. Hybridization with *E. oleifera* has increased the disease resistance of *E. guineensis*, while improving its biochemical and physiological characteristics [10,11].

These latter species, together with palms within the genera *Euterpe* and *Astrocaryum*, have not been extensively domesticated given that they tend to occur as solitary individuals or small groups within intact tropical forest and in seral stages leading to mature stand canopy closure. Nonetheless, these species are cultivated as food crops (including açai berry and hearts of palm); the small quantities of oil extracted from their edible fruits are used in salads or as ingredients in soaps and cosmetics [12].

Resulting land use changes have exerted effects especially on the capacity of ecosystems to sequester and store carbon in the plants of which they are composed. In the case of oil palms, this capacity may increase or decrease, depending upon the ecosystems that they frequently replace [9,13,14]. In mature forests, oil palms generally cause a loss in the quantity of carbon that is stored, whereas they favor an increase in storage in fallow land and savannah areas [6,14,15]. As a result, this variation in storage can have repercussions for the climatic and environmental equilibria in the affected tropical regions [1]. An important challenge is the ability to exploit and develop agricultural areas, such as agro-industrial plantations, without compromising or damaging the ecological integrity of the broader continuous forest (e.g., [5]). Such actions would help to accentuate the trend towards decelerating deforestation rates that has been noted by the Food and Agriculture Organization (FAO) [1]. Indeed, the loss of tropical forest area declined by 42% to 5.5 Mha year^{-1} during the period 2010–2015, from the estimated 9.5 Mha year^{-1} during the 1990s [2].

To improve attempts in land use redevelopment, it is important to have a better understanding of forest and agricultural biomass stocks, together with their respective spatial and temporal dynamics. The current study developed allometric models to estimate the total aboveground biomass of oil palms in a particular region, namely the Congo Basin. We aimed at determining which components of the total aboveground biomass were most effective and efficient in constructing these allometric equations. Total aboveground biomass is the total dry mass of aboveground organic material that is present in different plant compartments, including the stem, branches, leaves, stumps and bark [16]. The total biomass represents an important carbon storage reservoir within the plant and also constitutes the part that is most vulnerable to human activities and natural perturbations, regardless of whether these effects are acute or chronic.

A number of studies have been conducted to gather data on the total dry aboveground biomass of oil palms in targeted tropical areas of Africa, Amazonia and Asia [4,17–19]. The resulting allometric equations rely upon characteristic attributes of oil palms (and palms, in general), such as diameter, height, wood infra-density (stem dry mass vs. fresh volume), dry mass fraction and the number of leaves or age to estimate the biomass at different stages of tree development (young, mid-mature, mature). These attributes are frequently used as simple [4,20] or as composite [21–23] explanatory variables. In Benin, Thenkabail et al. [20] estimated the aboveground biomass of young oil palm trees (1- to 5-years-old) using stem height. In another study conducted in Benin, Aholoukpè et al. [24] proposed a simple equation for estimating the biomass of oil palm fronds using the dry mass of the rachis, i.e., the axis of the compound leaf. Khalid et al. [4] predicted the biomass for mid-mature

(23-years-old) Malaysian oil palm plantations by considering the total height as the explanatory variable. Saldarriaga et al. [18] developed equations in which the squared value of the DBH (diameter at breast height, measured at 1.3 m), together with the stem height, were used to estimate the plant biomass. In this case, the plants were relatively young, with DBHs ranging between 1 and 10 cm. The same variable combination expressed as $DBH^2 \times$ stem height was used by Hughes et al. [25] in southern Mexico to estimate biomass for wild palms (*Astrocaryum mexicanum* Liebm. ex Mart.) with DBHs < 10 cm. Cole and Ewel [26] also considered the same variable combination for four economically valuable forest species, but also included the leaf count in the estimation of the aboveground biomass in açai palm (*Euterpe oleracea* Mart.) plantations (DBH < 20 cm) in the Atlantic Lowlands of Costa Rica. Goodman et al. [27] studied the allometric relationships of nine species in the Arecaceae, including *Attalea phalerata* Mart. ex Sprung, which is a source of vegetable oil [28]. These authors substituted the dry-matter fraction for the leaf count in palm plantations covering all the stages of development (DBH between 4 and 50 cm) in Amazonian Peru, while Da Silva et al. [22] considered the stem infra-density in the case of young (DBH 3–13 cm) forest açai or *açaí-solitário* (*Euterpe precatoria* Mart.) in Amazonian Brazil.

All of these studies have provided various explanatory variables for estimating the aboveground biomass of palms worldwide, including those that produce oil. The performances of these different variables have yet to be compared in the same study. Furthermore, the allometric equations of these studies have not been compared with local allometric models for estimating the aboveground biomass of oil palms that were established in the Congo Basin (West Equatorial Africa). The oil palms that were used in these studies were consistently young, with a few cases of semi-mature and mature individuals. However, there is little research on mature individuals (>30-years-old).

Research estimating oil palm biomass is relatively sparse for the Congo Basin. Considering the gradual but relentless establishment of oil palm plantations, it is crucial that estimation methods be developed and comparative analyses be conducted relative to other tropical regions. Thus, the current study sought (1) to evaluate the attributes that were most relevant for characterizing the aboveground biomass of oil palm plantations in the Congo Basin, and (2) to develop the necessary allometric equations and compare their results with those that have been obtained from other tropical areas. This study was based upon field measurements that were acquired through destructive sampling in agro-industrial oil palm plantations operated by the Société Olam Palm Gabon.

2. Materials and Methods

2.1. Study Area

The oil palms were sampled in a plantation operated by the Société Olam Palm Gabon, which is located in the Makouké district, Moyen Ogooué Province, central Gabon. The region is characterized by a hot and humid equatorial climate, with two rainy seasons and two dry seasons. The plantation adjoins the Ogooué River, which is the largest river in Gabon. The temperatures in the area range from 27 °C to 38 °C and precipitation reaches 1800 to 2000 mm annually. The plantation is located on a ferralitic Cambisol-dominant soil. Established in 1981, it is the oldest plantation of oil palm in Gabon. It currently covers 5700 ha, with an average density of 134 oil palm trees per hectare. Mature oil palms represent an area of about 1500 ha. The plantation is divided into blocks. One of these blocks was the study area, located between longitudes 10°24′27″ E and 10°24′57″ E, and latitudes 0°30′06″ S and 0°30′16″ S (Figure 1). The block considered covered an area of 25 ha (1000 m × 250 m). The palm trees inside the block were mature and were planted in both dry and flooded areas (Figure 1).

In 2019, the plantation obtained "Roundtable on Sustainable Palm Oil" (RSPO) certification for its efforts to protect the environment and encourage sustainable development by complying with global standards for sustainable palm oil. The RSPO certification committed Olam Palm Gabon to respecting several principles and criteria for managing the plantation, including among others (1) reducing deforestation, (2) encouraging the responsible use of agro-chemical products in the production system and (3) avoiding conflicts between local communities and plantation owners. The implementation of all these actions contributed to the sustainable exploitation of oil palm trees. The company's commitment to not convert primary forests, areas of high conservation value, peatlands as well as land belonging to local communities into plantations, considerably limits deforestation.

Although located in the heart of the tropical forest in the Congo Basin, the area where the stuied plantation was positioned cannot be considered as fully representative of the conditions prevailing throughout the whole basin, which is a very large region. Nevertheless, it reflected certain similar natural aspects, notably in terms of tropical forest, biodiversity, climate, soil type and hydrography.

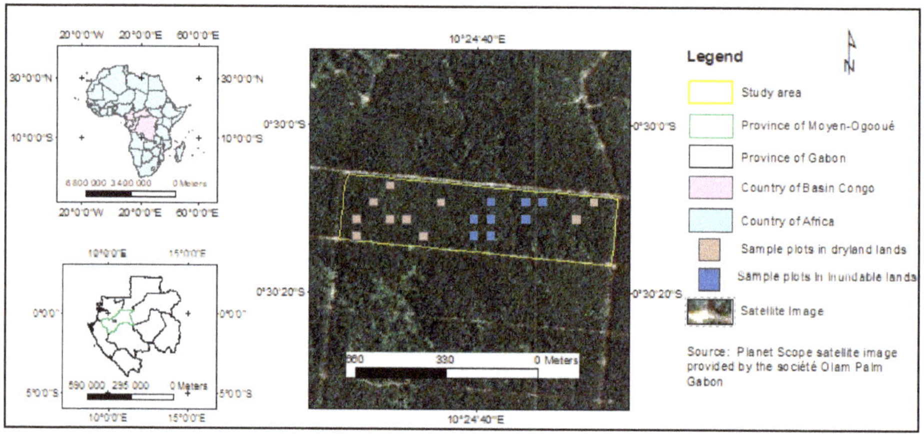

Figure 1. Study area.

2.2. Data Collection

2.2.1. Field Data Measurement

The sampling was conducted in a 35-year old plantation. In this area, eighteen oil palms were felled during the rainy season in October 2017 to obtain information on their aboveground biomass. Each felled palm tree was randomly selected within a 30.8 m × 30.8 m sample plot across the plantation block. The sample plots were randomly located in the study block. Eight of them were located in flood prone areas, while ten were on drylands (Figure 1). The individuals were healthy and not deformed by disease. The DBH of each selected palm, together with its stem height (H_T) and total height (H_{TOT}), were recorded using a standard measuring tape (DBH in cm and heights in m). The stem height was measured from the stump to the first branch, while the total height was taken from the stump to the top of the crown. The total number of leaves (N_F) on each palm was also counted. Following these measurements, the palm stem was sectioned into logs at fixed 0.5 m increments. The diameters and heights of the palm trees in the flood prone and neighboring areas were smaller (less than 52 cm for DBH and less than 16.4 m for H_{TOT}) than those located on the dryland (more than or equal to 52 cm for DBH and 15.1 m for H_{TOT}). The basic statistics of the measurements that were taken in the field are summarized in Table 1.

Once the measurements were taken and the stem was sectioned, each oil palm was separated into different components, namely the stem elements and the leaves, which included the petioles, rachis, leaflets and the fruits. The fresh mass of each component was immediately recorded using a one-ton scale. Subsamples were then taken from each component. Stem samples were taken from the first three cut logs, starting with the stump from the middle, and then from the crownward end of the stem. Samples were taken from the butt or bottom end of each log, in the form of a right angled triangle that varied in thickness from 2.7 to 4.8 cm, and from 12 to 48 cm for the lengths of the sides forming the right angle of the triangle. The same samples represented one-quarter of the large end of the stem section. Following weighing, all samples were inserted into numbered freezer bags for laboratory determinations. A summary of all the data on oil palms is presented in Table A1.

Table 1. Summary of the field measurements for 18 felled oil palms: n is the number of oil palms; DBH, H_T, H_{TOT} and N_F are respectively the diameter at breast height (cm, measured 1.3 m above the ground surface), the stem height (m), the total height (m) and the leaf number per tree.

Parameter	Minimum	Maximum	Mean	SE	%SE
DBH	48.8	57.9	53.1	0.71	1.34
H_T	6.65	10.0	8.46	0.22	2.60
H_{TOT}	14.5	18.2	15.97	0.22	1.38
N_F	27	39	33.27	0.92	2.77

SE is the standard error for each parameter.

2.2.2. Laboratory Measurements

Subsamples that were taken from the components of the 18 individuals palms (stems, petioles, rachis, leaflets and fruit) were dried under ambient air conditions and then placed in drying oven (or steamer) at 105 °C (except for the leaflets, 65 °C) to obtain wet-to-dry conversion factors. Six oil palms were selected to obtain dried fruit masses, but not all of the individuals bore fruit. The individual components were dried to constant mass (3 consecutive days) [27,29,30] and weighed on a 5 kg balance. Thus, the dry mass fraction (DMF) of each sample per component (stems, petioles, rachis, fruit and leaflets) was calculated as the ratio of the dry mass that was recorded in the laboratory to the corresponding fresh mass that was obtained in the field for each oil palm. A mean DMF value was determined for each component. Table 2 summarizes the means that were obtained for each component and for the whole palm tree.

Table 2. Summary of the infra-density, dry mass fractions and the average total dry mass of the palm components for 18 individuals that were felled in Makouké, central Gabon.

Components	Minimum	Maximum	Mean	SE	% SE
Descriptive Statistical Parameters for Dry Mass Fractions (DMF)					
Stem	0.253	0.347	0.301	0.006	2.020
Petiole	0.134	0.245	0.194	0.007	3.805
Fruit	0.156	0.221	0.190	0.009	5.059
Rachis	0.233	0.335	0.277	0.006	2.386
Leaflet	0.198	0.386	0.322	0.010	3.215
Whole oil palm	0.281	0.290	0.285	6.10^{-4}	0.220
Descriptive Statistical Parameters for Infra-Density (g·cm⁻³)					
Stem	0.25	0.3279	0.2930	0.0048	1.639
Descriptive Statistical Parameters for Total Dry Mass of Oil Palm Compartments (kg)					
Stem	199.19	419.46	302.77	13.66	4.51
Petiole	20.89	46.31	33.28	1.64	4.92
Fruit	14	82.5	58.57	10.54	17,99
Rachis	29.38	83.16	56.50	3.31	5.86
Leaflet	13.29	29.57	21.42	1.03	4.83
Leaf (Petioles, Fruit, Rachis + Leaflets)	77.15	148.79	114.93	5.19	4.52
Stem + Leaf	288.72	556.41	417.69	17.78	4.26

Finally, the fresh masses that were obtained in the field for each component of an oil palm were multiplied by the corresponding average DMF to obtain their respective dry masses. From these corrected values, it was then possible to determine the total dry mass of an individual oil palm. The dry mass: total fresh mass ratios of the palms allowed us to estimate the DMF for each of the 18 palms. The total dry masses of the different compartments of the 18 oil palms are presented in Table 2. In the laboratory, the infra-density (ϱ) of the oil palm stem tissue was determined according to the protocol of Rondeux [31], and Bauwens and Fayolle [32]. The mean infra-density of the oil palm stems is also presented in Table 2.

2.3. Establishment and Validation Allometric Models

The data analyses were performed using the XLSTAT software (https://www.xlstat.com/fr/). Scatterplots were created to better understand the distributions of the data. Consequently, outliers were identified and checks were performed to detect the possible sources of error. Only data that were correctly identified and reported were retained for the purposes of this study. Using data from the 18 oil palm trees, two classes of DBH (48–54 cm and 54–58 cm) were established to determine the proportions of biomass. The DBH class of 48-54 cm had 11 oil palms and that of 54–58 cm had 7 oil palms.

The basic expression that was employed in this research for creating the allometric equations took the following form [33]:

$$y = aX^b, \tag{1}$$

where y is the dependent variable (dry aboveground biomass), X is the product of one or more independent variables (e.g., DBH) and a and b are empirically estimated scaling factors. Typically, the $\log_e$-transformed form of the equation is used to linearize the expression, while at the same time homogenizing the variance, which increases the validity of statistical tests that are being used [34–36]. The equation can be rewritten as

$$\ln(y) = \ln(a) + b\ln(X), \tag{2}$$

The independent variables that were considered here are DBH, H_T, H_{TOT}, ϱ and N_F. To obtain unbiased estimations with log-transformed models, the bias caused by the conversion of ln (y) to the original non-transformed scale y, should be corrected. The correction factor (CF) was used to make this correction [26,27,30], such as CF = exp(root mean square error (RMSE) ^{2}/2), where RMSE is the mean square error of the regression equation. The original untransformed scale of y could be obtained by $y = (CF \times a)X^b$ [30]. To develop the equations, 60% of the oil palms were randomly selected and used (i.e., 11 of 18 palms). The remaining 40% (7 oil palms) were set aside for the independent validation of the results. Both the data for development and validation were randomly located over drylands and flood prone areas.

The performance of the established models was evaluated using different metrics. These included the coefficient of determination (r^2), the residual standard error (σ), the Akaike information criterion (AIC), the relative error (ER), the relative percentage error (%ER), and the root mean square error (RMSE) and its percentage (%RMSE). Similar metrics have been used in previous studies [30,37,38]. The expressions for calculating ER and RMSE are as follows:

$$ER = \frac{\sum_{i=n}^{n} [(\bar{y}_i - y_i)/y_i]}{n}, \tag{3}$$

$$RMSE = \sqrt{\frac{\sum_{i=1}^{n} (y_i - \bar{y}_i)^2}{n}}, \tag{4}$$

where n is the number of observations, y_i is the observed value for palm i, and $\bar{y}_i$ is its predicted value. The relative RMSE (%RMSE) was calculated as a percentage by dividing the RMSE by the observed mean [38]. The relative percent error (%ER) was obtained by multiplying ER by 100. The interpretation of the metrics differed when attempting to characterize the best performance. The higher the r^2, the more robust the equation was considered. In contrast, the lower the AIC value, the better the model fit. In all cases, the errors (ER, %ER, σ, RMSE, %RMSE) should be as small as possible.

However, Kuyah et al. [39] and Yang et al. [30] have recommended giving more weight to the bias and RMSE rather than to an adjusted r^2 or AIC in deciding the final optimal model [30].

The field data that were obtained from seven randomly selected sample plots were used to validate the equations that we had developed. The same metrics were considered in the validation, except for the standard residual error. Error distributions were established to better understand the predictive performance of the models.

2.4. Comparisons with Existing Biomass Allometric Models

Previous studies by different authors have established allometric equations for estimating the dry aboveground biomass of oil palms in several tropical regions of the world, i.e., Africa, Amazonia and Asia–Oceania. Table 3 summarizes the previously published equations, which were considered here for comparison. The work covered at least four different oil palm species. Depending on the study, the equations were available for the total above-ground biomass or for specific compartments (such as stem or leaves). The data from the seven validation sample plots were used to verify the applicability of each model to our study area and to compare their performance with the equations developed in this research. Our aim was to determine whether an existing biomass model for oil palm which was developed elsewhere could be used in the Congo Basin. The evaluations were made by quantifying the errors (ER, %ER, RMSE and %RMSE) for each model relative to the data that were used.

Table 3. Existing biomass models that were considered. B = total aboveground biomass (kg); B_F = total aboveground fresh biomass of an oil palm (kg); B_{Stem} = stem biomass (kg); N_F = number of leaves; B_{FSR} = leaf biomass without rachis (kg); B_{Rachis} = rachis biomass (kg); DBH = diameter at breast height (in cm, measured 1.3 m above ground surface); H_{Tcm} = stem height of a palm (cm); CF = correction factor; n = number of palms that were sampled. r^2 = coefficient of determination. The other variables have been previously defined in the text.

Source	Geographic Region	Palm Species	Existing Biomass Model (kg tree^{-1})	CF	r^2	n
Khalid et al. [4]	Malaysia	*Elaeis guineensis*	$B_F = 725 + 197 \times H_{TOT}$		0.96	7
Thenkabail et al. [20]	Benin	*Elaeis guineensis*	$B_F = 1.5729 \times H_{Tcm} - 8.2835$ $B = 0.3747 \times H_{Tcm} + 3.6334$		0,97 0.98	7 7
Hughes et al. [25]	Mexico	*Astrocaryum mexicanum*	$B = \exp(3.6272 + 0.5768 \times \ln(DBH^2 H_T))\ CF/10^6$	1.02	0.73	15
Saldarriaga et al. [18]	Colombia and Venezuela	Common	$B = \exp(-6.3789 - 0.877 \times \ln(1/DBH^2) +$ $2.151 \times \ln(H_T))$		0.89	19
Goodman et al. [27]	Amazonia (Peru)	Common	$B = 0.0950 \times (DMF \times DBH^2 H_T)$		0.99	106
Da Silva et al. [22]	Brazil	*Euterpe precatoria*	$B = 0.167 \times (DBH^2 H_{T\varrho})^{0.883}$ $B_{Stem} = \exp(0.1212 + 0.90 \times \ln(DBH^2 H_{T\varrho}))$ $B_{Leaf} = \exp(0.0065 + 0.69 \times \ln(DBH^2 H_T N_F))$		0.98 [1] 0.98 [1] 0.94 [1]	20 20 20
Cole and Ewel [26]	Tropical zone (Costa Rica)	*Euterpe oleraceae*	$B_{Stem} = 0.0314 \times (DBH^2 H_T)^{0.917} \times CF$ $B_{FSR} = 0.0237 \times (DBH^2 H_T N_F)^{0.512} \times CF$ $B_{Rachis} = 0.0458 \times (DBH^2 H_T N_F)^{0.388} \times CF$	1.04 1.036 1.036	0.95 0.94 0.90	156 182 187

[1] Da Silva [22] used adjusted values of the coefficient of determination (R_{adj}^2) rather than r^2.

3. Results

3.1. Distribution of Biomass Proportions

Average proportions are shown in Figure 2 for the aboveground oil palm biomass per compartment as a function of the DBH class. The aboveground biomass of the 18 oil palms that were sampled was concentrated mainly in the stems (72.51%). Leaf biomass (including petioles, rachises, fruits and leaflets) represented on average 27.50% of the total aboveground biomass. Average biomass proportions of rachises (13.53%) were much higher than those of petioles (7.95%) and fruits and leaflets (6.02%). The difference between the biomass proportions in each compartment by diameter class

(48–54 cm vs. 54–58 cm) was analyzed by simple linear regression. Proportions did not significantly differ between the DBH classes ($r^2 = 0.999$, $p < 0.0001$).

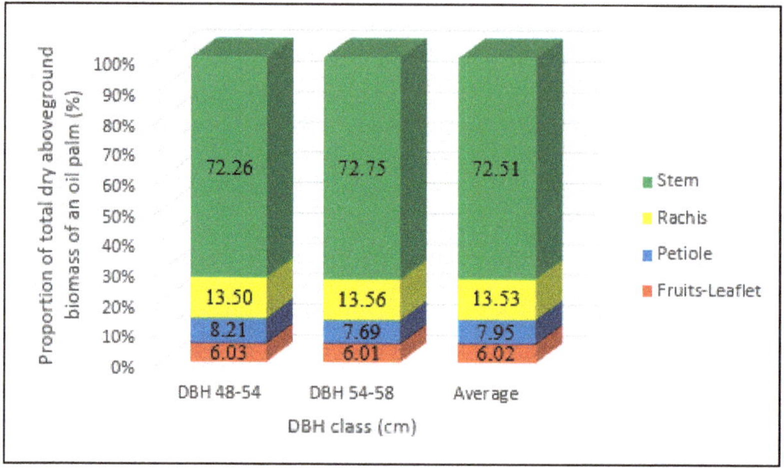

Figure 2. Mean proportions of the aboveground biomass for the 18 oil palms, Makouké, central Gabon.

3.2. Relationships between Variables

Before creating the biomass equations, we first examined the interrelationships between the variables under consideration. Clearly, the question was whether variation in ϱ, H_{TOT}, H_T and N_F could be explained by the diameter (DBH). As shown in Table 4 and Figure 3, the respective allometric relationships were significant ($p < 0.05$) between the dependent variables ϱ, H_{TOT}, H_T and N_F, vs. the independent variable DBH, with a moderate to strong r^2 (0.538 to 0.806). The %RMSE was < 3% (Table 4). On the one hand, the strongest relationship was obtained between the stem height and the DBH ($r^2 = 0.806$; $p = 0.0001$). On the other hand, the weakest relationship (albeit, statistically significant) with the DBH was obtained with the total tree height ($r^2 = 0.538$; $p = 0.010$).

3.3. Allometric Biomass Models That Were Developed

Allometric models of the aboveground oil palm biomass were developed for the different compartments, i.e., stem, leaves and the total aerial biomass. Several equations were tested to determine which were the best models; we referred to these as the local models. All the established local allometric models for estimating the aboveground biomass provided low errors overall (Table 5: %RMSE < 4%; $\sigma < 1$ kg for mean biomass = 417.7 kg; $r^2 \geq 0.564$, $p < 0.05$; %ER < 1.3%). Model 6, which was based upon DBH, yielded the highest $r^2 = 0.959$ ($p < 0.0001$) and the lowest errors (%RMSE = 0.54%, %ER = 0.003%) compared to all the other local models (7 and 8) using individual explanatory variables (ϱ, H_T and H_{TOT}) (Table 5).

Table 4. Criteria for evaluating the allometric relationships between the DBH and the dependent variables using data from 11 oil palms in Makouké, central Gabon. Values of the coefficients a and b of the models are given; σ is the residual standard error (in kg); p is the p-value of the model. CF is the correction factor for the log-transformed equation. Residual standard errors (σ, in kg), Akaike Information Criterion (AIC), relative error (ER), relative percentage error (%ER), root-mean-square error (RMSE, in kg) and its percentage (%RMSE) are shown for each equation.

Model	a	b	r^2	σ	AIC	CF	p	ER	%ER	RMSE	%RMSE
Model 1: $\ln(\varrho) = a + b \times \ln(DBH)$	−5.057	0.967	0.674	0.037	−73.343	1.0006	0.302	75×10^{-5}	0.075	0.034	2.793
Model 2: $\ln(N_F) = a + b \times \ln(DBH)$	−3.892	1.868	0.804	0.051	−63.390	1.0011	0.0001	17.5×10^{-5}	0.017	0.046	1.327
Model 3: $\ln(H_T) = a + b \times \ln(DBH)$	−4.342	1.608	0.806	0.044	−66.843	1.0008	0.0001	34.6×10^{-5}	0.034	0.039	1.869
Model 4: $\ln(H_{TOT}) = a + b \times \ln(DBH)$	−0.179	0.746	0.538	0.038	−69.769	1.0006	0.010	15.4×10^{-5}	0.015	0.034	1.258

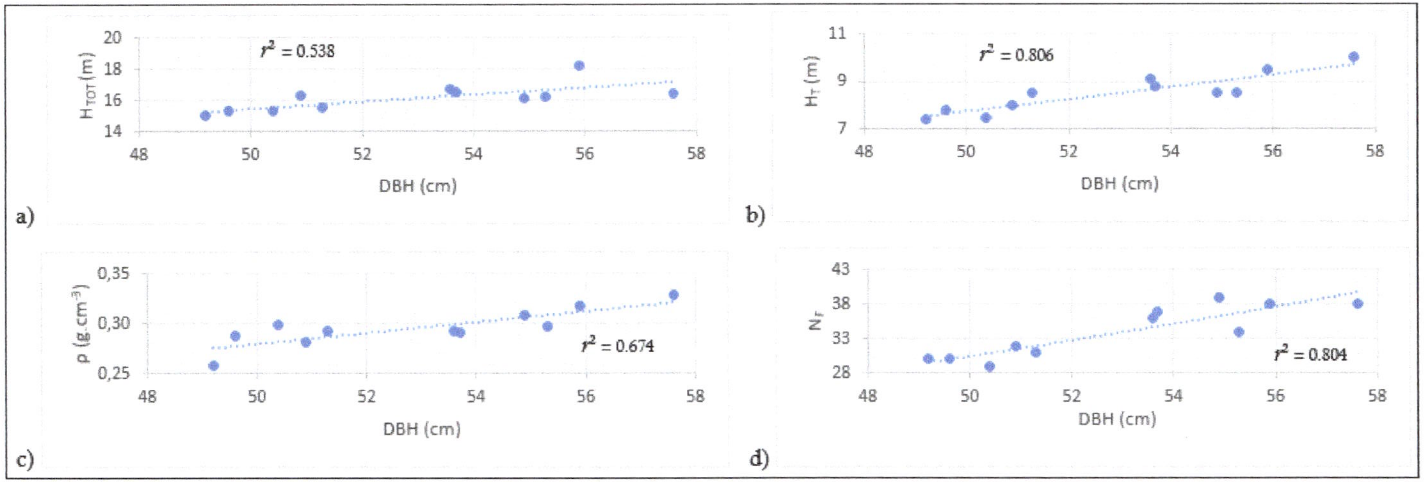

Figure 3. Allometric relationships between the variables and the DBH of the 11 oil palms used in this study to develop local biomass models: (**a**) relationship between total height vs. DBH; (**b**) relationship between stem height vs. DBH; (**c**) relationship between infra-density vs. DBH; and (**d**) relationship between number of leaves vs. DBH.

Table 5. Local allometric biomass models that were developed in this study. B is the total dry aboveground oil palm biomass. B_{Stem}, B_{Leaf}, B_{FSR} and B_{Rachis} are respectively stem, leaf (including petioles, rachis and leaflets), rachis-free leaf and rachis biomasses. P is the p-value of the model. Residual standard error (σ, in kg), the correction factor (CF), the ER and the RMSE are shown for each equation.

	Model	a	b	r^2	σ	AIC	P	CF	ER	%ER	RMSE	%RMSE
	Allometric Equations Using Infra-Density (ϱ) or DBH as the Predictor											
Model 5	$\ln(B) = a + b\ln(\varrho)$	8.755	2.223	0.685	0.099	−48.985	0.002	1.0041	22.1×10^{-5}	0.022	0.089	1.488
Model 6	$\ln(B) = a + b\ln(DBH)$	−6.256	3.100	0.959	0.035	−71.480	<0.0001	1.0005	2.8×10^{-5}	0.002	0.032	0.535
	Equations using height as the predictor											
Model 7	$\ln(B) = a + b\ln(H_T)$	2.616	1.604	0.824	0.074	−55.383	0.0001	1.0022	12.3×10^{-5}	0.012	0.067	1.112
Model 8	$\ln(B) = a + b\ln(H_{TOT})$	−0.443	2.333	0.562	0.117	−45.350	0.008	1.0056	30.4×10^{-5}	0.030	0.106	1.755
	Allometric Equations Using DBH and Height as Composite Predictors											
Model 9	$\ln(B) = a + b\ln(DBH^2 H_T)$	−2.335	0.832	0.942	0.042	−57.606	<0.0001	1.0007	4×10^{-5}	0.004	0.038	0.638
	Allometric Equations Using DBH, Height and Infra-Density as Composite Predictors											
Model 10	$\ln(B) = b\ln(DBH^2 H_T\, \varrho)$		0.683	0.999	0.0439	−58.560	0.0001	1.0008	-2.1×10^{-5}	−0.002	0.040	0.669
Model 11	$\ln(B) = a + b\ln(DBH^2 H_T\, \varrho)$	0.277	0.651	0.938	0.043	−66.938	<0.0001	1.0008	4.3×10^{-5}	0.004	0.039	0.658
	Allometric Equations Using DBH, H_T, ϱ or N_F as Composite Variables to Estimate Aboveground Biomass from Its Components (Stems, Rachises, Leaves with/without Rachises)											
Model 12	$\ln(B_{Stem}) = a + b\ln(DBH)$	−6.776	3.147	0.930	0.048	−64,933	<0.0001	1.0010	5.6×10^{-5}	0.005	0.043	0.762
	$\ln(B_{Leaf}) = a + b(DBH)$	−7.188	3.014	0.679	0.115	−45,605	0.002	1.0055	51.4×10^{-5}	0.051	0.104	2.197
	$\ln(B_{Stem}) = a + b\ln(DBH^2 H_T)$	−2.831	0.848	0.921	0.051	−63,594	<0.0001	1.0010	6.4×10^{-5}	0.006	0.046	0.810
Model 13	$\ln(B_{FSR}) = a + b\ln(DBH^2 H_T N_F)$	−3.124	0.530	0.564	0.146	−40,430	0.008	1.0088	115.1×10^{-5}	0.115	0.132	3.257
	$\ln(B_{Rachis}) = a + b\ln(DBH^2 H_T N_F)$	−4.041	0.597	0.702	0.122	−44,332	0.001	1.0062	74.9×10^{-5}	0.074	0.111	2.724
Model 14	$\ln(B_{Stem}) = b\ln(DBH^2 H_T\, \varrho)$		0.645	0.999	0.037	−71.844	<0.0001	1.0006	11×10^{-5}	0.011	0.034	0.610
	$\ln(B_{Leaf}) = b\ln(DBH^2 H_T N_F)$		0.351	0.999	0.103	−46.406	<0.0001	1.0061	109.6×10^{-5}	0.109	0.110	2.320
Model 15	$\ln(B_{Stem}) = a + b\ln(DBH^2 H_T\, \varrho)$	−0.295	0.678	0.958	0.037	−70.429	<0.0001	1.0006	3.4×10^{-5}	0.003	0.033	0.594
	$\ln(B_{Leaf}) = a + b\ln(DBH^2 H_T N_F)$	−2.852	0.561	0.747	0.103	−48.205	0.001	1.0043	40.4×10^{-5}	0.040	0.093	1.952

The local models that were constructed from stem heights (%RMSE < 1.12; r^2 > 0.8; $p \leq$ 0.0001; AIC < −55.4) were more efficient than those designed using the total heights (%RMSE < 1.8; r^2 > 0.5; $p \leq$ 0.008; AIC < −45.4). The allometric models of the aboveground biomass using composite variables (DBH^{2}H$_T$ or DBH^{2}H$_T$ ϱ) performed in a manner that was relatively similar to those solely based upon DBH (Table 5).

The allometric relationships between stem and leaf biomass with DBH, as an independent variable (Model 12), were significant with r^2 values of 0.930 and 0.679, respectively. The errors associated with these relationships were relatively small (%RMSE = 0.76%; RMSE = 0.04 kg for stems, %RMSE = 0.10%; RMSE = 2.19 kg for leaves) (Table 5). All of the allometric models estimating stem and leaf biomasses (13 and 14), with the exception of Model 15, exhibited evaluation performances that were close to those of Model 12 (Table 5). Errors for Model 15 in predicting stem and leaf biomass were lower (RMSE < 0.094 kg; %RMSE < 1.96%) than those for Models 12, 13 and 14 (RMSE < 0.14 kg; %RMSE < 3.28%). In summary, according to the results that we obtained, the total aboveground biomass of oil palm was best correlated with the DBH compared to the stem or leaf biomass (Figure 4).

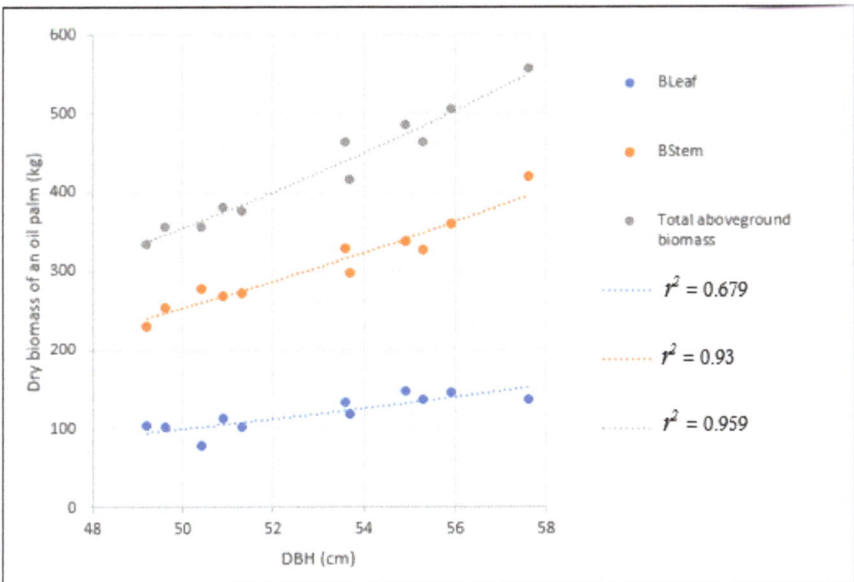

Figure 4. Allometric relationships between the biomass components of 11 oil palms and their corresponding DBH.

3.4. Validation of Local Allometric Models

Seven palm trees were used to validate the relationships between the individual variables that were considered (ϱ, H$_{TOT}$, H$_T$ and N$_F$) and the DBH, together with the allometric relationships that were obtained with log sections. Validation results are reported in Table 6. All the variables that were considered were significantly related to DBH ($r^2 \geq$ 0.66; $p \leq$ 0.026), with relatively small errors (%RMSE $\leq$ 9.6%; %ER $\leq$ 7.5%). The relationships that were obtained for the total height (Model 4) and infra-density (Model 1) appeared to be the most robust following validation (Table 6).

Table 6. Validation of the allometric relationships between the individual explanatory variables (ϱ, H_{TOT}, H_T and N_F) and the DBH (for estimates of a and b, see Table 4).

Model	r^2	AIC	p	ER	%ER	RMSE	%RMSE
Model 1: $\ln(\varrho) = a + b \times \ln(DBH)$	0.787	−60.077	0.008	0.034	3.407	0.014	4.845
Model 2: $\ln(N_F) = a + b \times \ln(DBH)$	0.750	13.371	0.012	0.075	7.552	3.098	9.638
Model 3: $\ln(H_T) = a + b \times \ln(DBH)$	0.660	−3.111	0.026	0.0006	0.068	0.645	7.697
Model 4: $\ln(H_{TOT}) = a + b \times \ln(DBH)$	0.927	−15.660	0.001	0.001	0.136	0.583	3.712

Different local allometric biomass estimation models that were proposed in Section 3.3 were validated using the data that were independently collected from the seven sample plots. The same performance evaluation metrics were considered. Table 7 summarizes the validation results that were obtained. Adding the stem height and infra-density or the stem height and leaf count to the model using the DBH alone as the predictor improved the predictions, as shown in the results. The introduction of infra-density (ϱ) to the models that were based upon the combination (DBH^2H_T) contributed to the improvement of all the validation criteria of these allometric models. As an example, the root mean square error decreased from 8.2% to 5.7% when moving from Model 9 (excluding infra-density) to Model 11 (including infra-density). In the same vein, taking into account the leaf number (N_F) in allometric models using DBH^2H_T improved the estimates of leaf biomass and by extension, the entire palm tree. The results of Allometric Models 14 and 15 clearly showed these improvements compared to models in which the leaves were not considered.

Model 9 yielded the highest r^2 (0.939, $p < 0.0001$), the smallest AIC (45.3) and the lowest %RMSE (8.2%) among all the allometric models using structural parameters that were measured directly on oil palm (DBH and H_T) (Table 7). With the addition of a variable that is not directly measurable, such as infra-density (ϱ), Allometric Model 11 slightly improves upon Allometric Model 9. Allometric Model 15 includes both infra-density and leaf number (a parameter usually not available). This model exhibited the best performance in this study, with a relative RMSE of 5% (Table 7). By combining the aboveground biomass of the stems ($DBH^2H_T\varrho$) and leaves ($DBH^2H_TN_F$), Model 15 stands out as the best of the local biomass allometric models that were developed in the study (Table 7). The expressions of these three (3) models are described below:

$$\text{Allometric Model 9}: \ B \ = \ 1.0007 * \exp\left[-2.335 + 0.832 \times \ln\left(DBH^2H_T\right)\right], \tag{5}$$

$$\text{Allometric Model 11}: \ B \ = \ 1.0008 * \exp\left[0.277 + 0.65 \times 1\ln\left(DBH^2H_T\rho\right)\right], \tag{6}$$

$$\text{Allometric Model 15}: \ B \ = \ B_{Stem} + B_{Leaf} = 1.0005 * \exp\left[-0.295 + 0.678 \times \ln(DBH^2H_T\rho)\right] \\ +1.0043 * \exp\left[-2.852 + 0.561 \times \ln\left(DBH^2H_TN_F\right)\right] \tag{7}$$

Table 7. Validation of the local allometric models of oil palm biomass; estimates for a and b are available in Table 5.

Model	r^2	AIC	P	ER	%ER	RMSE	%RMSE
Allometric Equations Using a Single Explanatory Variable, i.e., Infra-Density or DBH							
Model 6: $\ln(B) = a + b\ln(DBH)$	0.887	49.601	0.002	0.091	9.109	45.386	11.253
Model 5: $\ln(B) = a + b\ln(\varrho)$	0.757	54.962	0.011	0.010	1.079	38.143	9.457
Allometric Equations Using Height as an Explanatory Variable							
Model 8: $\ln(B) = a + b\ln(H_{TOT})$	0.730	55.712	0.014	0.012	1.242	41.954	10.402
Model 7: $\ln(B) = a + b\ln(H_T)$	0.810	53.234	0.006	0.042	4.157	38.854	9.633
Allometric Equations Using DBH and Height as Compound Explanatory Variables							
Model 9: $\ln(B) = a + b\ln(DBH^2H_T)$	0.939	45.305	0.0003	0.065	6.501	33.027	8.188
Allometric Equations Using DBH, Height, and ϱ as Compound Explanatory Variables							
Model 10: $\ln(B) = b\ln(DBH^2H_T\,\varrho)$	0.961	42.153	0.0001	0.048	4.815	26.786	6.641
Model 11: $\ln(B) = a + b\ln(DBH^2H_T\,\varrho)$	0.961	42.206	0.0001	0.042	4.247	23.339	5.786
Allometric Equations Using Biomass Components (Stems, Rachises, Leaves with/without Rachis)							
Model 12: $\ln(B) = [\ln(B_{Stem}) + \ln(B_{Leaf})] = [a_1 + b_1\ln(DBH) + a_2 + b_2\ln(DBH)]$	0.887	49.605	0.002	0.089	8,916	44.856	11.121
Model 13: $\ln(B) = [\ln(B_{Stem}) + \ln(B_{FSR}) + \ln(B_{Rachis})] = [a_1 + b_1\ln(DBH^2H_T) + a_2 + b_2\ln(DBH^2H_TN_F) + a_3 + b_3\ln(DBH^2H_TN_F)]$	0.956	42.950	0.0001	0.052	5.268	27.325	6.774
Model 14: $\ln(B) = [\ln(B_{Stem}) + \ln(B_{Leaf})] = [b_1\ln(DBH^2H_T\varrho) + b_2\ln(DBH^2H_TN_F)]$	0.969	40.519	< 0.0001	0.044	4.420	21.352	5.294
Model 15: $[B_{Stem} + B_{Leaf}] = [a_1 + b_1\ln(DBH^2H_T\varrho) + a_2 + b_2\ln(DBH^2H_TN_F)]$	0.972	39.922	< 0.0001	0.036	3.684	20.692	5.130

3.5. Validation of Existing Allometric Biomass Models

Nine existing models (Table 3) were evaluated using the data from the seven palm trees. The results are compiled in Table 8. Among the existing allometric models only using stem height as the predictor, the allometric equation Thenk2004a that was proposed by Thenkabail et al. [20] produced the lowest error (%RMSE = 13.7%). Although the values were lower, these results were relatively close to those provided by Allometric Model 7 that was proposed in this study (%RMSE = 9.63%). The other two existing allometric models that were based upon height had errors > 20% (Table 8). In the category of allometric models using the DBH and height together, the two existing models of Hughes et al. [25] and Saldarriaga et al. [18] could not be directly used in the study area because they produced very large estimation errors (%RMSE > 100%), unlike Model 9 in this study (%RMSE of about 8%). The allometric equation DaSilva2015a, which was devised by Da Silva et al. [22], stood out among the models that were based upon DBH, height, infra-density or the dry mass fraction (DMF) as composite explanatory variables. It produced a small error (%RMSE = 9.3%), which was close to that of the Local Allometric Model 11 which was developed in this study (%RMSE = 5.8%). Figure 5a illustrates the proximity of the two allometric models. Finally, in the category of allometric models estimating the total aerial biomass of the palm from the biomass of its components (stems, leaves or rachis) considering the DBH, H_T, and ϱ and N_F as composite explanatory variables, the two existing models produced errors of less than 24% (Table 8). The allometric equation DaSilva2015b from Da Silva et al. [22] provided an %RMSE of 10.9%. However, this error was almost double that produced by Allometric Model 15 in this study. Figure 5b illustrates the results of the five existing models that produced the lowest errors in our study area. The results of the three best local allometric models that were proposed in this study are also shown for comparison purposes. Dispersion is greater in the estimated biomasses for larger diameters (>52 cm). The allometric equations ColEwe2006 [26] and Thenk2004b [20] systematically underestimated the biomass in the area. The associated errors were generally > 20%. Other existing allometric models in Figure 5b produced relative errors < 15%, although they were not developed specifically for the region. Of course, local models were more efficient with relative errors that were generally < 10%.

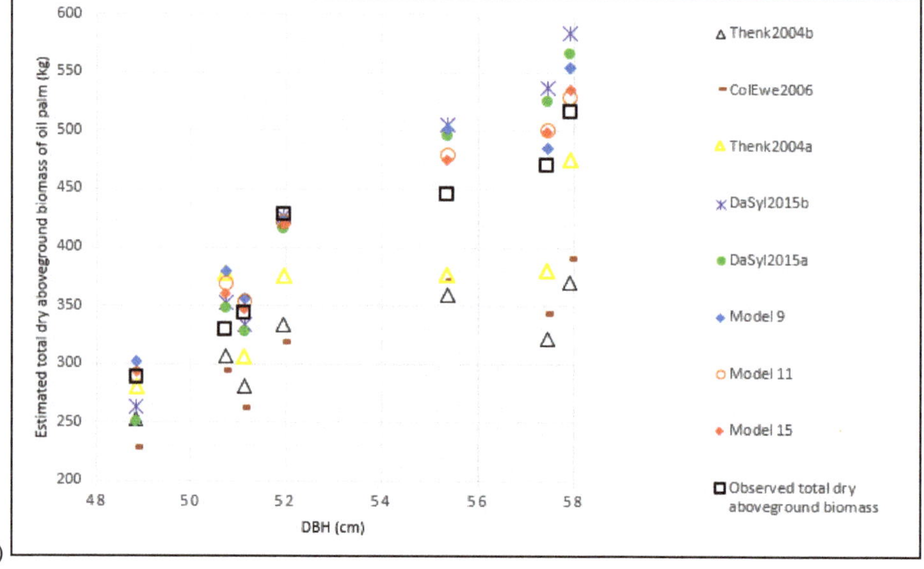

Figure 5. *Cont.*

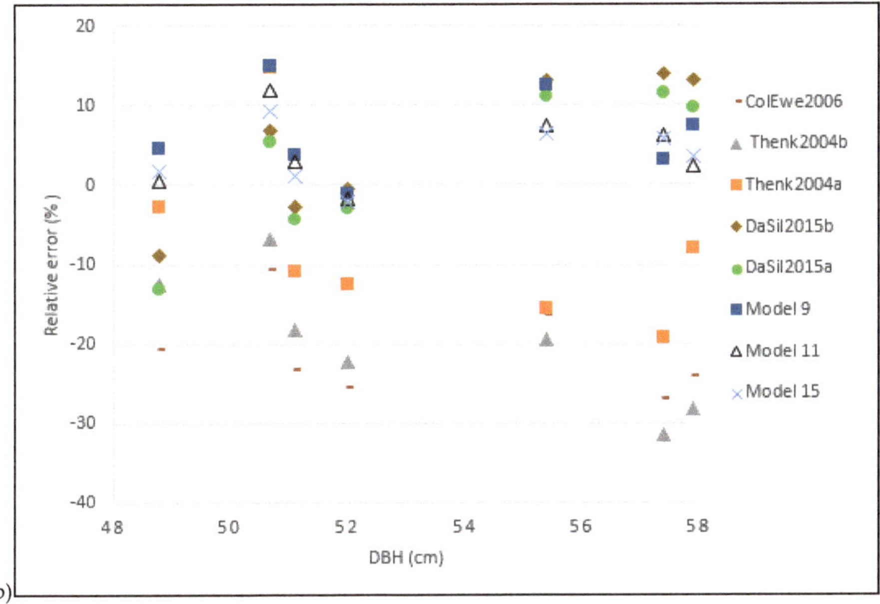

Figure 5. Comparison of the selected existing allometric models with the best allometric models that were proposed in this study: (**a**) biomass variation according to DBH; (**b**) relative error variation according to DBH.

Table 8. Comparison of the existing allometric biomass models to the corresponding local models that were developed in this study.

Reference	Name	ER	%ER	RMSE	%RMSE
Allometric Equations Using Height as an Explanatory Variable					
Khalid et al. [4]	Khal1999	1.725	172.583	669.968	166.109
Thenkabail et al. [20] (Dry biomass model)	Thenk2004b	−0.198	−19.838	96.752	23.988
Thenkabail et al. [20] (Fresh biomass model)	Thenk2004a	−0.077	−7.752	55.317	13.715
This study	**Model 7**	**0.042**	**4.157**	**38.854**	**9.633**
Allometric Equations Using DBH and Height as Compound Explanatory Variables					
Saldarriaga et al. [18]	Sald1988	−0.999	−99.999	410.677	101.821
Hughes et al. [25]	Flyn1999	−0.999	−99.996	410.664	101.818
This study	**Model 9**	**0.065**	**6.501**	**33.027**	**8.188**
Allometric Equations Using DBH, Height and Infra-Density or Dry Mass Fraction as Composite Explanatory Variables					
Goodman et al. [27]	Good2013	−0.994	−99.408	408.402	101.257
Da Silva et al. [22] (Not compartmentalized allometric biomass model)	DaSil2015a	0.024	2.413	37.699	9.347
This study	**Model 11**	**0.042**	**4.247**	**23.339**	**5.786**
Allometric Equations Estimating Aboveground Biomass (B) from Biomass Components (Stems, Leaves or Rachis) Using DBH, H_T and ϱ or N_F					
Cole and Ewel [26]	ColEwe2006	−0.211	−21.122	92.841	23.018
Da Silva et al. [22] (Compartmentalized allometric biomass model)	DaSil2015b	0.050	5.007	44.157	10.948
This study	**Model 15**	**0.036**	**3.684**	**20.692**	**5.130**

4. Discussion

4.1. Interpretation of Biomass Distribution

The distribution of the aboveground oil palm biomass among the components showed that on average the stems produced most of the biomass (about 73%) compared to the leaves (rachis + petiole + fruits/leaflets), which produced a total of about 27%. The palms that were considered in this study were generally older (>30 years) than those that were measured in most previous studies. They were practically at the stage of maximum maturity in the case of plantations. Nevertheless, other studies have reported distributions of relatively similar proportions for younger age classes. For example, Da Silva et al. [22] also noted the high proportion of stem biomass (86.4%) to leaf biomass (13.6%) in Brazil for oil palms with DBHs between 3.9 and 12.7 cm. The results that were obtained by Cole and Ewel [26] in Costa Rica were even more comparable than those obtained in this study, i.e., about 78% for stems and 22% for leaves for 13-year-old oil palms with DBH < 20 cm. According to the various results, the proportions of oil palm compartment biomasses (stem vs. leaf) varied only slightly according to age and diameter in the tropical regions that were considered. This lack of variability could likely be attributed to the maintenance that was practiced in these plantations. In natural environments, palm growth may be less homogeneous, depending upon site conditions, which in turn could affect the proportions of biomass among tree compartments. The plantation considered in this study was composed of old trees of 35 years old. This constitutes a limitation, as it is not representative of young and medium age plantations which can be found in the Congo Basin and elsewhere.

4.2. Evaluation of Local Allometric Biomass Equations

In this study, the stem diameter was measured at 1.3 m above the ground surface (DBH), consistent with other studies estimating the aboveground biomass of oil palms [22,23,27]. Several allometric relationships were established to estimate the oil palm biomass in this research (see Table 5).

Our results show an improvement in allometric relationships between biomass and DBH when height or infra-density are taken into account (Table 7). The integrative variable DBH^2H_T (Model 9) is effective in estimating palm biomass on the study site, as has been the case in other tropical areas [18,25,36]. The cylindrical shape of the oil palm stem, geometrically characterized by the combination DBH^2H_T, could explain the strong relationship with biomass. Indeed, the latter is essentially concentrated in the stem (Figure 2). Estimating biomass using DBH^2H_T is an alternative, non-destructive method in different tropical oil palm-producing regions.

The infra-density (ϱ) of wood varies according to the type of species, plantation density and growing conditions [30]. The average value of ϱ obtained for the palm stems that were considered in this study (>30-years) equals 0.293 g·cm^{-3}. The estimate was within the range from 0.21 to 0.41 g·cm^{-3} that was defined by Supriadi et al. [40] for *Elais guineensis*-type palms. The relationship between biomass and infra-density appears to be very significant (see Allometric Model 5, Table 7). Thus, the weighting of the composite variable DBH^2H_T with infra-density resulted in considerable improvement in palm biomass prediction, as demonstrated by the very small error (%RMSE = 5.8%) that was obtained with Allometric Model 11 (Table 7). The combination of the three variables (DBH, H_T, ϱ) has also provided significant results in previous work [21,22].

To consider the contributions of all the components of the palm, we integrated the biomass that was contained in the leaves with that of the stem. The resulting Allometric Model 15 stands out as the best performing of all models that were proposed in this study, with a %RMSE of 5.1% (Tables 5 and 7). Previous work in other regions has reported similarly convincing results and demonstrated the importance of considering the contributions of various components into account ([22,26]; see Table 3). Nevertheless, a close look at the results shows that leaf inclusion did not appreciably improve biomass estimates, compared to the results that were based solely upon $DBH^2H_T\varrho$ (Model 11) or DBH^2H_T (Allometric Model 9). Despite incurring larger errors, Allometric Model 9 remains an interesting alternative to Models 11 and 15 in the absence of infra-density or leaf number data.

Although strong relationships were developed in the study, it should be mentioned that the low number of samples used remains a limitation. The study was based on a destructive approach. Thus, only eighteen trees could be felled to acquire the data both for the development of the equations and for their validation. A larger number of samples associated with a wider range of DBH and height values is therefore recommended for future studies.

4.3. Comparison of Local Models to Existing Allometric Biomass Models

Several allometric relationships have been proposed for estimating oil palm biomass elsewhere in Africa, Asia–Oceania and the Neotropics (Central and South America). The current study sought to understand whether some of these relationships were directly applicable to our site in the Congo Basin. As expected, several allometric models produced large errors (%RMSE > 100%), together with unrealistic variation in predicted biomass, especially for the highest DBHs (52 to 58 cm). This was the case for the equations of Khalid et al. [4], Saldarriaga et al. [18], Goodman et al. [27] and Hughes et al. [25] that were established for Malaysia, Colombia and Venezuela as well as Peru and Mexico, respectively (Table 8). The very high errors could be explained by the very different site conditions and the great disparity in the experimental data that were available for constructing the equations. Nonetheless, the allometric equations of Thenkabail et al. [20] (Thenk2004a and b), Cole and Ewel [26] (ColEwe2006) and those of Da Silva et al. [22] seem to be very applicable to the Congo Basin. Building upon research that was conducted by Yang et al. [30], we investigated errors that were associated with these five existing allometric relationships, together with our three best local models. Figure 6 shows the performance of these different models as a function of the two selected classes of DBH (class 1: 48–52 cm; class 2: 52–58 cm). Among the different allometric models shown, the errors appeared larger for the ColEwe2006 model [26] and Thenk2004b model [20] in both DBH classes, but remained < 25%. The biomass errors observed with the allometric model of Thenkabail et al. [20] could be due to the small stem heights that were used in their study (28–195 cm). Those of Cole and Ewel [26] could possibly be explained by the short to tall oil palms that were considered in the development of their model (1.3 to 20 m). The Thenk2004b allometric model that was developed in Benin and the two models of Da Silva et al. [22] that were developed in Brazil yielded errors < 15%. In particular, the allometric model DaSilv2015a produced results that were close to those of the models proposed in this study, especially in the first DBH class. Indeed, the allometric models of Da Silva et al. [22] were constructed from data comparable to those collected in our study ($n = 20$; $H_T = 8.8$ m; $\varrho = 0.3306$ g·cm^{-3}). This could be the cause of their strong performance, especially in class 1 DBH. The DBHs (3.9 and 12.7 cm) that were used to establish the allometric models of Da Silva et al. [22] could have caused the slightly larger errors that were observed in class 2.

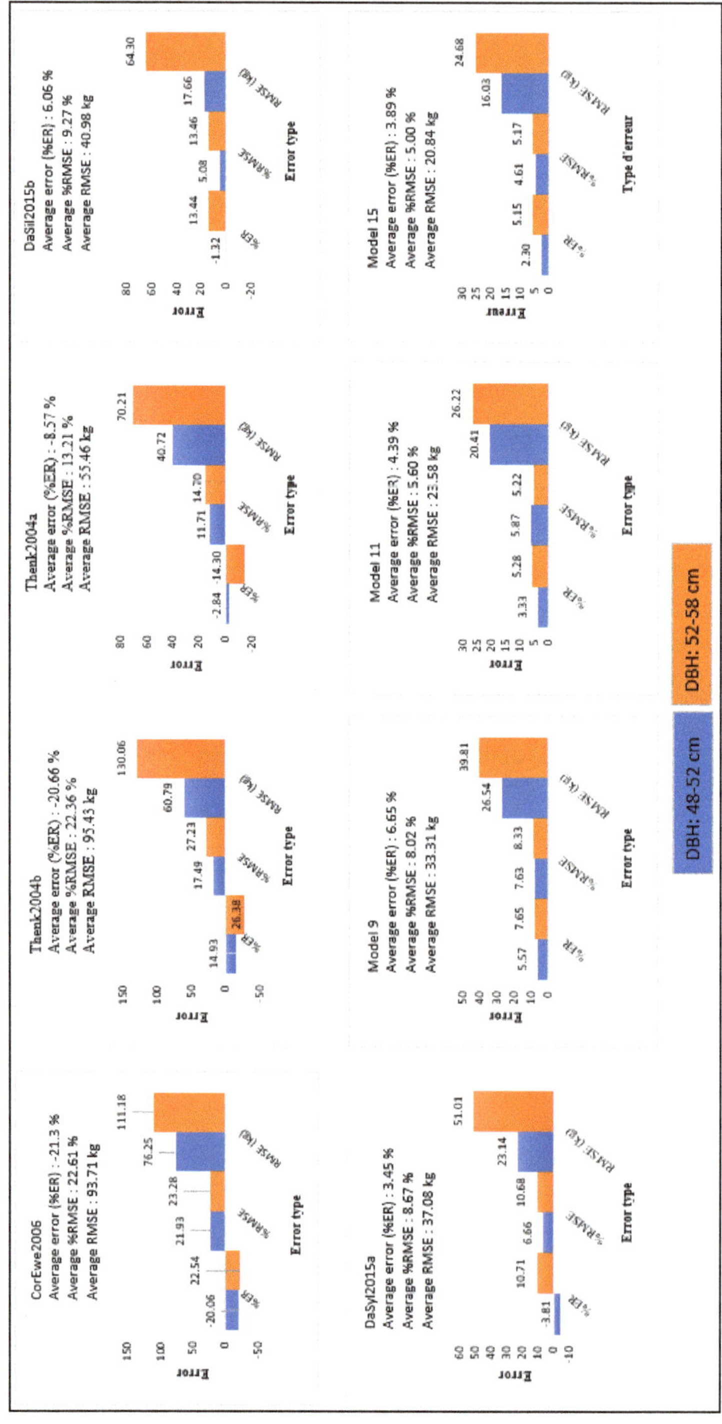

Figure 6. Errors in the allometric models when considering the two DBH classes (48–52 cm, 52–58 cm).

5. Conclusions

The present study estimated aboveground biomass using the structural parameters of oil palms that were acquired in the Congo Basin using destructive sampling. About three-quarters of the palm biomass was concentrated in the stem. Several allometric equations that were based on diameter at breast height, height, infra-density, number of leaves, or some combination of these different variables, were developed. The composite variable that was based upon the combination (DBH^2H_T) emerged as the most interesting and perhaps useful explanatory variable for estimating oil palm biomass in the current study. It was the basis of the three best models that were obtained. The best of the three (Allometric Model 15) integrates the contributions of leaves and is characterized by a low error (%RMSE about 5%). The second high-performance allometric model, which weights DBH^2H_T by infra-density, also produces a low error of about 6%. The third allometric model (Model 9), which was based solely upon DBH^2H_T, was the most practical alternative, given its relatively small error (about 8%) and the fact that information on infra-density and palm leaf number is not always available. The study shows that some allometric equations developed in other regions could have been used to estimate the palm biomass in the site that we selected in the Congo Basin, but with slightly larger errors than those of the three proposed allometric models. However, several existing models were not applicable because of the large errors they produced in the site, due to differences in palm oil species, age or site conditions.

One of the main limitations in this study resides in the reduced number of samples used to develop and then validate the allometric equations. Only the data from eighteen oil palms were available for this study. Increasing the number of samples in future works would allow the development of probably more robust equations. Such equations are essential to assess the carbon produced by oil palms and understand the impact of the establishment of agro-industrial plantations in tropical forest areas in the context of climate change, while helping their sustainable management. Combinations of data that were acquired from experimental sites in different tropical regions across the globe could make it possible to envisage the development of regional and pan-tropical allometric relationships for estimating the aboveground biomass of oil palms.

Author Contributions: Conceptualization, P.M. and K.G.; methodology, P.M., A.N., K.G.; software, P.M.; validation, P.M. and Kalifa Goïta.; formal analysis, P.M.; investigation, P.M.; resources, P.M., A.N. and A.P.M.B.; data curation, P.M.; writing—original draft preparation, P.M.; writing—review and editing, P.M., K.G., A.N., A.P.M.B.; visualization, P.M.; supervision, K.G.; project administration, P.M.; funding acquisition, P.M. and K.G. All authors have read and agreed to the published version of the manuscript.

Funding: This study was funded by the Programme Canadien de Bourses de la Francophonie (Government of Canada, Department of Foreign Affairs, Trade and Development, Canadian Partnership Branch) and by the Natural Sciences and Engineering Research Council of Canada (NSERC).

Acknowledgments: The authors acknowledge the financial support of the Programme Canadien de Bourses de la Francophonie (Government of Canada, Department of Foreign Affairs, Trade and Development, Canadian Partnership Branch) and the Natural Sciences and Engineering Research Council of Canada (NSERC).The following corporation and government institutions contributed substantially to the data measurement campaign: Olam Palm Gabon; Institut de Recherche en Ecologie Tropicale (IRET); Direction Générale des Eaux et Forêts and the Projet de Développement d'un Système d'Inventaire des Ressources Forestières Nationales contribuant à la Gestion Durable des Forêts (PDSIRFN); the Ecole Nationale des Eaux et Forêts (ENEF); and the Institut de Recherche Agronomiques et Forestières (IRAF). We are grateful to those individuals who participated in field surveys, and in drying and weighing samples, especially Jacques Mouloungou, Bruno Nkoumakali, Inès Nelly Moussavou, and Dieudonné Ndzengboro.

Conflicts of Interest: The authors declare no conflict of interest. The funders had no role in the design of the study; in the collection, analyses, or interpretation of data; in the writing of the manuscript, or in the decision to publish the results.

Appendix A

Table A1. Field and laboratory data that were used to develop aboveground biomass models for oil palm.

Plot Number	B_F (Total Fresh Aboveground Biomass of Oil Palm) (kg)	DBH (Diameter at Breast Height, 1.3 m) (cm)	H_{TOT} (Total Height) (m)	H_T (Stem Height) (m)	N_F (Number of Leaves Per Palm)	DMF (Dry Mass Fraction) Stem Mean	Dmf (Mean Dry Mass Fraction) Of Oil Palm	ϱ (Mean Infra-Density of Oil Palm Stem) (g·cm^{-3})	B_{Rachis} (Dry Rachis Biomass) (kg)	B_{FSR} (Dry leaf Biomass without Rachis) (kg)	B_{Leaf} (Dry Leaf Biomass of Oil Palm: Petioles, Fruits, Rachises and Leaflets) (kg)	B_{Stem} (Dry Stem Biomass of Oil Palm) (kg)	B (Total Dry Aboveground Biomass) (kg)
1	1336.80	50.9	16.3	8.0	32	0.2914	0.2857	0.2819	59.8752	52.9108	112.786	269.1137	381.8997
2	1643.30	53.6	16.7	9.1	36	0.3085	0.2818	0.2921	63.0907	70.4268	133.5175	329.4937	463.0112
3	1950.20	57.6	16.4	10.0	38	0.3081	0.2853	0.3279	66.9161	70.0291	136.9452	419.4599	556.4051
4	1176.95	49.2	15.0	7.4	30	0.3105	0.2836	0.2587	50.7830	53.1305	103.9135	229.9270	333.8405
5	1259.35	49.6	15.3	7.8	30	0.3314	0.2832	0.2872	43.1878	59.7505	102.9383	253.6564	356.5947
6	1227.44	50.4	15.3	7.5	29	0.3008	0.2902	0.2993	42.3007	36.6779	78.9786	277.2167	356.1953
7	1462.40	53.7	16.5	8.8	37	0.3365	0.2846	0.2917	55.4954	63.7667	119.2621	296.9790	416.2411
8	1623.30	55.3	16.2	8.5	34	0.3343	0.2855	0.2972	69.4109	67.4892	136.9001	326.5954	463.4955
9	1710.05	54.9	16.1	8.5	39	0.3471	0.2843	0.3077	74.2342	74.5585	148.7927	337.3431	486.1358
10	1294.15	51.3	15.5	8.5	31	0.2972	0.2901	0.2927	55.4954	46.9775	102.4729	272.9780	375.4509
11	1763,1	55.9	18.2	9.5	38	0.3115	0.2869	0.3180	83.1600	63.0959	146.2559	359.6233	505.8792
12	1803.75	57.9	16.6	9.8	38	0.3099	0.2862	0.3030	45.6964	68.9378	114,6342	401,5572	516.1914
13	1156.25	50.7	15.3	8.1	27	0.2987	0.2852	0.2749	29.3832	47.7701	77.1533	252,6299	329.7832
14	1672.25	57.4	17.0	8.5	37	0.2857	0.2814	0.3260	62.5918	74.5959	137.1877	333.4184	470.6061
15	1543.70	55.4	16.6	9.5	32	0.2536	0.2889	0.2935	65.1420	52.9177	118.0597	327.9238	445.9835
16	1216.85	51.1	15.0	7.4	28	0.2650	0.2822	0.2776	41.2474	51.4161	92.6635	250.6978	343.3613
17	1021.80	48.8	14.5	6.65	30	0.2706	0.2826	0.2500	40.8038	48.7180	89.5218	199.1936	288.7154
18	1493.20	52.0	15.1	8.8	33	0.2727	0.2871	0.2950	68.2466	48.4361	116.6827	311.9835	428.6662

References

1. Food and Agriculture Organization. *Évaluation des Ressources Forestières Mondiales 2015: Comment les Forêts de la Planète Changent-Elles?* 2nd ed.; FAO: Rome, Italy, 2016; pp. 30–31. Available online: http://www.fao.org/3/a-i4793f.pdf (accessed on 28 April 2018).
2. Keenan, R.J.; Reams, G.A.; Achard, F.; de Freitas, J.V.; Grainger, A.; Lindquist, R. Dynamics of global forest area: Results from the FAO Global Forest Resources Assessment 2015. *For. Ecol. Manag.* **2015**, *352*, 9–20. [CrossRef]
3. Tuma, J.; Fleiss, S.; Eggleton, P.; Frouz, J.; Klimes, P.; Lewis, O.T.; Yusah, K.M.; Fayle, T.M. Logging of rainforest and conversion to oil palm reduces bioturbator diversity but not levels of bioturbation. *Appl. Soil Ecol.* **2019**, *144*, 123–133. [CrossRef]
4. Khalid, H.; Zin, Z.Z.; Anderson, J.M. Quantification of oil palm biomass and nutrient value in a mature plantation. I, Above-ground biomass. *J. Oil Palm Res.* **1999**, *11*, 23–32. Available online: http://jopr.mpob.gov.my/wp-content/uploads/2013/07/joprv11n1-p31.pdf (accessed on 20 October 2018).
5. Folefack, A.J.J.; Ngo Njiki, G.M.; Darr, D. Safeguarding forests from smallholder oil palm expansion by more intensive production? The case of Ngwei forest (Cameroon). *For. Policy Econ.* **2019**, *101*, 45–61. [CrossRef]
6. Olorunfemi, I.E.; Komolafe, A.A.; Fasinmirin, J.T.; Olufayo, A.A. Biomass carbon stocks of different land use management in the forest vegetative zone of Nigeria. *Acta Oecol.* **2019**, *95*, 45–56. [CrossRef]
7. Megevand, C.; Mosnier, A.; Hourticq, J.; Sanders, K.; Doetinchem, N.; Streck, C. *Deforestation Trends in the Congo Basin: Reconciling Economic Growth and Forest Protection*; The World Bank, Directions in Development—Environment and Sustainable Development: Washington, DC, USA, 2013; pp. 4–118. [CrossRef]
8. Glinskis, E.A.; Gutiérrez-Vélez, V.H. Quantifying and understanding land cover changes by large and small oil palm expansion regimes in the Peruvian Amazon. *Land Use Policy* **2019**, *80*, 95–106. [CrossRef]
9. De Almeida, A.S.; Vieira, I.C.G.; Ferraz, S.F.B. Long-term assessment of oil palm expansion and landscape change in the eastern Brazilian Amazon. *Land Use Policy* **2020**, *90*, 104321. [CrossRef]
10. Rivera, Y.D.; Moreno, A.L.; Romero, H.M. Biochemical and physiological characterization of oil palm interspecific hybrids (*Elaeis oleifera* x *Elaeis guineensis*) grown in hydroponics. *Acta Biol. Colomb.* **2013**, *18*, 465–472. Available online: https://www.researchgate.net/publication/257938174 (accessed on 11 January 2020).
11. Meléndez, M.R.; Ponce, W.P. Pollination in the oil palms *Elaeis guineensis*, *E. oleifera* and their hybrids (OxG), in tropical America. *Pesqui. Agropecu. Trop.* **2016**, *46*, 102–110. [CrossRef]
12. Pacheco-Palencia, L.A.; Mertens-Talcott, S. Chemical composition, antioxidant properties, and thermal stability of a phytochemical enriched oil from Acai (*Euterpe oleracea* Mart.). *J. Agric. Food Chem.* **2008**, *56*, 4631–4636. [CrossRef]
13. Morel, A.C.; Fisher, J.B.; Malhi, Y. Evaluating the potential to monitor aboveground biomass in forest and oil palm in Sabah, Malaysia, for 2000–2008 with Landsat ETM+ and ALOSPALSAR. *Int. J. Remote Sens.* **2012**, *33*, 3614–3639. [CrossRef]
14. Kho, L.-K.; Rudbeck Jepsen, M. Carbon stock of oil palm plantations and tropical forests in Malaysia: A review. *Singap. J. Trop. Geogr.* **2015**, *36*, 249–266. [CrossRef]
15. Lahteenoja, O.; Ruokolainen, K.; Schulman, L.; Oinonen, M. Amazonian peatlands: An ignored C sink and potential source. *Glob. Change Biol.* **2009**, *15*, 2311–2320. [CrossRef]
16. Neumann, M.; Moreno, A.; Mues, V.; Härkönen, S.; Mura, M.; Bouriaud, O.; Lang, M.; Achten, W.M.J.; Thivolle-Cazat, A.; Bronisz, K.; et al. Comparison of carbon estimation methods for European forests. *For. Ecol. Manag.* **2016**, *361*, 397–420. [CrossRef]
17. Tinker, P.B.H.; Smilde, K.W. Dry-matter production and nutrient content of plantation oil palms in Nigeria. II. Nutrient Content. *Plant. Soil* **1963**, *19*, 350–363. Available online: https://www.jstor.org/stable/42933246 (accessed on 15 May 2018). [CrossRef]
18. Saldarriaga, J.G.; West, D.C.; Tharp, M.L.; Uhl, C. Long-term chronosequence of forest succession in the upper Rio Negro of Colombia and Venezuela. *J. Ecol.* **1988**, *76*, 938–958. [CrossRef]

19. Yuen, J.Q.; Fung, T.; Ziegler, A.D. Review of allometric equations for major land covers in SE Asia: Uncertainty and implications for above- and below-ground carbon estimates. *For. Ecol. Manag.* **2016**, *360*, 323–340. [CrossRef]

20. Thenkabail, P.S.; Stucky, N.; Griscom, B.W.; Ashton, M.S.; Diels, J.; van der Meer, B.; Enclona, E. Biomass estimations and carbon stock calculations in the oil palm plantations of African derived savannas using IKONOS data. *Int. J. Remote* **2004**, *25*, 5447–5472. [CrossRef]

21. Corley, R.H.V.; Tinker, P.B.H. *The Oil Palm*, 4th ed.; Blackwell Science: Hoboken, NJ, USA, 2003; pp. 89–103.

22. Da Silva, F.; Suwa, R.; Kajimoto, T.; Ishizuka, M.; Higuchi, N.; Kunert, N. Allometric equations for estimating biomass of *Euterpe precatoria*, the most abundant palm species in the Amazon. *Forests* **2015**, *6*, 450–463. [CrossRef]

23. Kotowska, M.M.; Leuschner, C.; Triadiati, T.; Meriem, S.; Hertel, D. Quantifying above- and belowground biomass carbon loss with forest conversion in tropical lowlands of Sumatra (Indonesia). *Glob. Change Biol.* **2015**, *21*, 3620–3634. [CrossRef]

24. Aholoukpè, H.; Dubos, B.; Flori, A.; Deleporte, P.; Amadji, G.; Chotte, J.L.; Blavet, D. Estimating aboveground biomass of oil palm: Allometric equations for estimating frond biomass. *For. Ecol. Manag.* **2013**, *292*, 122–129. [CrossRef]

25. Hughes, R.F.; Kauffman, J.B.; Jaramillo, V.J. Biomass, carbon, and nutrient dynamics of secondary forests in a humid tropical region of Mexico. *Ecology* **1999**, *80*, 1892–1907. [CrossRef]

26. Cole, T.G.; Ewel, J.J. Allometric equations for four valuable tropical tree species. *For. Ecol. Manag.* **2006**, *229*, 351–360. [CrossRef]

27. Goodman, R.C.; Phillips, O.L.; Del Castillo Torres, D.; Freitas, L.; Cortese, S.T.; Monteagudo, A.; Baker, T.R. Amazon palm biomass and allometry. *For. Ecol. Manag.* **2013**, *310*, 994–1004. [CrossRef]

28. Moraes, R.M.; Borchsenius, F.; Blicher-Mathiesen, U. Notes on the biology and uses of the Motacú palm (*Attalea phalerata*, Arecaceae) from Bolivia. *Econ. Bot.* **1996**, *50*, 423–428. Available online: www.jstor.org/stable/4255886 (accessed on 10 January 2020).

29. Sunaryathy, P.-I.; Suhasman Kanniah, K.-D.; Tan, K.-P. Estimating aboveground biomass of oil palm trees by using the destructive method. *World J. Agric. Res.* **2015**, *3*, 17–19. [CrossRef]

30. Yang, X.; Blagodatsky, S.; Liu, F.; Beckschäfer, P.; Xu, J.; Cadisch, G. Rubber tree allometry, biomass partitioning and carbon stocks in mountainous landscapes of sub-tropical China. *For. Ecol. Manag.* **2017**, *404*, 84–99. [CrossRef]

31. Rondeux, J. *La Mesure des Arbres et des Peuplements Forestiers*, 2nd ed.; Les Presses Agronomiques de Gembloux: Gembloux, Belgium, 1999; pp. 97–108.

32. Bauwens, S.; Fayolle, A. Protocole de Collecte des Données sur le Terrain et au Laboratoire Nécessaires pour Quantifier la Biomasse Aérienne des Arbres et pour l'établissement d'équations Allométriques. *Nature +* **2014**, *40*. [CrossRef]

33. Gould, S.J. Allometry and size in ontogeny and phylogeny. *Biol. Rev.* **1966**, *41*, 587–640. [CrossRef]

34. Brown, S.; Gillespie, A.J.R.; Lugo, A.E. Biomass estimation methods for tropical forests with applications to forest inventory data. *For. Sci.* **1989**, *35*, 881–902. [CrossRef]

35. Onyekwelu, J.-C. Above-ground biomass production and biomass equations for even-aged *Gmelina arborea* (ROXB) plantations in southwestern Nigeria. *Biomass Bioenerg.* **2004**, *26*, 39–46. [CrossRef]

36. Chave, J.; Andalo, C.; Brown, S.; Cairns, M.A.; Chambers, J.Q.; Eamus, D.; Fölster, H.; Fromard, F.; Higuchi, N.; Kira, T.; et al. Tree allometry and improved estimation of carbon stocks and balance in tropical forests. *Oecologia* **2005**, *145*, 87–99. [CrossRef] [PubMed]

37. Ngomanda, A.; Engone Obiang, L.N.; Lebamba, J.; Moundounga Mavouroulou, Q.; Gomat, H.; Mankou, G.S.; Loumeto, J.; Midoko Iponga, D.; Kossi Ditsouga, F.; Zinga Koumba, R.; et al. Site-specific versus pantropical allometric equations: Which option to estimate the biomass of a moist central African forest? *For. Ecol. Manag.* **2014**, *312*, 1–9. [CrossRef]

38. Hansen, E.H.; Gobakken, T.; Bollandsås, O.M.; Zahabu, E.; Næsset, E. Modeling aboveground biomass in dense tropical submontane rainforest using airborne laser scanner data. *Remote Sens.* **2015**, *7*, 788–807. [CrossRef]

39. Kuyah, S.; Dietz, J.; Muthuri, C.; Jamnadass, R.; Mwangi, P.; Coe, R.; Neufeldt, H. Allometric equations for estimating biomass in agricultural landscapes: I. Aboveground biomass. *Agric. Ecosyst. Environ.* **2012**, *158*, 216–224. [CrossRef]

40. Supriadi, A.; Rachman, O.; Sarwono, E. Characteristics and sawing properties of oil-palm (*Elaeis guineensis* Jacq.) wood logs. *Buletin Penelitian Hasil Hutan* **1999**, *17*, 1–20. [CrossRef]

 *forests*

Article

Biomass Allocation into Woody Parts and Foliage in Young Common Aspen (*Populus tremula* L.)—Trees and a Stand-Level Study in the Western Carpathians

Bohdan Konôpka [1,2], Jozef Pajtík [1], Vladimír Šebeň [1,*], Peter Surový [2] and Katarína Merganičová [2,3]

[1]   Forest Research Institute, National Forest Centre, T.G. Masaryka 22, SK-960 01 Zvolen, Slovakia; bohdan.konopka@nlcsk.org (B.K.); jozef.pajtik@nlcsk.org (J.P.)
[2]   Faculty of Forestry and Wood Sciences, Czech University of Life Sciences Prague, Kamýcká 129, CZ-165 000 Prague, Czech Republic; psurovy@gmail.com (P.S.); merganicova@tuzvo.sk (K.M.)
[3]   Forestry Faculty, Technical University Zvolen, T.G. Masaryka 24, SK-960 01 Zvolen, Slovakia
*   Correspondence: seben@nlcsk.org

Received: 19 February 2020; Accepted: 16 April 2020; Published: 20 April 2020

Abstract: Our research of common aspen (*Populus tremula* L.) focused on the forested mountainous area in central Slovakia. Forest stands (specifically 27 plots from 9 sites) with ages between 2 and 15 years were included in measurements and sampling. Whole tree biomass of aspen individuals was destructively sampled, separated into tree components (leaves, branches, stem, and roots), and then dried and weighed. Subsamples of fresh leaves from three crown parts (upper, middle, and lower) were scanned, dried, and weighed. Allometric biomass models with stem base diameter as an independent variable were derived for individual tree components. Basic foliage traits, i.e., leaf mass, leaf area, and specific leaf area, were modelled with regard to tree size and leaf position within the crown. Moreover, biomass stock of the woody parts and foliage as well as the leaf area index were modelled using mean stand diameter as an independent variable. Foliage traits changed with both tree size and crown part. Biomass models showed that foliage contribution to total tree biomass decreased with tree size. The total foliage area of a tree increased with tree size, reaching its maximum value of about 12 m^2 for a tree with a diameter of 120 mm. Leaf area index increased with mean stand diameter, reaching a maximum value of 13.5 m^2 m^{-2}. Since no data for biomass allocation for common aspen had been available at either the tree or stand levels, our findings might serve for both theoretical (e.g., modelling of growth processes) and practical (forestry and agro-forestry stakeholders) purposes.

Keywords: tree components; biomass models; stem base diameter; specific leaf area; leaf area index

---

1. Introduction

Knowledge on the patterns of biomass partitioning of a variety of species is of high importance for carbon reporting [1], tree physiology, plant ecology, and process-based modelling [2], and has also applications for forestry management [3]. Although generalised biomass models can be used to estimate biomass stock in forests stands [4], locally fitted models are recommended by the IPCC to minimise the bias of estimations [5]. In addition, young stands are characterised by fast changes in biomass allocation in individual tree compartments, due to which biomass equations developed for older stands are not appropriate [6–8].

Considering a range of forest tree species, a lot of works focused on biomass allocation and foliage traits with regard to a number of different factors, e.g., competition in a stand [9,10]. Few authors [11–13] used the ratio between foliage dry mass and total dry plant biomass (leaf mass ratio; LMR) or between foliage area and total plant dry biomass (leaf area ratio; LAR) to link them to ecological and production processes. Similarly, specific leaf area (ratio between leaf area and dry leaf mass; SLA) has often been studied in plant ecology as one of the adaptation strategy indicators [14]. Canopy leaf area of forests serves as a dominant physical control over primary production, transpiration, energy exchange, and other physiological attributes related to a variety of ecosystem processes; hence, it is a substantial element of ecological studies [15]. Here, the leaf area index (LAI, defined as the amount of leaf area in the canopy per unit ground area) is the main parameter of the canopy [16]. None of these indicators has been studied in common aspen either at the tree or the stand level.

Common aspen, or also named Eurasian aspen (*Populus tremula* L.), is one of the most widely distributed tree species around the world [17]. It is a light-demanding broadleaved softwood tree species that is native to boreal zones as well as to cooler parts, such as the temperate zones of Europe and Asia [18]. It is one of the so-called pioneer species that occupy post-disturbance areas after windbreaks, fire events, or clear cutting [17,19]. Although in forest stands dominated by other species it is usually scattered, it can create homogeneous stands at post-disturbance areas [20].

The aspen grows very fast, especially during the first 15–20 years, i.e., in the period when crown competition increases [20]. Even though the commercial importance of aspen is limited, the species is often considered as a relevant part of forest ecosystems due to its fundamental importance for other plant and animal species [21,22]. This species creates conditions for existence of a variety of herbivorous, saprophytic invertebrates, fungi and lichens, birds, etc. [20]. It is one of the most attractive tree species for forage for large herbivores [23]. Hence, its presence increases the carrying capacity of red deer hunting grounds and can ensure the biological protection of other economically important tree species [24].

Moreover, the species is considered to be soil ameliorating since its foliage litter contains high concentrations of calcium, potassium, and magnesium [19]. It is also an important species for regulating the microclimate and for enhancing the structural and biological diversity of open agricultural landscapes in the temperate zone of Europe [22,25].

The share of aspen in Slovak forests is small, since from the point of its contribution to the wood stock and forest area it was ranked 18th and 16th, respectively, out of all tree species (approximately 50) recorded in the national forestry inventory [26]. According to the data from the last National Forest Inventory of Slovakia, aspen was recorded at 8% of the inventory plots [26]. Although it occurred at elevations from 200 to 1300 m a.s.l., aspen was most frequent at elevations between 450 and 550 m a.s.l. [10]. Despite its good ecological values, it is not an important tree species for any biotopes of European and national significance, nor is it a main tree species of any biotope. In Slovakia and Czechia, aspen (similarly to many other softwood broadleaved tree species) occur more frequently in stands growing on former agricultural land than on forest lands [27].

Since common aspen is fast growing and tolerant to extreme or nutrient-poor ecological conditions, it might be a prospective species to produce biomass for energetic purposes or for pulp and paper production [28]. Common aspen and its hybrids (especially *P. tremula* × *P. tremuloides*) are frequently planted in Nordic and Baltic countries [29–31] and used for pulp production. The increasing demand of woody biomass by the energy sector as well as by the forest industry will increase the need for alternative wood sources. Applying "short rotation forestry" based on fast-growing tree species, including those from the *Populus* genus, is a promising alternative [32]. While a number of papers have focused on the biomass of a variety of tree species, nearly no information on the biomass characteristics for common aspen is available and absolutely no works studied its biomass allocation patterns or component traits in terms of ecological conditions or biological aspects. Our review of the European literature focusing on common aspen indicated neglected interest in this species.

Therefore, the main aim of our paper was to quantify total tree biomass and its allocation to components in common aspen at both the tree and stand levels. A further aim of the paper was to quantify foliage traits (especially SLA) and stand canopy status (LAI) with respect to tree or stand size, specifically stem base diameter.

2. Material and Methods

2.1. Site and Stand Description

Our research focused on the forested mountainous area of central Slovakia belonging to the Western Carpathians. In general, the forest composition of the Western Carpathians, which is a part of the Carpathian Mountain Range, changes with altitude. At the lowest altitudes, oaks (mainly *Quercus robur* L. and *Q. petraeae* (Matt.) Liebl.) dominate, while at the middle altitudes European beech (*Fagus sylvatica*) is the dominant species. At higher altitudes (over approx. 900 m a.s.l.), coniferous species such as Norway spruce (*Picea abies* (L.) H. Karst.), Silver fir (*Abies alba* Mill.), Scots pine (*Pinus sylvetris* L.), and European larch (*Larix decidua* Mill.) prevail.

A preliminary selection of forest stands containing common aspen was conducted using a database of Programs of Forest Management by Stand Units in Slovakia (available on: http://gis.nlcsk.org/lgis/) based on specific information on tree species composition and stand age. The main criteria for forest selection were (1) the share of the target tree species, i.e., aspen, had to be equal to or greater than 90%, and (2) a stand age of maximum 20 years. Afterwards, we examined preselected forest stands in the field, where we checked the origin and the actual contribution of aspen to tree species composition in stands. The final selection of nine forest stands (Figure 1) was performed considering exclusively natural regeneration and nearly a 100% share of aspen. The youngest selected stand was 2 years old and the oldest stand was 15 years old. All stands were dense with no large gaps in the forest canopy.

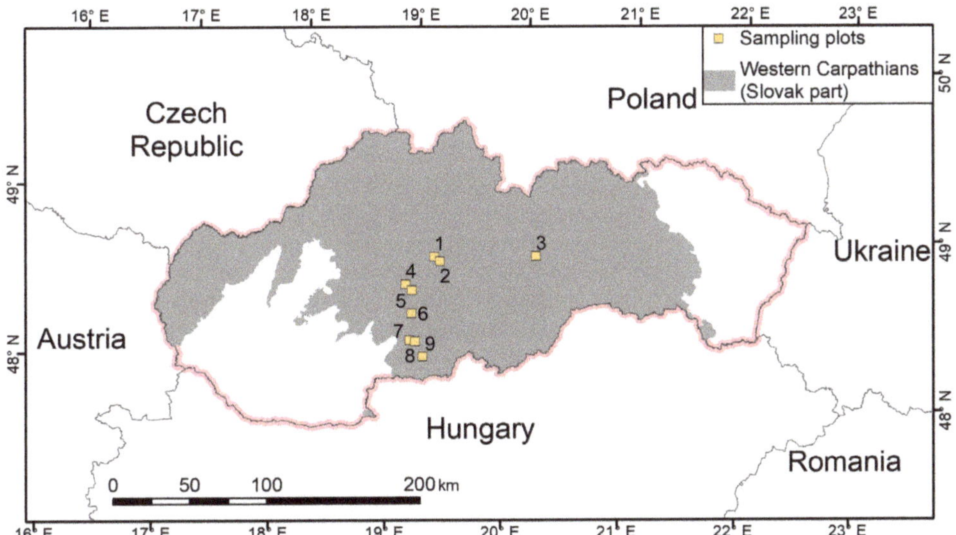

Figure 1. Localization of common aspen stands selected for biomass sampling (namely: 1—Podkonice I, 2—Podkonice II, 3—Telgart, 4—Kasova Lehota, 5—Straze, 6—Dobra Niva, 7—Suchan I, 8—Suchan II, 9—Opava).

The altitudinal range of the selected stands was between 335 m and 610 m a.s.l. (Table 1, i.e., they occurred in the forest vegetation zone with natural prevalence of oak and beech). In the selected region and altitudes, annual mean temperatures are between 5.0 °C and 8.5 °C, annual precipitation totals fluctuate between 600 and 900 mm, and the growing season usually lasts from 130 to 175 days. Soils are represented by Cambisols, Luvisols, and Rendzina (Table 1).

Table 1. Site characteristics of the forest stands selected for common aspen sampling and measurements.

Locality Code	Name of Locality	Altitude (m a.s.l.)	Longitude (°)	Latitude (°)	Exposition (°)	Slope (%)	Soil Group
1	Podkonice I	550	48.7932	19.2672	228	15	Rendzina
2	Podkonice II	545	48.7931	19.2671	230	16	Rendzina
3	Telgart	870	48.8359	20.1711	61	9	Cambisol
4	Kasova Lehotka	610	48.6204	19.0288	231	121	Cambisol
5	Straze	335	48.5864	19.0897	213	13	Luvisols
6	Dobra Niva	365	48.4522	19.1003	50	7	Cambisol
7	Suchan I	540	48.2897	19.1022	355	14	Luvisols
8	Suchan II	540	48.2896	19.1023	357	13	Luvisols
9	Opava	525	48.1998	19.2235	227	22	Cambisol

2.2. Field and Laboratory Works

Field measurements and tree samplings were performed in the second half of the growing season of 2018. In each selected stand, three circular plots were established with a distance between the plots of at least 10 m. The radius of the plots varied between 1.5 m and 3.0 m, as it depended on stand density and was chosen to cover at least 30 trees. Diameters at stem base (hereinafter as d_0) and tree heights were measured for all trees at the plots. In total, 27 circular plots and 971 trees were measured. Mean tree and stand characteristics for the nine forest stands are presented in Table 2. Plot-level data were used for subsequent calculations at a stand level using input data recorded at the tree level (destructive tree sampling).

Afterwards, 20 aspen trees were selected from each stand (altogether 180 individuals based on stratified random sampling) to cover all bio-sociological classes (i.e., dominant, co-dominant, sub-dominant, and suppressed). The root system of each tree was excavated to include all roots with a diameter of at least 1 mm. In addition, 5 leaves were sampled from the lower, middle, and upper thirds of the tree crowns, i.e., 15 leaves were sampled from each tree. Leaf subsamples were packed in paper envelopes identified with plot and tree codes, and the position of the leaves in the crown.

Excavated trees were cross-cut at a stem base to separate the belowground (root system) and aboveground parts. Subsequently, branches with foliage were cut off from the main stem. Stem length (tree height) and diameter d_0 were measured. Each tree component, i.e., root system, stems, and branches with foliage, were packed in paper bags identified with plot and tree codes. All samples were transported to the laboratory. Subsamples of leaves were scanned and their one-side leaf surface area (LA hereinafter) was calculated using the Easy Leaf Area software [33]. Each leaf was oven-dried (under 95 °C for 24 h) and weighed with a precision of 0.001 g. Individual leaf mass and area were used for the calculation of SLA at a specific crown part (upper, middle, and lower).

Leaves were manually trimmed back from the branches. Consequently, roots, stems, branches, and foliage were stored in a warm (about 28 °C), dry, and well-ventilated room. After several weeks, individual tree components were placed in the oven and dried under a temperature of 95 °C for a determined period; specifically, foliage for 48 h and woody parts (i.e., roots, stems, and branches) for 72 h to obtain dry matter. After drying, all components were weighed with a precision of 0.1 g.

Table 2. Stand characteristics of common aspen (mean values and standard deviations were calculated from three circular plots).

Locality Code	Name of Locality	Mean Diameter d_0 (mm)	S.D.	Mean Height (m)	S.D.	Mean Canopy Closure (%)	S.D.	Mean Basal Area (m^2 per are)	S.D.	Mean Number of Trees (per are)	S.D.
1	Podkonice I	30.4	1.3	3.29	0.24	77	15	0.347	0.125	469	139
2	Podkonice II	11.8	1.9	1.45	0.49	58	16	0.133	0.017	1241	240
3	Telgart	26.1	16	2.20	1.00	60	17	0.114	0.063	380	388
4	Kasova Lehotka	57.9	7.3	6.40	0.50	95	5	0.613	0.244	227	63
5	Straze	29.2	1.8	4.09	0.61	73	31	0.289	0.175	432	247
6	Dobra Niva	54.4	4.3	5.82	0.66	100	0	0.493	0.117	210	20
7	Suchan I	30.8	10.3	2.99	0.64	83	15	0.176	0.057	303	37
8	Suchan II	12.7	1.5	1.22	0.21	50	0	0.122	0.014	976	160
9	Opava	108.8	11.2	9.55	0.83	70	33	1.109	0.839	123	89

2.3. Calculations and Modelling

From the data we derived models that quantified the biomass in the foliage and woody parts of aspen trees, as well as other variables representing leaf growth (*LA, SLA, LAI, LAR, LMR*). Models were derived for several levels:

- The level of individual leaves (leaf mass w_f, *LA* and *SLA*)
- The tree level (foliage mass w_f, mass of woody parts w_{wp}, *LA, LAR, LMR*)
- The stand level (foliage mass w_f, mass of woody parts w_{wp}, *LAI*).

An allometric function with two regression coefficients (b_1 and b_2),

$$Y = b_1 d_0^{b_2},\tag{1}$$

was used to describe the relationship of the particular dependent variable (represented by Y) to the d_0 stem base diameter. In the models for the leaf and tree levels, the tree stem base diameter d_0 was used, while in the stand level models a mean stand diameter at the stem base d_{0g} was used. The frequently used diameter at breast height was not applicable in our analyses, because some of the selected trees did not reach a height of 1.3 m. Almost all constructed equations were derived in a basic power form. The equations of foliage and wood biomass were derived as linearised logarithmic allometric equations that were subsequently transformed back to the exponential form. This approach was applied to ensure a methodological link to previous works that quantified the biomass of components in young trees [34]. The shape of the allometric function after the logarithmic transformation and its back transformation is

$$Y = e^{(b_1 + b_2 \cdot \ln d_0)} \cdot \lambda\tag{2}$$

where λ is a correction factor.

A logarithmic version of the equation is frequently used to exclude heteroscedasticity of residues, which is always present in the calculation of biomass components. Using a logarithmic equation enables the estimation of parameters with linear regressions, which fulfils the assumption of constant variance of residues. Although in the last years non-linear regression methods have been developing, the opinions on these two methodological approaches differ between papers, e.g., [35–37].

Models at a level of individual leaves were derived for leaves at different crown parts (upper, middle, and lower) using measured data of d_0 (mm), w_f (g), and *LA*. *SLA* was calculated using the following formula

$$SLA = \frac{LA}{w_f}\tag{3}$$

Since only d_0 diameter and w_f and w_{wp} mass were measured for a tree, leaf area *LA* of the whole tree had to be calculated otherwise. In our work we derived a new model from the measured foliage mass of a particular tree and a tree *LA* calculated as an arithmetical average from the LA values of nine leaves taken from three crown parts of the particular tree using the equation

$$LA = w_f \frac{\sum\limits_{i=1}^{n}\left(\frac{LA_i}{w_{f_i}}\right)}{n}\tag{4}$$

where:

LA—leaf area of a tree (cm^2);
LA_i—leaf area of *i*th sampled leaf (cm^2);
w_f—mass of all tree leaves (g);
w_{f_i}—mass of *i*th sampled leaf (g);
n—number of leaves sampled from one tree (i.e., 15).

For the models at a stand level, mean stand diameter d_{0g} (mm) was calculated as a quadratic mean of all tree diameters using the formula

$$d_{0g} = \sqrt{\frac{\sum_{i=1}^{n} d_i^2}{n}} \qquad (5)$$

Leaf area index *LAI* was calculated as the *LAI* under the assumed full canopy coverage ($LAI_{100\%}$) using the relationship

$$LAI_{100\%} = \frac{100 \sum_{i=1}^{n} LA_i}{Sc} \qquad (6)$$

where:

LA_i—leaf area of *i*th tree at a plot (m^2 m^{-2});
S—plot area in m^2;
C—crown canopy coverage at a plot in %;
n—number of trees at a plot.

Similarly, variables in other models (volume of woody parts, and foliage volume) were also converted to 100% crown canopy.

Finally, a generalised linear mixed model (GLMM) was created to evaluate the combined influence of d_0 and position of the leaves in the crown on leaf traits. The model has the following form:

$$y = \alpha + \beta \times \ln(d_0) + \gamma_1 \times M + \gamma_2 \times U \qquad (7)$$

where:

y—a dependent variable, namely leaf mass, leaf area, or specific leaf area;
d_0—diameter at stem base;
M—a dummy variable that represents the middle crown part, i.e., if the leaf was taken from the middle part, the value is 1, otherwise it is 0;
U—a dummy variable that represents the upper crown part;
α, β, γ_1 and γ_2—the regression coefficients of the model.

All statistical analyses were performed in Statistica 10.0 and R software version 3.5.1 [38].

3. Results

Heights and stem base diameters of the sampled trees varied between 0.4 m and 10.5 m and from 3.3 mm to 100.9 mm, respectively (Figure 2; Table 3). The relationship between d_0 diameter and tree height was best described by a fractional power relationship, specifically, $h = \frac{d_0^2}{8.221 + 7.077 d_0 + 0.021 d_0^2}$, which explained almost 90% of the variability. This relationship may help users who prefer using tree height as an independent variable in biomass models, to convert our models that are based on d_0 diameter.

The values of the measured leaves revealed high variation in all the assessed leaf characteristics, i.e., leaf mass (from 0.019 to 0.513 g), leaf area (between 3.79 and 50.81 cm^2) as well as SLA (71.33 to 374.24 cm^2 g^{-1}; Table 4). Leaf mass of individual leaves differed along the vertical crown profile, with the heaviest leaves in the upper crown part and the lightest ones in the bottom part of the crown (Figure 3a). At the same time, leaf area increased with tree size as represented by d_0 diameter (Figure 3b), although differences between the crown parts were less evident than in leaf mass. The influence of leaf position in the crown and tree size on SLA was very evident (Figure 3c). The largest SLA values were found for small trees and lower crown parts. Moreover, differences in SLA between the crown parts increased with tree size.

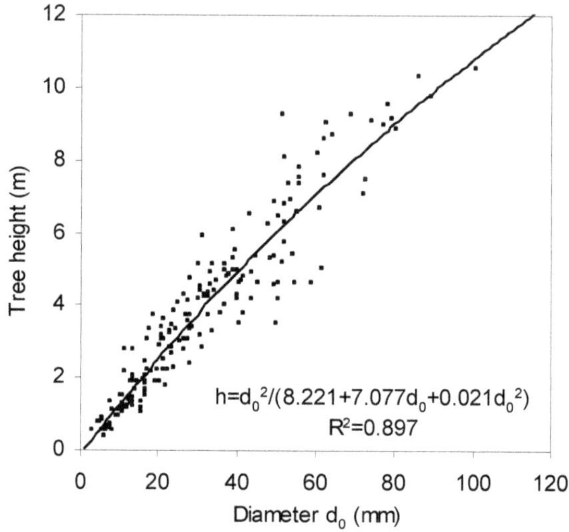

Figure 2. Relationship between d_0 stem base diameter and tree height derived from the sampled common aspen trees (values of SE were 11.812, 0.693, and 0.009, and the *p*-values 0.487, < 0.001, and 0.017 for parameters b$_1$, b$_2$, and b$_3$, respectively; MSE = 0.612).

Table 3. Descriptive statistics of the sampled common aspen trees.

Tree Parameter (unit)	Mean	S.D.	Min.	Max.	25th Percentile	75th Percentile
Diameter d_0 (mm)	31.4	20.0	3.3	100.9	15.5	43.1
Height (m)	3.81	2.42	0.40	10.54	1.87	4.96
Roots biomass (g)	262	376	1	2190	23	310
Stem biomass (g)	700	1293	1	8352	30	674
Branches biomass (g)	251	460	0	2770	32,813	284
Foliage biomass (g)	94	131	1	737	33,878	129
Woody parts biomass (g)	1397	2448	12,145	14,995	69	1391
Whole tree biomass (g)	1456	2530	45,809	15,651	83	1538

Table 4. Statistical characteristics of leaf traits of the sampled common aspen trees.

Leaf Trait (unit)	Crown Part	Mean	S.D.	Min.	Max.	25th Percentile	75th Percentile
Leaf mass (g)	Upper	0.171	0.079	0.029	0.513	0.114	0.222
	Middle	0.129	0.063	0.029	0.347	0.080	0.168
	Lower	0.101	0.058	0.019	0.329	0.054	0.130
Leaf area (cm^2)	Upper	21.50	7.29	6.33	44.63	16.32	25.50
	Middle	19.71	7.73	8.20	50.27	14.11	23.84
	Lower	17.42	8.10	3.79	50.81	11.98	21.76
SLA (cm^2 g^{-1})	Upper	138.88	42.68	71.33	308.17	108.16	159.40
	Middle	167.31	47.01	91.50	374.24	131.90	193.70
	Lower	190.78	49.85	97.25	355.21	152.06	227.78

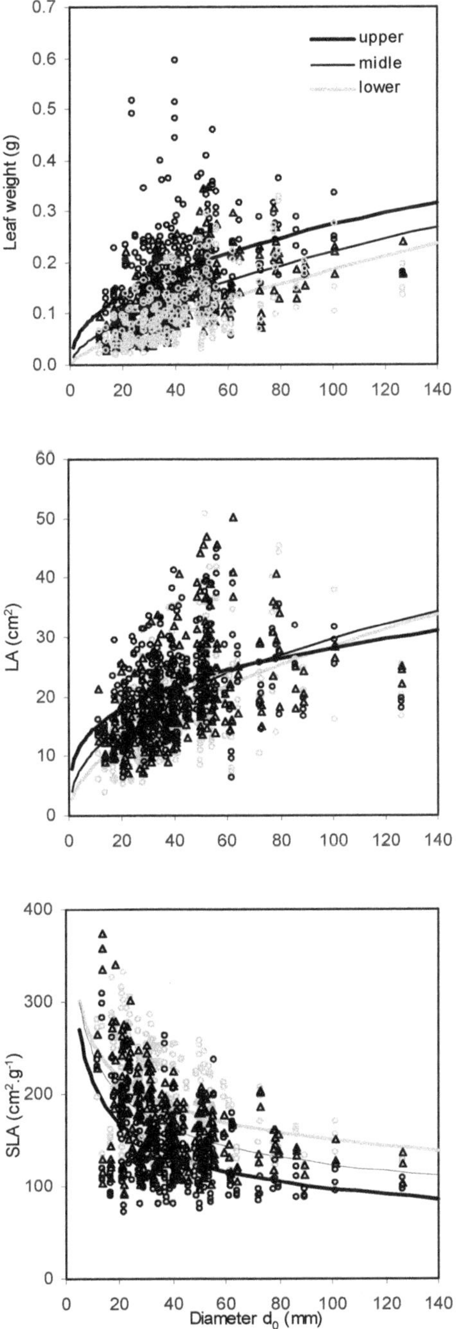

Figure 3. Relationship between d_0 stem base diameter and individual leaf mass (upper plate), leaf area (middle plate), or specific leaf area (lower plate) in the sampled common aspen trees with respect to foliage position along the tree crown.

Our mixed models (Table 5) also revealed a significant influence of the crown part from which the leaves were sampled. The models confirmed that both leaf mass and leaf area increase from the bottom to the top of the crown, i.e., leaves in the upper crown part have larger values of area and mass than in the middle part, which has larger and heavier leaves than the lower part. In the case of specific leaf area, we found an opposite order, i.e., lower > middle > upper, which means that the specific leaf area of the bottom leaves is greater than the SLA of the middle and upper ones (leaves on the top of the crown can be considered as those with minimum SLA, given the diameter at the base is equal for the samples).

Table 5. Results of the generalised linear mixed models derived for leaf traits based on Equation (7).

Leaf Trait (unit)	GLMM Component	Estimate	Standard Error	t-Value	p-Value	Significance
	Intercept	−0.195	0.0167	−11.695	0.000	***
	$Ln(d_0)$	0.082	0.0045	18.119	0.000	***
Leaf mass (g)	Middle crown part	0.082	0.0050	5.665	0.000	***
	Upper crown part	0.079	0.0050	15.797	0.000	***
	$R^2 = 0.625, AIC = -2606$					
	Intercept	−12.863	1.764	−7.292	0.000	***
	$Ln(d_0)$	8.391	0.478	17.560	0.000	***
Leaf area (cm^2)	Middle crown part	2.301	0.526	4.374	0.000	***
	Upper crown part	4.020	0.529	4.374	0.000	***
	$R^2 = 0.725, AIC = 6468$					
	Intercept	359.55	10.866	33.088	0.000	***
	$Ln(d_0)$	−46.77	2.944	−15.889	0.000	***
SLA (cm^2 g^{-1})	Middle crown part	−23.48	3.247	−7.233	0.000	***
	Upper crown part	−52.51	3.244	−16.183	0.000	***
	$R^2 = 0.654, AIC = 10089$					

Note: *** significant at 99.9% significance level.

Allometric biomass models for individual tree components as well as for whole-tree biomass showed their close dependences on d_0 diameter (in all cases $p < 0.001$; see Table 6). Fitted values showed that while aspen trees with a diameter of 50 mm had about 2.5 kg of woody biomass and 0.2 kg of foliage biomass, individuals with a diameter of 100 mm had 16.4 kg of biomass in woody parts and 0.7 kg in foliage (Figure 4). This means that biomass in woody parts increased with tree size faster (in the d_0 interval 50–100 mm as much as 6.6 times) than in foliage (in the diameter interval 50–100 mm only by 3.5 times). If foliage area at a tree level is considered, increase in the d_0 interval from 50 mm to 100 mm was from 2.6 m^2 to 8.7 m^2 (tripled value; Figure 5). As for LAR and LMR, both indicators decreased with tree size (Figure 6; Table 7). This fast decrease was typical in very small trees (with d_0 below approximately 20 mm); then the trend became milder.

Table 6. Biomass models of individual tree components in common aspen using stem base diameter as an independent variable, showing their regression coefficients (b$_1$, b$_2$), standard errors (SE), p-values (P), coefficients of determination (R^2), mean squared errors (MSE), logarithmic transformation bias (λ), and standard deviation (SD) (see Equation (2)).

Biomass of Tree	b$_1$	S.E.	P	b$_2$	S.E.	P	R^2	MSE	λ	S.D.
Components (g)										
Roots (A)	−3.315	0.130	<0.001	2.410	0.039	<0.001	0.954	0.145	1.078	0.474
Stem (B)	−3.612	0.107	<0.001	2.795	0.033	<0.001	0.976	0.098	1.048	0.323
Branches (C)	−5.683	0.196	<0.001	3.010	0.060	<0.001	0.935	0.306	1.116	0.710
Leaves (D)	−2.907	0.225	<0.001	2.020	0.069	<0.001	0.829	0.429	1.191	0.634
Woody parts (A+B+C)	−2.760	0.093	<0.001	2.699	0.028	<0.001	0.982	0.071	1.036	0.293
Whole tree (A+B+C+D)	−2.379	0.092	<0.001	2.618	0.028	<0.001	0.981	0.070	1.036	0.293

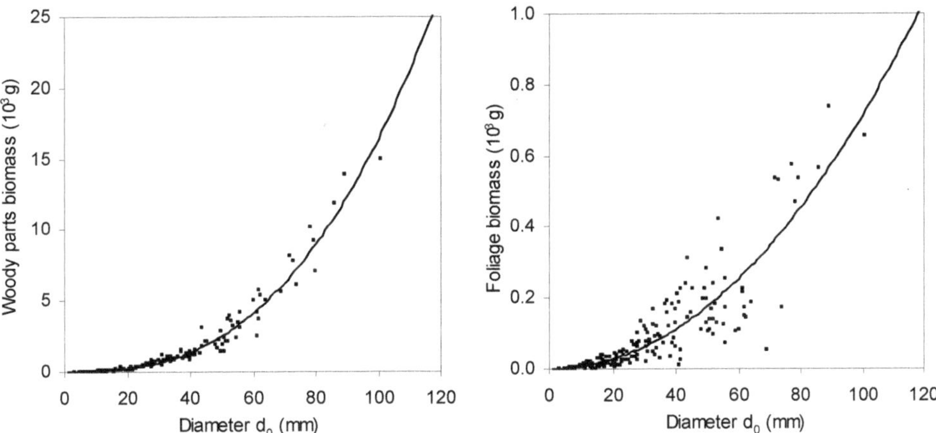

Figure 4. Biomass of woody parts (left plate) and foliage (right plate) in the sampled common aspen trees plotted against the d_0 stem base diameter (see also Table 6).

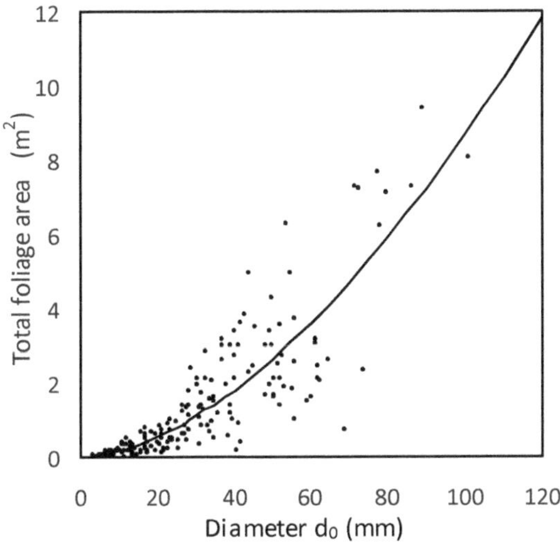

Figure 5. Total tree foliage area of the sampled common aspen trees against the d_0 stem base diameter (see also Table 7).

Table 7. Models for calculating tree foliage area, leaf area ratio (LAR), and leaf mass ratio (LMR) in common aspen using stem base diameter as an independent variable, showing their regression coefficients (b_1, b_2), standard errors (S.E.), p-values (P), coefficients of determination (R^2), and mean squared errors (MSE) (see Equation (1)).

Tree Characteristics (unit)	b_1	S.E.	P	b_2	S.E.	P	R^2	MSE
Tree foliage area (cm^2)	0.0031	0.0012	0.010	1.723	0.093	<0.001	0.734	0.834
LAR ($cm^2\ kg^{-1}$)	308.4	37.2	<0.001	−0.869	0.051	<0.001	0.634	165.93
LMR ($kg\ kg^{-1}$)	0.657	0.083	<0.001	−0.585	0.048	<0.001	0.458	1.297

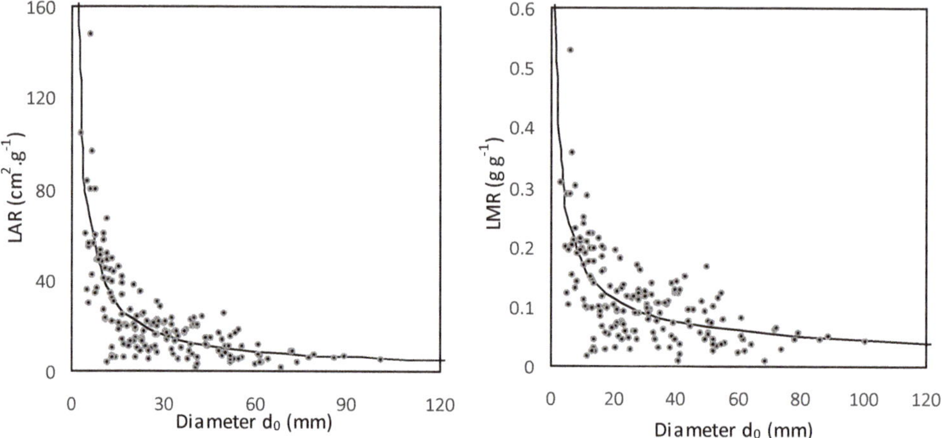

Figure 6. Leaf area ratio (left plate) and leaf mass ratio (right plate) plotted against the d_0 stem base diameter of the sampled common aspen trees (see also Table 3).

Finally, quantities of woody parts and foliage were estimated at the stand level. The fitted curves (Figure 7 and Table 8) showed that the increase in stand biomass in woody parts with an increasing d_0 was greater than the increase in foliage biomass. For instance, while an aspen stand with a mean diameter d_0 of 50 mm had about 8 kg of biomass in woody parts per m^2, a stand with a d_0 of 100 mm contained about 32 kg of woody biomass per m^2. This represents a fourfold increase. On the other hand, the difference between the aspen stands with mean diameters d_0 of 50 mm and 100 mm was only doubled (0.5 kg m^{-2} versus 1.2 kg m^{-2}). Values of LAI increased with d_0 first fast, and then more slowly (Figure 8, Table 8). For instance, while a stand with d_0 equal to 50 mm had an LAI of 7.3 m^2 m^{-2}, the LAI of a stand with a d_0 of 100 mm was 10.9 m^2 m^{-2}. The greatest value of LAI equal to 13.5 was found in the aspen forest with a mean stand diameter d_0 of 120 mm.

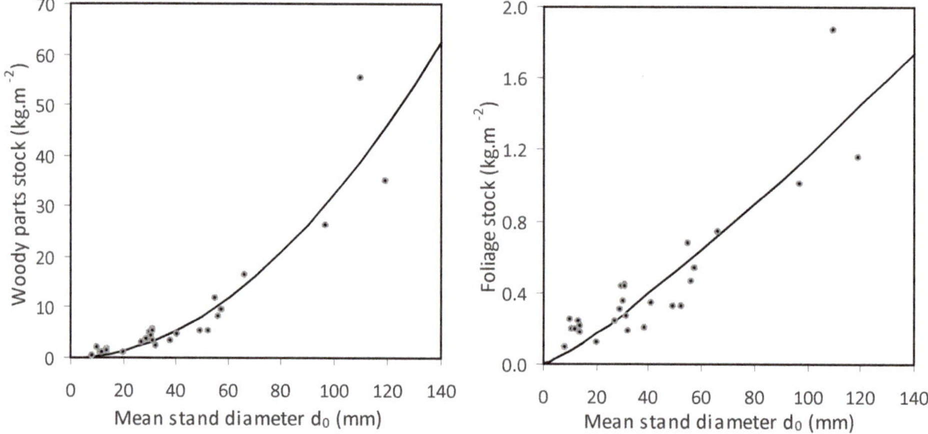

Figure 7. Stand biomass of woody parts (left plate) and foliage (right plate) in common aspen fully-stocked stands plotted against the d_0 mean stand diameter at the stem base (see also Table 8).

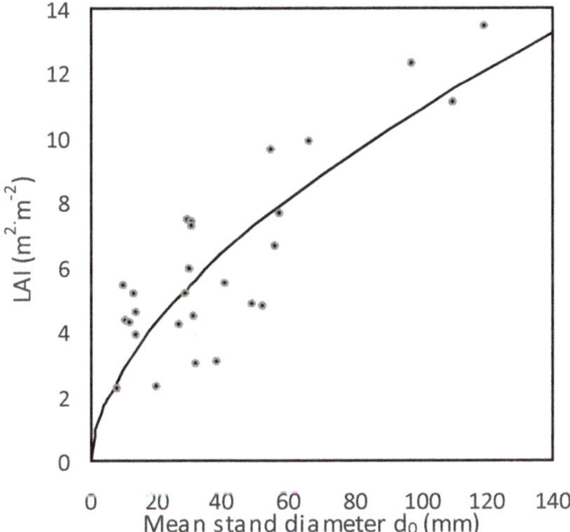

Figure 8. Leaf area index of fully-stocked common aspen stands plotted against the d_0 mean stand diameter at the stem base (see also Table 8).

Table 8. Models for calculating the biomass of woody parts, foliage, and leaf area index in common aspen stands using stem base diameter as an independent variable, showing their regression coefficients (b_1, b_2), standard errors (S.E.), p-values (P), coefficients of determination (R^2), and mean squared errors (MSE) (see Equation (1)).

Stand Characteristics (unit)	b_1	S.E.	P	b_2	S.E.	P	R^2	MSE
Woody parts stock (kg m^{-2})	0.0039	0.0036	0.291	1.959	0.201	<0.001	0.882	18.497
Foliage stock (kg m^{-2})	0.005	0.003	0.082	1.184	0.125	<0.001	0.810	0.029
Leaf area index (m^2 m^{-2})	0.929	0.305	0.005	0.522	0.082	<0.001	0.616	3.160

4. Discussion

4.1. Foliage- and Tree-Level Traits

Foliage traits are important characteristics that specify the rate of absorbing photosynthetically active radiation used in photosynthesis (leaf surface area), the metabolic costs of leaf construction (leaf mass), and the efficiency with which the leaf captures light relative to the biomass invested in the plant biomass (leaf mass ratio) or leaf (specific leaf area). Their values obtained from field measurements are required not only for the assessment of the trade-off between ecosystem mass and energy balance, but also for accurate parametrisation and predictions of forest growth by process-based models [39]. Increasing mass and area of single leaves with tree size was typical for young common aspen trees, while the opposite situation was found for SLA. Besides tree size, the position of the leaves along the vertical crown profile appeared to be important for foliage properties, more clearly for leaf mass and SLA than leaf area.

Our study indicated that while smaller trees had larger (as for area) leaves in the upper crown part and smallest in the lower crown part, the situation in bigger trees was reversed. Larger leaves in the upper crown part were also found for European beech trees with a d_0 diameter below 60 mm [40]. Similarly, [41] showed that shade leaves in young beech stands had prevailingly lower leaf areas than sunlit leaves. A reverse situation was recorded in old beech stands [42,43]. The increase of SLA from the top to the bottom of the crown was demonstrated also in other works [40,42,44,45]. It seems that SLA is a better indicator of foliage plasticity to environmental factors, especially light conditions [42,46], than the area of an individual leaf.

Our results showed that the SLA of common aspen leaves fluctuated between approx. 100 and 300 cm^2 g^{-1}. Other published values of SLA for common aspen varied from 39 to 182 cm^2 g^{-1} [4], i.e., they were lower than most of our values (Figure 3c) although they represented stands of similar ages (from 5 to 24 years). The results from European beech stands showed SLA values similar to our results, ranging from 80 to 480 cm^2 g^{-1} [41,42,44,47,48].

The values of LMR for small aspen trees were high and sharply decreased with tree size. The LMR of aspen trees with a d_0 diameter of about 20 mm was below 0.1, while the LMR of trees with a d_0 of 80 mm reached a value of only about 0.05. Our previous study [49] for *Fagus sylvatica* and *Quercus petraea* showed that the LMR of trees with a d_0 between 20 and 80 mm decreased from 0.2 to 0.1 and from 0.15 to 0.10, respectively. These values indicate that in the investigated diameter interval, the contribution of foliage to total biomass of *Fagus sylvatica* and *Quercus petraea* was double the foliage contribution in aspen. This suggests a high growth efficiency of common aspen in tree biomass production [50].

Our height–diameter model for young common aspen trees showed a steep, almost a linear increase in tree height with increasing d_0 (Figure 2). Since we have previously constructed height–diameter models for other tree species [51], we were able to analyse inter-specific differences. We compared aspen versus seven other broadleaved tree species, i.e., *Fagus sylvatica*, *Carpinus betulus*, *Quercus petraea*, *Acer pseudoplatanus*, *Fraxinus excelsior*, *Salix caprea*, and *Sorbus aucuparia*. The comparison showed that aspen trees were higher than the trees of other species with the same d_0 diameter. For instance, while aspen with a d_0 equal to 80 mm was approximately 9 m tall, the height of *Acer pseudoplatanus* was only 8 m, *Carpinus betulus* was 7 m, and all other species were less than 7 m tall. This indicates that in comparison with other species young aspen trees invest more carbohydrates into height increment than into radial stem increment. From the point of inter-specific competition for light, that might be its advantage in young mixed stands. In this context, for instance, [31] pointed out the very high competitive ability of aspen, especially on fertile soils with fresh moisture conditions.

The calculated mean woody biomass value of the investigated common aspen trees was lower than the published values from young hybrid aspen stands (see e.g., [31,52,53]) or other poplar cultivars [54] of similar ages, while the range of our values was usually greater than in other studies. This was expected since our trees originated from natural regeneration, while the other studies dealt with trees from plantations established for biomass production.

4.2. Stand Level Traits

Mean woody stand biomass stock of young common aspen at our plots was 8.20 ± 12.30 kg/m^2. The value does not significantly differ from other published values, e.g., [31] reported an average value of aboveground woody biomass for his hybrid aspen stands equal to 13.5 ± 5.3 kg/m^2. Similar to tree biomass, the variability in stand biomass between plots was large (Figure 7). A wide range of aboveground stand biomass values was also reported in other studies, e.g., the aboveground biomass of hybrid aspen stands in Sweden ranged from 1.43 to 21.9 kg/m^2 [31,55], or young common aspen stands aged below 20 years included in the Eurasian database [56,57] were characterised by aboveground biomass values ranging from 1.6 to 10.99 kg/m^2. Faster biomass accumulation has been documented at fertile sites with good soil conditions [30].

Stand biomass stock of woody tree parts in young common aspen forests exponentially increased with increasing stand dimensions (mean stand diameter d_0), while foliage biomass stock increased almost linearly (Figure 7). Hence, the ratio between woody part biomass and foliage biomass in an aspen forest with a mean stand stem base diameter of 50 mm was approximately 20, the same ratio was almost 30 if the mean stand diameter was 100 mm. These changes with stand development reflect not only the physiological (increasing growth efficiency) but also production–ecological aspects of forests, especially their carbon sequestration potential. Woody parts represent long-term carbon cycling components, while leaves store carbon only for a short time. Approximately 1/20 of total tree biomass annually falls off on the ground in young aspen forests with a mean stand diameter of 50 mm, while in the stand with a diameter of 100 mm it is only 1/30 (here we obviously neglected tree mortality as a source of carbon transfer from biomass to necromass). Foliage litter was estimated to be about 0.5 kg per m^2 and 1.1 kg per m^2 in stands with a d_0 mean diameters of 50 mm and 100 mm, respectively. A similar decrease in foliage contribution to total tree biomass has been recorded in beech [58] or birch stands [1].

The LAI of our aspen stands fluctuated between 2 m^2 m^{-2} and 14 m^2 m^{-2}. The maximum estimated value of nearly 14 m^2 m^{-2} is greater than the values published in most of the other papers covering a variety of forest tree species (often European beech) [46,47,59–63]. The prevailing part of the other papers presented LAI values up to 10 m^2 m^{-2}. In fact, the results of most papers originated from older growth stages than in our case. The few papers that showed LAI values near 14 m^2 m^{-2} [64–67] originated from young growth stages. This might suggest that forest stands reach their maximum LAI in the young stages and after that the values decrease with stand development, which has already been documented for Norway spruce [65]. Our results bring new findings about biomass allocation and foliage traits of common aspen at both the tree and stand levels. Since this kind of study has been missing for common aspen, our results might not be compared or synthesized with other knowledge. To obtain biomass information about less abundant and/or currently commercially unimportant species, foresters often use equations adopted from the already published sources, or those developed for other species. However, such an approach may cause a significant bias in the obtained estimates, since growing conditions and tree properties vary across the regions and country-specific models also differ from each other [67]. Comparison of different models for birch showed that non-local models may overestimate total aboveground biomass of thin trees with a diameter at breast height below 4 cm [68]. Moreover, equations derived 20 years before or more may no longer be valid due to the recent changes in environmental conditions [69]. Hence, model updates are required even for thoroughly studied commercial species.

5. Conclusions

This study brings novel findings for a very productive tree species with modest ecological demands, but competitive to other plant species (both weeds and trees). Therefore, it might be planted on sites which are not attractive and/or suitable for traditionally important commercial tree species (e.g., European beech and Norway spruce). Moreover, the species could be very useful especially for energy production. A rapid increase in bioenergy demands has been predicted for most future scenarios, considering especially ecological (e.g., global carbon balance) and technological issues [70]. Biomass of short-living tree species may significantly contribute to bioenergy production and carbon sequestration [71]. Therefore, we believe that our results might be significant for both theoretical (e.g., biomass production and partitioning modelling) and practical (forestry and agro-forestry stakeholders) purposes.

Author Contributions: Conceptualization, B.K.; data curation, J.P. and V.Š.; funding acquisition, B.K.; investigation, B.K., J.P. and V.Š.; methodology, B.K. and P.S.; visualisation, J.P. and V.Š; supervision, B.K.; writing—original draft, B.K., J.P., V.Š., P.S. and K.M.; writing—review and editing, B.K. and K.M. All authors have read and agreed to the published version of the manuscript.

Funding: This research was funded by grant "EVA4.0", No. CZ.02.1.01/0.0/0.0/16_019/0000803 financed by OP RDE, also by the projects APVV-0584-12 and APVV-18-0086 from the Slovak Research and Development Agency as well as by the project "Research and innovation for supporting competitiveness of the Slovak forestry sector" (SLOV-LES) financed by the Ministry of Land Management and Rural Development of the Slovak Republic.

Conflicts of Interest: The authors declare no conflict of interest.

References

1. Jagodzinski, A.M.; Dyderski, M.; Horodecki, P. Tree- and Stand-Level Biomass Estimation in a *Larix decidua* Mill. Chronosequence. *Forests* **2018**, *9*, 587. [CrossRef]
2. Merganičová, K.; Merganič, J.; Lehtonen, A.; Vacchiano, G.; Sever, M.; Augustynczik, A.D.; Grote, R.; Kyselová, I.; Mäkelä, A.; Yousefpour, R.; et al. Forest carbon allocation modeling climate change. *Tree Physiol.* **2019**, *39*, 1937–1960. [CrossRef] [PubMed]

3. Mensah, S.; Kaki, R.L.G.; Siefert, T. Patterns of biomass allocation between foliage and woody structure: The effects of tree size and specific functional traits. *Ann. For. Res.* **2016**, *59*, 49–60. [CrossRef]

4. Forrester, D.I.; Tachauer, I.H.H.; Annighoefer, P.; Barbeito, I.; Pretzsch, H.; Ruiz-Peinado, R.; Stark, H.; Vacchiano, G.; Zlatanov, T.; Charaborty, T.; et al. Generalized biomass and leaf area allometric equations for European tree species incorporating stand structure, tree age and climate. *For. Ecol. Manag.* **2017**, *396*, 160–175. [CrossRef]

5. Eggleston, S.; Buedia, L.; Miwa, K.; Ngara, T.; Tanabe, K. *IPCC Guidelines for National Greenhouse Gas Inventories*; Institute for Global Environmental Strategy: Hayama, Japan, 2006.

6. Jagodzinski, A.M.; Zasada, M.; Bronisz, K.; Bronisz, A.; Bijak, S. Biomass conversion and expansion factors for a chronosequence of young naturally regenerated silver birch (*Betula pendula* Roth) stands growing on post-agricultural sites. *For. Ecol. Manag.* **2017**, *384*, 208–220. [CrossRef]

7. Jagodzinski, A.M.; Dyderski, M.; Gesikiewicz, K.; Horodecki, P.; Cysewska, A.; Wierczynska, S.; Maciejczyk, K. How do tree stand parameters affect young Scots pine biomass?—Allometric equations and biomass conversion and expansion factors. *For. Ecol. Manag.* **2018**, *409*, 74–83. [CrossRef]

8. Jagodzinski, A.M.; Dyderski, M.K.; Gęsikiewicz, K.; Horodecki, P. Effects of stand features on aboveground biomass and biomass conversion and expansion factors based on a *Pinus sylvestris* L. chronosequence in Western Poland. *Eur. J. For. Res.* **2019**, *138*, 673–683. [CrossRef]

9. Hommel, R.; Siegwolf, R.T.W.; Zavadlav, S.; Arend, M.; Schaub, M.; Galiano, L.; Haeni, M.; Kayler, Z.; Gessler, A. Impact of interspecific competition and drought on the allocation of new assimilates in trees. *Plant Biol.* **2016**, *18*, 785–796. [CrossRef]

10. Yang, X.Z.; Zhang, W.H.; He, Q.Y. Effects of intraspecific competition on growth, architecture and biomass allocation of *Quercus liaotungensis*. *J. Plant Interact.* **2019**, *14*, 284–294. [CrossRef]

11. Pickup, M.; Westoby, M.; Basden, A. Dry mass costs of deploying leaf area in relation to leaf size. *Funct. Ecol.* **2005**, *19*, 88–97. [CrossRef]

12. Shipley, B. Net assimilation rate, specific leaf area and leaf mass ratio: Which is most closely correlated with relative growth rate? A meta-analysis. *Funct. Ecol.* **2006**, *20*, 565–574. [CrossRef]

13. Milla, R.; Reich, P.B.; Niinemets, U.; Castro-Diez, P. Environmental and developmental controls on specific leaf area are little modified by leaf allometry. *Funct. Ecol.* **2008**, *22*, 565–576. [CrossRef]

14. Tomlinson, K.W.; Poorter, L.; Sterck, F.J. Leaf adaptations of ever-green and deciduous trees of semi-arid and humid savannas on three continents. *J. Ecol.* **2013**, *101*, 430–440. [CrossRef]

15. Savoy, P.; MacKay, D.S. Modeling the seasonal dynamics of leaf area index based on environmental constraints to canopy development. *Agric. For. Meteorol.* **2015**, *200*, 46–56. [CrossRef]

16. Asaadi, A.; Arora, V.K.; Melton, J.R.; Bartlett, P. An improved parameterization of leaf area index (LAI) seasonality in the Canadian Land Surface Scheme (CLASS) and Canadian Terrestrial Ecosystem Model (CTEM) modelling framework. *Biogeosciences* **2018**, *15*, 6885–6907. [CrossRef]

17. Worrell, R. European aspen (*Populus tremula* L.): A review with particular reference to Scotland, I. Distribution, ecology and genetic variation. *Forestry* **1995**, *68*, 93–105. [CrossRef]

18. Pagan, J.; Randuška, D. *Atlas drevín 1. (Pôvodné dreviny)*; Obzor: Bratislava, Slovakia, 1987; 360p.

19. Kacálek, D.; Mauer, O.; Podrázský, V.; Slodičák, M.; Holušová, K.; Špulák, O.; Souček, J.; Jurásek, A.; Leugner, J.; Dušek, D. *Soil Improving and Stabilizing Functions of Forest Trees*; Lesnická práce Ltd.: Prague, Czech Republic, 2017; 300p.

20. San-Miguel-Ayanz, J.; de Rigo, D.; Caudullo, G.; Durrant, T.H.; Mauri, A. *European Atlas of Forest Tree Species*; Publication Office of European Union: Luxembourg, 2016; 200p.

21. Myking, T.; Bohler, F.; Austrheim, G.; Solberg, E.J. Life history strategies of aspen (*Populus tremula* L.) and browsing effects: A literature review. *Forestry* **2011**, *84*, 61–71. [CrossRef]

22. Baum, C.; Eckhardt, K.-U.; Hahn, J.; Weih, M.; Dimitriou, I.; Leinweber, P. Impact of poplar on soil organic matter quality and microbial communities in arable soils. *Plant Soil Environ.* **2013**, *59*, 95–100. [CrossRef]

23. Bergström, R.; Hjeljord, O. Moose and vegetation interactions in northwestern Europe and Poland. *Swed. Wildl. Res. Suppl.* **1987**, *1*, 213–228.

24. Konôpka, B.; Pajtík, J.; Bošeľa, M.; Šebeň, V.; Shipley, L.A. Modeling forage potential for red deer (*Cervus elaphus*): A tree level approach. *Eur. J. For. Res.* **2020**. [CrossRef]

25. Tognetti, R.; Cocozza, C.; Marchetti, M. Shaping the multifunctional tree: The use of Salicaceae in environmental restoration. *iForest* **2013**, *6*, 37–47. [CrossRef]

26. Šebeň, V. Národná inventarizácia a monitoring lesov Slovenskej republiky 2015–2016. *Lesn. Štud.* **2017**, *65*, 255.

27. Šebeň, V.; Kučera, M.; Meragničová, K.; Konôpka, B. The current state of non-forest land in the Czech Republic and Slovakia—Forest cover estimates based on the national inventory data. *Cent. Eur. For. J.* **2018**, *64*, 207–222.

28. Výbohová, E.; Kučerová, V. The changes of selected characteristics of cellulose at water prehydrolysis of poplar (*Populus tremula* L.) wood. *Acta Fac. Xylogiae Zvolen* **2013**, *55*, 77–86.

29. MacKenzie, N. *Ecology, Conservation and Management of Aspen. A Literature Review*; Scottish Native Woods: Aberfeldy, UK, 2010; 41p.

30. Tullus, A.; Rytter, L.; Tullus, T.; Weih, M.; Tullus, H. Short-rotation forestry with hybrid aspen (*Populus tremula* L. × *P. tremuloides* Michx.) in Northern Europe. *Scand. J. For. Res.* **2012**, *27*, 10–29. [CrossRef]

31. Johansson, T. Biomass production of hybrid aspen growing on former farm land in Sweden. *J. For. Res.* **2013**, *24*, 237–246. [CrossRef]

32. Fahlvik, N.; Rytter, L.; Stener, L.G. Production of hybrid aspen on agricultural land during one rotation in southern Sweden. *J. For. Res.* **2019**. [CrossRef]

33. Easlon, H.M.; Bloom, A.J. Easy Leaf Area: Automated digital image analysis for rapid and accurate measurement of leaf area. *Appl. Plant Sci.* **2014**, *2*, 1400033. [CrossRef]

34. Pajtík, J.; Konôpka, B.; Lukac, M. Biomass functions and expansion factors in young Norway spruce (*Picea abies* [L.] Karst) trees. *For. Ecol. Manag.* **2008**, *256*, 1096–1103. [CrossRef]

35. Cienciala, E.; Černý, M.; Tatarinov, F.; Apltauer, J.; Exnerová, Z. Biomass function applicable to Scots pine. *Trees* **2016**, *20*, 483–495. [CrossRef]

36. Lai, J.; Yang, B.; Lin, D.; Kerkhoff, A.J.; Ma, K. The allometry of coarse root biomass: Long-transformed linear regression or nonlinear regression? *PLoS ONE* **2013**, *8*, e77007. [CrossRef] [PubMed]

37. Mascaro, J.; Litton, C.M.; Hughes, R.F.; Uowolo, A.; Schnitzer, S.A. Is logarithmic transformation necessary in allometry? Ten, one-hundred, one-thousand-time yes. *Biol. J. Linn. Soc. Lond.* **2014**, *111*, 230–233. [CrossRef]

38. R Development Core Team. *R: A Language and Environment for Statistical Computing*; R foundation for Statistical Computing: Vienna, Austria, 2012; Available online: http://www.R-project.org/ (accessed on 10 February 2012).

39. Marshall, J.D.; Monserud, R.A. Foliage height influences specific leaf area of three conifer species. *Can. J. For. Res.* **2003**, *33*, 164–170. [CrossRef]

40. Konôpka, B.; Pajtík, J. Similar foliage area but contrasting foliage biomass between young beech and spruce stands. *Cent. Eur. For. J.* **2014**, *60*, 205–213.

41. Closa, I.; Irigoyen, J.J.; Goicoechea, N. Microclimatic conditions determined by stem density influence leaf anatomy and leaf physiology of beech (*Fagus sylvatica* L.) growing within stands that naturally regenerate from clear-cutting. *Trees* **2010**, *24*, 1029–1043. [CrossRef]

42. Barna, M. Adaptation of European beech (*Fagus sylvatica* L.) to different ecological conditions: Leaf size variation. *Pol. J. Ecol.* **2004**, *52*, 35–45.

43. Scartazza, A.; Di Baccio, D.; Bertolotto, P.; Gavrichova, O.; Matteucci, G. Investigating the European beech (*Fagus sylvatica* L.) leaf characteristics along the vertical canopy profile: Leaf structure, photosynthetic capacity, light energy dissipation and photoprotection mechanisms. *Tree Physiol.* **2016**, *36*, 1060–1076. [CrossRef]

44. Bartelink, H.H. Allometric relationships for biomass and leaf area of beech (*Fagus sylvatica* L). *Ann. Sci. For.* **1997**, *54*, 39–50. [CrossRef]

45. Petritan, A.M.; von Lupke, B.; Petritan, I.C. Influence of light availability on growth, leaf morphology and plant architecture of beech (*Fagus sylvatica* L.) maple (*Acer psuedoplatanus* L.) and ash (*Fraxinus excelsior* L.) saplings. *Eur. J. For. Res.* **2009**, *128*, 61–74. [CrossRef]

46. Meier, I.C.; Leuschner, C. Leaf size and leaf area index in *Fagus sylvatica* forests: Competing effects of precipitation, temperature, and nitrogen availability. *Ecosystems* **2008**, *11*, 655–669. [CrossRef]

47. Bouriaud, O.; Soudani, K.; Bréda, N. Leaf area index from litter collection: Impact of specific leaf area variability within a beech stand. *Can. J. Remote Sens.* **2003**, *29*, 371–380. [CrossRef]

48. Leuschner, C.; Voss, S.; Foetzki, A.; Clases, Y. Variation in leaf area index and stand leaf mass of European beech across gradients of soil acidity and precipitation. *Plant Ecol.* **2006**, *186*, 247–258. [CrossRef]

49. Pajtík, J.; Konôpka, B.; Lukac, M. Individual biomass factors for beech, oak and pine in Slovakia: A comparative study in young naturally regenerated stands. *Trees* **2011**, *25*, 277–288. [CrossRef]

50. Gersonde, R.; O'Hara, K.L. Comparative tree growth efficiency in Sierra Nevada mixed-conifer forests. *For. Ecol. Manag.* **2005**, *219*, 95–108. [CrossRef]

51. Pajtík, J.; Konôpka, B.; Šebeň, V. *Mathematical Biomass Models for Young Individuals of Forest Tree Species in the Region of the Western Carpathians*; National Forest Centre: Zvolen, Slovakia, 2018; 89p.

52. Tullus, A.; Tullus, H.; Soo, T.; Pärn, L. Above-ground biomass characteristics of young hybrid aspen (*Populus tremula* L. × *P. tremuloides* Michx.) plantations on former agricultural land in Estonia. *Biomass Bioenergy* **2009**, *33*, 1617–1625. [CrossRef]

53. Jansons, A.; Rieksts-Riekstins, J.; Senhofa, S.; Katrevics, J.; Lazdina, D.; Sisenis, L. Above-ground Biomass Equations of Populus Hybrids in Latvia. *Balt. For.* **2017**, *23*, 507–514.

54. Niemczyk, M.; Wojda, T.; Kaliszewski, A. Biomass productivity of selected poplar (*Populus* spp.) cultivars in short rotations in northern Poland. *N. Z. J. For. Sci.* **2016**, *46*, 22. [CrossRef]

55. Johansson, T. Biomass equations for determining fractions of European aspen growing on abandoned farmland and some practical implications. *Biomass Bioenergy* **1999**, *17*, 471–480. [CrossRef]

56. Usoltsev, V.A. *Eurasian Forest Biomass and Primary Production Data*; Ural Branch of Russian Academy of Sciences: Yekaterinburg, Russia, 2010; 570p.

57. Usoltsev, V.A. *Forest Biomass and Primary Production Database for Eurasia: CD-Version*, 2nd ed.; Enlarged and Reharmonized; Ural State Forest Engineering University: Yekaterinburg, Russia, 2013.

58. Šebeň, V.; Konôpka, B.; Pajtík, J. Quantifying carbon in dead and living trees; a case study in young beech and spruce stand over 9 years. *Cent. Eur. For. J.* **2017**, *63*, 133–141. [CrossRef]

59. Gond, V.; de Pury, D.G.G.; Veroustraete, F.; Ceulemans, R. Seasonal variation in leaf area index, leaf chlorophyll, and water content: Scaling-up to estimate fPAR and carbon balance in a multilayer, multispecies temperate forest. *Tree Physiol.* **1999**, *19*, 673–679. [CrossRef]

60. Gower, S.T.; Kucharik, C.J.; Norman, J.N. Direct and indirect estimation of leaf area index, f_{APAR} and net primary production of terrestrial ecosystems. *Remote Sens. Environ.* **1999**, *70*, 29–51. [CrossRef]

61. Bréda, N.J.J. Ground-based measurements of leaf area index: A review of methods, instruments and current controversies. *J. Exp. Bot.* **2003**, *54*, 2403–2417. [CrossRef] [PubMed]

62. Thimonier, A.; Sedivy, I.; Schleppi, P. Estimating leaf area index in different types of mature forest stands in Switzerland: A comparison of methods. *Eur. J. For. Res.* **2010**, *129*, 543–562. [CrossRef]

63. Ariza-Carricondo, C.; Di Mauro, F.; De Beck, M.O.; Roland, M.; Gielen, B.; Vitale, D.; Ceulemans, R.; Papale, D. A comparison of different methods for assessing leaf area index in four canopy types. *Cent. Eur. For. J.* **2019**, *65*, 67–80. [CrossRef]

64. Tadaki, Y. Studies on the production structure of forest. XVII. Vertical change of specific leaf area in forest canopy. *J. Jpn. For. Soc.* **1970**, *52*, 263–268.

65. Pokorný, R.; Stojnič, S. Leaf area index in Norway spruce stands in relation to age and defoliation. *Beskydy* **2012**, *5*, 173–180. [CrossRef]

66. Konôpka, B.; Pajtík, J.; Marušák, R.; Bošeľa, M.; Lukac, M. Specific leaf area and leaf area index in developing stands of *Fagus sylvatica* L. and *Picea abies* Karst. *For. Ecol. Manag.* **2016**, *364*, 52–59. [CrossRef]

67. Neumann, M.; Moreno, A.; Mues, V.; Härkönen, S.; Mura, M.; Bouriaud, O.; Lang, M.; Achten, W.M.J.; Thivolle-Cazat, A.; Bronisz, K.; et al. Comparison of carbon estimation methods for European forests. *For. Ecol. Manag.* **2016**, *361*, 397–420. [CrossRef]

68. Bronisz, K.; Strub, M.; Cieszewski, C.; Bijak, S.; Bronisz, A.; Tomusiak, R.; Wojtan, R.; Zasada, M. Empirical equations for estimating aboveground biomass of *Betula pendula* growing on former farmland in central Poland. *Silva Fenn.* **2016**, *50*, 1559. [CrossRef]

69. Lang, M.; Lilleleht, A.; Neumann, M.; Bronisz, K.; Rolim, S.G.; Seedre, M.; Uri, V.; Kiviste, A. Estimation of above-ground biomass in forest stands from regression on their basal area and height. *For. Stud.* **2016**, *64*, 70–92. [CrossRef]

70. Lauri, P.; Forsell, N.; Gusti, M.; Korsuo, A.; Havlik, P.; Obersteiner, M. Global Woody Biomass Harvest Volumes and Forest Area Use under Different SSP-RCP Scenarios. *J. For. Econ.* **2019**, *34*, 285–309. [CrossRef]

71. Jha, K.K. Biomass production and carbon balance in two hybrid poplar (*Populus euramericana*) plantations raised with and without agriculture in southern France. *J. For. Res.* **2018**, *29*, 1689–1701. [CrossRef]

Article

Net Primary Productivity of *Pinus massoniana* Dependence on Climate, Soil and Forest Characteristics

Xin Huang [1], Chunbo Huang [2], Mingjun Teng [1], Zhixiang Zhou [1,*] and Pengcheng Wang [1]

[1] College of Horticulture and Forestry Sciences/Hubei Engineering Technology Research Center for Forestry Information, Huazhong Agricultural University, Wuhan 430070, China; hxhanson@163.com (X.H.); tengmingjun@hotmail.com (M.T.); wangpc@mail.hzau.edu.cn (P.W.)

[2] School of Geography and Information Engineering, China University of Geosciences, Wuhan 430074, China; poahcb@hotmail.com

* Correspondence: whzhouzx@mail.hzau.edu.cn; Tel.: +86-027-8728-4232

Received: 4 January 2020; Accepted: 2 April 2020; Published: 4 April 2020

Abstract: Understanding the spatial variation of forest productivity and its driving factors on a large regional scale can help reveal the response mechanism of tree growth to climate change, and is an important prerequisite for efficient forest management and studying regional and global carbon cycles. *Pinus massoniana* Lamb. is a major planted tree species in southern China, playing an important role in the development of forestry due to its high economic and ecological benefits. Here, we establish a biomass database for *P. massoniana*, including stems, branches, leaves, roots, aboveground organs and total tree, by collecting the published literature, to increase our understanding of net primary productivity (NPP) geographical trends for each tree component and their influencing factors across the entire geographical distribution of the species in southern China. *P. massoniana* NPP ranges from 1.04 to 13.13 $Mg·ha^{-1}·year^{-1}$, with a mean value of 5.65 $Mg·ha^{-1}·year^{-1}$. The NPP of both tree components (i.e., stem, branch, leaf, root, aboveground organs, and total tree) show no clear relationships with longitude and elevation, but an inverse relationship with latitude ($p < 0.01$). Linear mixed-effects models (LMMs) are employed to analyze the effect of environmental factors and stand characteristics on *P. massoniana* NPP. LMM results reveal that the NPP of different tree components have different sensitivities to environmental and stand variables. Appropriate temperature and soil nutrients (particularly soil available phosphorus) are beneficial to biomass accumulation of this species. It is worth noting that the high temperature in July and August (HTWM) is a significant climate stressor across the species geographical distribution and is not restricted to marginal populations in the low latitude area. Temperature was a key environmental factor behind the inverse latitudinal trends of *P. massoniana* NPP, because it showed a higher sensitivity than other factors. In the context of climate warming and nitrogen (N) deposition, the inhibition effect caused by high temperatures and the lack or imbalance of soil nutrients, particularly soil phosphorus, should be paid more attention in the future. These findings advance our understanding about the factors influencing the productivity of each *P. massoniana* tree component across the full geographical distribution of the species, and are therefore valuable for forecasting climate-induced variation in forest productivity.

Keywords: net primary productivity; *Pinus massoniana*; geographical gradient; environmental factors; stand characteristics; regional scale

---

1. Introduction

Forest ecosystems are a major component of the terrestrial ecosystem worldwide and play an irreplaceable role in regulating global carbon balances and mitigating atmospheric concentrations of greenhouse gases, as well as in biodiversity and water conservation [1,2]. Forest biomass is the total amount of organic matter formed by CO_2 sequestration in the process of photosynthesis per unit area, and its accumulation rate is usually used as an indicator of forest productivity [3]. Therefore, forest biomass and productivity are important for measuring carbon sequestration ability and assessing carbon balance of forest ecosystems, and play an important role in global carbon cycle research [4]. Since the industrial revolution in the mid-20th century, the rapid development of human activities and modern industry, especially the burning of fossil fuels, massive deforestation, and grassland reclamation, have had a tremendous impact on the global ecosystem [5]. Therefore, studies on the environmental response of forest productivity in the context of global climate change, such as atmospheric nitrogen (N) deposition, elevated CO_2 concentration, and climate warming, have received increasing attention because of their great practical significance for reducing carbon emissions and mitigating global warming [6,7].

The production, allocation, and turnover of forest carbon on a large regional scale has received considerable attention in the past and is becoming increasingly important [8]. The accurate assessment of forest productivity on a large regional scale and the understanding of its influencing factors can provide a theoretical basis for enhancing forest productivity and studying the terrestrial carbon cycle [9]. Because of the complex terrain and large environmental differences in China, the geographical distribution of forest biomass and productivity is diverse [10–12]. As an important component of forest resources in China, plantations play an increasingly prominent role in maintaining the global carbon balance and mitigating global warming [13]. As artificially regulated ecosystems, the carbon sequestration function of plantation ecosystems is directly or indirectly affected by tree species, afforestation strategies, and tending operations [14]. Most large-scale studies on biomass and productivity have been conducted for a variety of forest types in China, rather than for a particular tree species, although many of these forests are single-species plantations [15]. Research on the response of forest productivity to environmental factors at the tree species level can reveal species-specific ecophysiological characteristics. On this basis, more targeted forest management strategies (e.g., site selection for afforestation and selection of tree species for close-to-nature transformation of monospecific plantations) could be proposed for different tree species in different regions, especially for major planted tree species, to improve forest productivity and carbon storage [15,16].

In the past 20 years, China has implemented a series of ecological projects, such as the Grain-to-Green Program, to improve the situation of low forest coverage and poor forest resources, and has now become the country with the largest area of plantations [17]. *Pinus massoniana* Lamb. is widely distributed in subtropical areas of China. It is an important afforestation tree species in barren mountains owing to its strong adaptability, fast growth, high yield, and drought tolerance, and is also one of the most representative forest types in China [18]. As one of the major planted and native tree species in southern China, *P. massoniana* has made a great contribution to plantation development [13,15]. The eighth forest resources inventory of China published by the State Forestry Administration in 2014 showed that the total area and volume of *P. massoniana* were 1.0×10^7 ha and 5.9×10^8 m^3, respectively, accounting for 6.1% and 4.0% of the total area and volume of arbor forests in China. However, *P. massoniana* plantations are faced with the problems of their irrational structure, poor stability and degraded productivity. Therefore, seeking scientific ways of regulating its productivity to improve its ecological and economic benefits is a major concern. Studies on biomass and productivity of *P. massoniana* began in the early 1980s [19], and mainly focused on biomass estimation and its allocation patterns in different site conditions, stand ages and densities [20], allometric equations research [21] and the effect of species mixtures on biomass [22]. In general, the research objectives of the above-mentioned studies mainly focused on quantitative assessment of biomass, and few studies involved biomass accumulation rate, which is usually quantified by forest productivity [23]. Therefore, knowledge about productivity responses to environmental factors is underdeveloped. Moreover, although tree-ring chronologies have been widely used to explore the relationship between climate and *P. massoniana* tree growth [24], these experiments have been carried out in individual study sites on a local scale, and the variation in productivity and its environmental control on a regional scale is still unclear. In addition, biotic factors, such as tree age and density, also influence forest productivity [25]. The variable environmental factors (e.g., temperature, precipitation, and soil properties) and forest characteristics (e.g., stand age and density) within the *P. massoniana* distribution area offers an opportunity to examine its productivity distribution patterns and its influencing factors on a large regional scale.

Numerous quantitative evaluations in the form of meta-analyses have been carried out to explore forest productivity and its influencing factors [26,27]. A lot of field measurements of *P. massoniana* biomass and productivity estimations have been conducted at multiple sites over the past four decades. However, these data were mostly published in Chinese journals and reports and are not accessible to non-Chinese scientists [27]. A *P. massoniana* biomass database could promote the completion of the established biomass database for China's forests and contribute to generating a huge database worldwide that will aid in the validation of ecosystem models, improving our understanding of the global carbon cycle and accurately evaluating carbon storage [26].

In this study, we established a biomass database for *P. massoniana*, including stems, branches, leaves, roots, aboveground organs, and total tree values by collecting data from the published literature, to increase our understanding of the geographical trends of *P. massoniana* net primary productivity (NPP) and its influencing factors across the full geographical distribution of the species in southern China. Our objectives were: (1) to document the spatial distribution of *P. massoniana* NPP on a large regional scale, (2) to quantify the effects of environmental factors (climate and soil variables) and stand characteristics on NPP of each tree component (i.e., stem, branch, leaf, root, aboveground organs and total tree), and (3) to identify the key environmental factors causing spatial heterogeneity in *P. massoniana* NPP.

2. Materials and Methods

2.1. Data Collection and Treatments

Our *P. massoniana* biomass database was generated by collecting published data (Figure 1, Table S1). To compile a comprehensive database, we searched the Web of Science for English literature and China National Knowledge Internet for Chinese literature. To minimize the variability associated with comparing biomass/productivity estimates derived from different methodologies, we selected all references included in the database that (1) included data actually measured in field experiments on *P. massoniana* stands (not forest inventory data or remote sensing based studies), (2) did not contain severe anthropogenic disturbances, such as close-to-nature transformation, irrigation, and harvest and (3) only included measurements of monospecific stands of *P. massoniana*; data on mixed stands was excluded. *P. massoniana* growth is affected by competition with other tree species in mixed stands [22]. The different types of species interactions formed depending on tree species composition, stand age, afforestation density and site conditions bring uncertainties to the evaluation of functions and services of forest ecosystems, including productivity [28]. Moreover, we only selected studies that were (4) derived from stable growing communities (data from stands younger than three years old was excluded), and (5) contained at least one of the following biomass data at the stand level: stem, branch, leaf, root, sum of aboveground organs or total tree. Studies on allometric scaling among biomass components of *P. massoniana* have focused primarily on the individual tree level [21]. However, forest NPP can only be estimated at the stand level [29]. All biomass data were converted to common units (Mg·ha^{-1}) prior to analysis. We retrieved missing latitude or longitude information for 16 sites without such data from Google Earth according to the site name [27]. Elevation information not provided in the literature was obtained based on the longitude, latitude of the sampling sites and a 1 km resolution Digital Elevation Model obtained from Cold and Arid Regions Sciences Data Center at Lanzhou (http://westdc.westgis.ac.cn/). A total of 87 references were acquired after screening (Figure 1, Table S1).

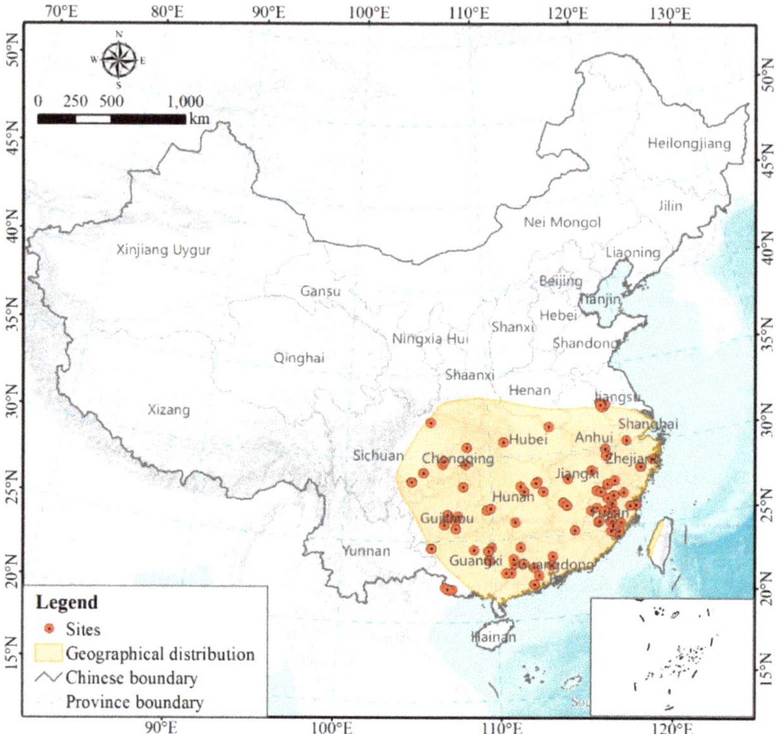

Figure 1. Geographical distribution of field study sites included in our database. Some plots are not visible as they are very close to each other and overlap. Geographical distribution range (shaded area) of *P. massoniana* adapted from Zhou (2001) [18].

Most of the literature only include biomass data and NPP data was rarely provided. Therefore, a uniform formula was used to estimate productivity based on biomass and stand age to enhance data comparability between different studies. NPP is the rate of production of biomass and organic compounds by the plant or ecosystem and consists of three components [23]:

$$NPP = \Delta Y + \Delta L + \Delta G, \tag{1}$$

where ΔY is the growth increment in a specified time interval (usually 1 year) estimated from temporal changes in forest biomass, ΔL is the loss part, consists of the death and litterfall production, and ΔG is the rate of grazing by herbivores. The full suite of components of NPP is rarely measured in forest ecosystems, owing to the difficulty of measuring ΔL and ΔG. Considering the large amount of missing data for these two components in the literature, the NPP of total tree (NPP_{tree}) includes aboveground and belowground NPP and is the sum of four compartments in this study:

$$NPP_{tree} = NPP_{stem} + NPP_{bra} + NPP_{leaf} + NPP_{root}, \tag{2}$$

where NPP_{stem}, NPP_{bra}, NPP_{leaf} and NPP_{root} are the annual net increments of stems (including bark), branches, leaves, and roots, respectively. Moreover, NPP of aboveground organs (NPP_{ag}) is the sum of NPPs from stems, branches, and leaves:

$$NPP_{ag} = NPP_{stem} + NPP_{bra} + NPP_{leaf}, \tag{3}$$

To simplify our analysis, in this study we did not consider carbon allocations to fruits, flowers and exudates.

2.2. Influencing Factors

2.2.1. Soil Data and Stand Characteristics

We collected soil data including alkali-hydrolysable nitrogen (AN, mg/kg), available phosphorus (AP, mg/kg), available potassium (AK, mg/kg), bulk density (BD, g/cm^3), soil organic matter (SOM, g/kg) and pH from the China Dataset of Soil Properties for Land Surface Modeling provided by Cold and Arid Regions Sciences Data Center at Lanzhou (http://westdc.westgis.ac.cn/). The data were obtained from the second national soil survey with a resolution of 1 km.

Stand characteristics used in this study involve stand density and age extracted from the literature.

2.2.2. Climatic Variables

Climatic factors used in this study included mean annual temperature (MAT, °C), mean annual precipitation (MAP, mm), mean high temperatures in warm months (HTWM, °C), and mean low temperatures in cold months (LTCM, °C). HTWM and LTCM were used to explore the effects of high and low temperatures on *P. massoniana* NPP, because the accumulation of forest biomass in different study sites is influenced by environmental factors over a relatively long period of time rather than a temporary time [30,31]. For the study area, July and August were warm months, and December was the cold month, as determined by frequency statistics of the months when the annual maximum and minimum temperatures occurred from 1981 to 2010 (Figure 2). Several study sites were located far from meteorological stations, and have different biomass accumulation stages. Therefore, spatially-interpolated climate data within the same observational period (1981–2015) was used. MAT and MAP data (1981–2015) were extracted from a China climate dataset (1 km resolution) provided by the Data Center for Resources and Environmental Sciences, Chinese Academy of Sciences (http://www.resdc.cn). Monthly HTWM and LTCM from 1981 to 2015 were obtained from China National Meteorological Information Center (http://data.cma.cn/) and interpolated into 1 km grid cells employing the kriging method. Spatial data was interpolated with ArcGIS 10.2 software.

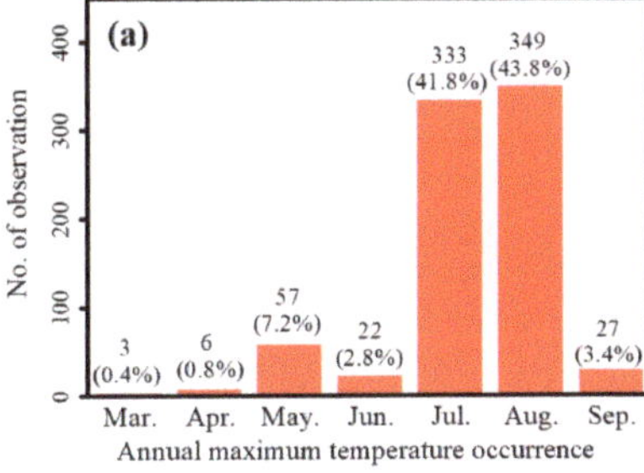
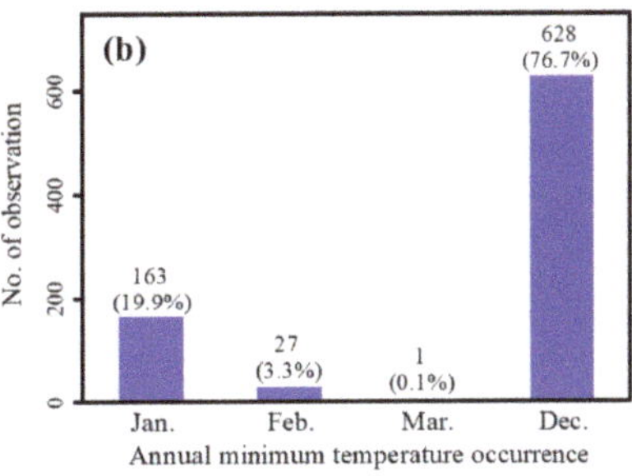

Figure 2. Frequency distributions of (**a**) annual maximum temperature and (**b**) annual minimum temperature occurrence time of 819 meteorological stations from 1981 to 2010 in southern China, where *P. massoniana* dominate. Data were obtained from meteorological stations in 11 provinces (Guizhou, Chongqing, Guangxi, Hunan, Fujian, Guangdong, Jiangxi, Hubei, Sichuan, Zhejiang and Anhui) provided by China National Meteorological Information Center (http://data.cma.cn/). The percentage of observations are presented in parentheses.

2.3. Statistical Analysis

We performed a descriptive statistical analysis to determine the distribution and variability of the NPP for *P. massoniana* forests, and fitted the distribution and variability curves with Gaussian functions. To stabilize heteroscedasticity and improve model *R*-square values in linear regression, all variables were transformed using the natural logarithm (i.e., $\ln(x_i)$) prior to analysis. Pearson correlation analyses were used to evaluate the relationship between the NPP of each *P. massoniana* component and site conditions (longitude, latitude, and elevation). The same approach was used to estimate the relationship between environmental factors and site conditions. All statistical analyses above were performed with SPSS (version 20, IBM Corp., Armonk, NY, USA).

The linear mixed-effects models (LMMs) were employed to analyze the effect of environmental factors and stand characteristics on *P. massoniana* NPP. To find the best models for our data we first built a full mixed-effects model containing all of the potential explanatory variables. Based on previous studies [25–27], soil variables (i.e., AN, AP, AK, BD, SOM, and pH), climate variables (i.e., MAT, MAP, HTWM, and LTCM), and stand characteristics (stand age and density), which were not highly correlated were checked by a correlation matrix and selected as potential explanatory variables. The twelve potential explanatory variables were set as fixed-effect terms and study sites were set as random-effect terms, to account for site-specific effects, such as forest management and microsite. Based on the full model, in each analysis, we constructed a set of candidate models that included different combinations of potential explanatory variables, in which we also included the null model with all fixed effects deleted [32]. Models were compared using the Akaike Information Criterion corrected for small samples (AICc). The best-performing model with the lowest AICc was selected as the final model. LMMs were performed using the package "lmerTest" [33] in R software (version 3.6.1) [34]. The model performance was evaluated by the "MuMIn" package [35], in which AICc, marginal R-square (variance explained by fixed factors) and conditional R-square (variance explained by both fixed and random factors) were calculated [36]. In addition, a regression analysis was performed to analyse the relationships between key environmental factors and NPP based on the results of LMMs.

3. Results

3.1. Variability of P. mansoniana Distribution and NPP

P. massoniana components exhibited large NPP variation across sampling sites (Figure 1, Table 1), ranging from 0.29 to 8.71, 0.10 to 2.11, 0.02 to 2.31, 0.07 to 2.11, 0.88 to 10.81, and 1.04 to 13.13 $\mathrm{Mg \cdot ha^{-1} \cdot year^{-1}}$ for stems, branches, leaves, roots, aboveground organs and total tree, respectively, with mean values of 3.51, 0.69, 0.34, 0.81, 4.53, 5.65 $\mathrm{Mg \cdot ha^{-1} \cdot year^{-1}}$, respectively (Figure 3).

Table 1. Net primary productivity (NPP) of *P. massoniana* stem, branch, leaf, root, aboveground organs, and total tree. Number of observations (N), mean value (Mean), maximum value (Max), minimum value (Min) and standard error (SE) were reported.

Component	N	NPP ($\mathrm{Mg \cdot ha^{-1} \cdot year^{-1}}$)			
		Mean	Max	Min	SE
Stem	172	3.51	8.71	0.29	0.13
Branch	168	0.69	2.11	0.10	0.03
Leaf	168	0.34	2.31	0.02	0.02
Root	148	0.81	2.11	0.07	0.03
Aboveground	185	4.53	10.81	0.88	0.15
Total tree	161	5.65	13.13	1.04	0.20

In this study, *P. massoniana* geographical distribution covered 11 provinces in southern China and the full geographical distribution of the species was well represented in our database (Figures 1 and 4a). *P. massoniana* study sites ranged from 25 to 1357 m of elevation, and most sample sites were set in the regions below 1000 m (Figure 4b). $\mathrm{NPP_{stem}}$, $\mathrm{NPP_{bra}}$, $\mathrm{NPP_{leaf}}$, $\mathrm{NPP_{root}}$, $\mathrm{NPP_{ag}}$, and $\mathrm{NPP_{tree}}$ showed no clear relationships with longitude and elevation, but significant latitudinal trends ($p < 0.01$, Table 2).

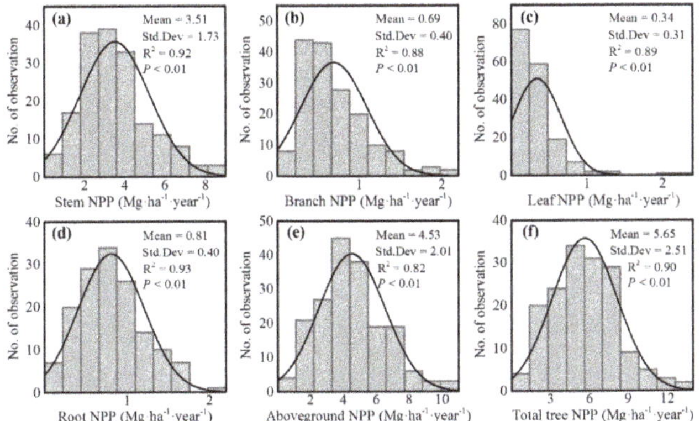

Figure 3. Frequency distributions of (**a**) stem, (**b**) branch, (**c**) leaf, (**d**) root, (**e**) aboveground organs and (**f**) total tree net primary productivity (NPP) of *P. massoniana* forests in south China. The mean and standard deviation values are presented. The curves were fitted by a Gaussian function.

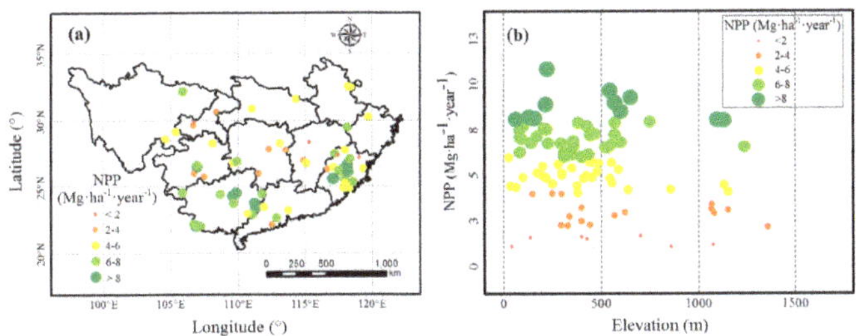

Figure 4. Net primary productivity (NPP) spatial distribution of *P. massoniana* forests in relation to longitude, latitude (**a**), and elevation (**b**).

Table 2. Pearson correlations between site conditions and *P. massoniana* stem, branch, leaf, root, aboveground, and total tree net primary productivity (NPP) on a logarithmic (ln) scale.

Site Conditions	NPP_{stem}	NPP_{bra}	NPP_{leaf}	NPP_{root}	NPP_{ag}	NPP_{tree}
Longitude (°E)	−0.112	−0.054	−0.007	−0.110	−0.100	−0.119
Latitude (°N)	−0.285 **	−0.369 **	−0.208 **	−0.251 **	−0.338 **	−0.344 **
Elevation (m)	−0.087	−0.019	0.102	−0.069	−0.061	−0.094

Note: (1) stem net primary productivity (NPP_{stem}); branch net primary productivity (NPP_{bra}); leaf net primary productivity (NPP_{leaf}); root net primary productivity (NPP_{root}); aboveground net primary productivity (NPP_{ag}); net primary productivity of total tree (NPP_{tree}); (2) ** represent $p < 0.01$.

3.2. Factors Influencing NPP of Different P. massoniana Components

P. massoniana NPP_{stem}, NPP_{bra}, NPP_{leaf}, NPP_{root}, NPP_{ag}, and NPP_{tree} exhibited different sensitivities to environmental factors and stand characteristics (Table 3). NPP_{stem} was significantly positively correlated with MAT, age, density, AP, SOM and BD, but significantly negatively correlated with HTWM. NPP_{bra} was significantly positively correlated with MAP and density. NPP_{leaf} was significantly positively correlated with MAP, but significantly negatively correlated with age. NPP_{root} was significantly positively correlated with BD, age, and density. NPP_{ag} significantly increased with MAT, age, and density. NPP_{tree} significantly increased with BD, stand age, and density, but significantly decreased with HTWM.

Table 3. Summary of the final model to test the effects of environmental factors and stand characteristics on *P. massoniana* stem, branch, leaf, root, aboveground, and total tree net primary productivity (NPP). Only the best-performing model, with the lowest the Akaike Information Criterion corrected for small samples (AICc), is shown (see Table S2 for the results of model performance).

NPP Component	Parameter	Variable									
		Intercept	MAT	HTWM	AGE	DENSITY	BD	MAP	AN	AP	SOM
Stem	Estimate	8.797	1.575	−5.099	0.562	0.216	1.170	—	—	0.424	0.370
	SE	5.746	0.466	1.745	0.086	0.067	0.585	—	—	0.128	0.171
	t-value	1.531	3.378	−2.922	6.520	3.233	1.999	—	—	3.311	2.160
	p-value	0.131	<0.01	<0.01	<0.001	<0.01	<0.05	—	—	<0.01	<0.05
Branch	Estimate	2.524	—	−3.997	—	0.148	0.913	1.294	—	—	—
	SE	7.119	—	2.472	—	0.059	0.654	0.455	—	—	—
	t-value	0.355	—	−1.617	—	2.533	1.397	2.844	—	—	—
	p-value	0.724	—	0.111	—	<0.05	0.168	<0.01	—	—	—
Leaf	Estimate	6.673	—	−4.488	−0.588	0.144	0.661	1.098	—	—	—
	SE	7.924	—	2.777	0.097	0.073	0.725	0.515	—	—	—
	t-value	0.842	—	−1.616	−6.045	1.964	0.912	2.133	—	—	—
	p-value	0.403	—	0.111	<0.001	0.052	0.366	<0.05	—	—	—
Root	Estimate	−1.479	1.173	−2.815	0.429	0.257	1.885	—	0.833	—	—
	SE	5.436	0.656	1.589	0.091	0.070	0.898	—	0.439	—	—
	t-value	−0.272	1.787	−1.772	4.699	3.680	2.098	—	1.894	—	—
	p-value	0.786	0.079	0.081	<0.001	<0.001	<0.05	—	0.063	—	—
Aboveground	Estimate	1.494	1.142	−1.666	0.333	0.182	—	—	—	—	—
	SE	3.510	0.412	0.998	0.077	0.064	—	—	—	—	—
	t-value	0.426	2.771	−1.669	4.333	2.852	—	—	—	—	—
	p-value	0.671	<0.01	0.099	<0.001	<0.01	—	—	—	—	—
Total tree	Estimate	1.920	0.989	−4.158	0.409	0.258	1.356	0.681	0.598	—	—
	SE	4.635	0.520	1.910	0.080	0.066	0.647	0.529	0.341	—	—
	t-value	0.414	1.900	−2.177	5.092	3.911	2.096	1.287	1.757	—	—
	p-value	0.680	0.062	<0.05	<0.001	<0.001	<0.05	0.202	0.083	—	—

Note: mean annual temperature (MAT); mean high temperatures in warm months (HTWM); stand age (AGE); stand density (DENSITY); bulk density (BD); mean annual precipitation (MAP); alkali-hydrolysable nitrogen (AN); available phosphorus (AP); soil organic matter (SOM).

3.3. Relationship between Site Conditions and Environmental Factors

P. massoniana NPPs showed significant trends in latitude, rather than in longitude and elevation (Table 2), therefore the trend of environmental variables in latitude was described minutely here, which was necessary to further explore the cause of formation of the spatial heterogeneity of *P. massoniana* NPP. MAT, LTCM, and MAP were significantly negatively correlated with latitude ($p < 0.01$), with correlation coefficients of −0.291, −0.413, and −0.256 on a logarithmic scale, respectively (Table 4). HTWM showed no clear relationship with latitude (Table 4). For soil properties, SOM was significantly negatively correlated with latitude ($p < 0.01$), with a correlation coefficient of −0.216 on a logarithmic scale (Table 5). AP and PH were significantly positively correlated with latitude ($p < 0.01$), with correlation coefficients of 0.223 and 0.216 on a logarithmic scale, respectively (Table 5). AN, AK, and BD showed no clear relationship with latitude (Table 5).

Table 4. Pearson correlations between site conditions and climate factors on a logarithmic (ln) scale.

Variables	MAT	LTCM	HTWM	MAP
Longitude (°E)	−0.073	−0.173 *	0.352 **	0.242 **
Latitude (°N)	−0.291 **	−0.413 **	−0.061	−0.256 **
Elevation (m)	−0.334 **	0.319 **	−0.391 **	−0.048

Note: (1) mean annual temperature (MAT); mean low temperatures in cold months (LTCM); mean high temperatures in warm months (HTWM); mean annual precipitation (MAP). (2) * and ** represent $p < 0.05$ and $p < 0.01$, respectively.

Table 5. Pearson correlations between site conditions and soil factors on a logarithmic (ln) scale.

Variables	AP	AN	AK	PH	SOM	BD
Longitude (°E)	−0.068	−0.175 *	−0.291 **	0.051	−0.341 **	0.152 *
Latitude (°N)	0.223 **	−0.093	−0.074	0.216 **	−0.216 **	−0.112
Elevation (m)	0.016	0.417 **	0.292 **	0.153 *	0.280 **	−0.213 **

Note: (1) bulk density (BD); alkali-hydrolysable nitrogen (AN); available phosphorus (AP); soil organic matter (SOM); available potassium (AK). (2) * and ** represent $p < 0.05$ and $p < 0.01$, respectively.

3.4. Relationship between Temperatures and P. massoniana NPP

P. massoniana NPP correlated highly with temperatures. As the MAT increased, NPP of both *P. massoniana* components (i.e., NPP_{stem}, NPP_{bra}, NPP_{leaf}, NPP_{root}, NPP_{ag}, and NPP_{tree}) increased signifcantly ($p < 0.05$, Figure 5a,d,g,j,m,p). In addition, LTCM was also significantly positively correlated with NPP_{stem}, NPP_{bra}, NPP_{leaf}, NPP_{ag}, and NPP_{tree} ($p < 0.05$, Figure 5b,e,h,n,q).

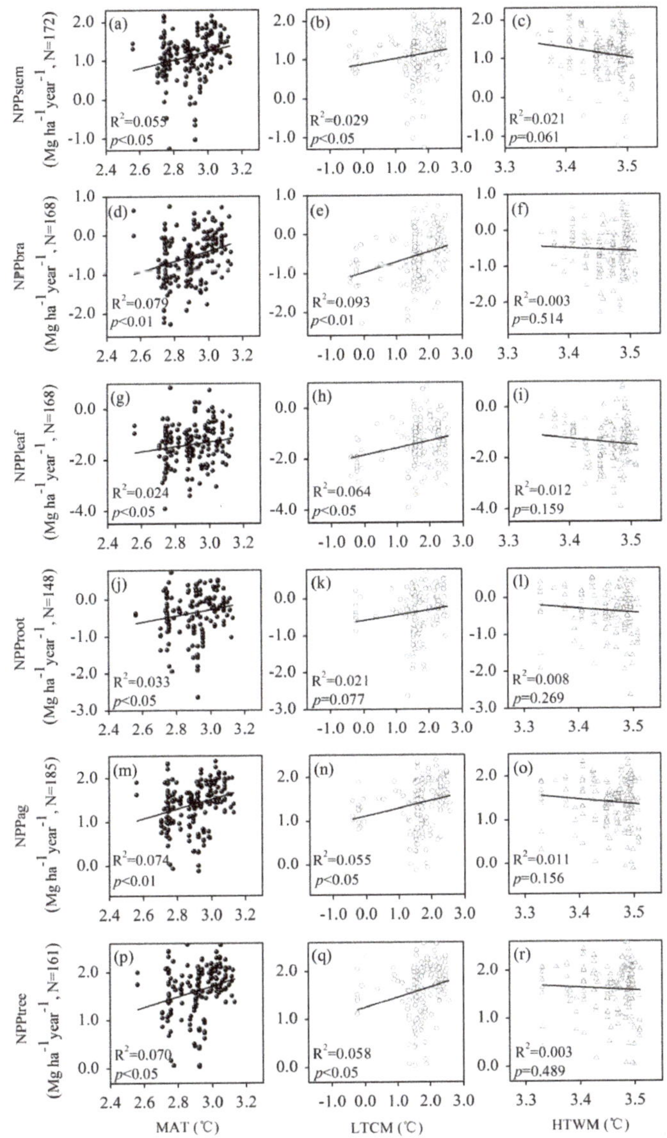

Figure 5. The relationships between *P. massoniana* net primary productivity (NPP) and MAT (**a,d,g,j,m,p**), LTCM (**b,e,h,k,n,q**), and HTWM (**c,f,i,l,o,r**) on a logarithmic (ln) scale. Mean annual temperature (MAT); mean low temperatures in cold months (LTCM); mean high temperatures in warm months (HTWM).

4. Discussion

4.1. Factors Influencing NPP of P. massoniana Forests

4.1.1. Climate Effects

P. massoniana NPP_{stem} and NPP_{ag} were significantly positively correlated with MAT (Table 3). This finding is consistent with previous studies [37–39], indicating that the warming effect can promote biomass accumulation of *P. massoniana*. Zhang et al. (2011) [30] showed that low air temperature from January to March was the major factor controlling the interannual variations of net carbon uptake, and cold springs greatly shortened the growing season length, reducing the carbon uptake period. Similarly, Zheng et al. (2012) [31] found that even a short-term freezing event during the transitional periods from dormancy to growth in early spring could result in disastrous damage to pine forests in temperate China. However, HTWM was significantly negatively correlated with NPP_{stem} and NPP_{tree} (Table 3), because higher temperatures in summer can increase plant evapotranspiration and respiration rates, leading to decreased biomass accumulation rates in subtropical forests [40]. Therefore, temperatures had an important effect on *P. massoniana* NPP. Large differences of NPP in sensitivity to temperature and precipitation were identified among various biomes and regions [41,42]. Regression analysis showed that NPP of both *P. massoniana* components increased significantly as MAT increased (Figure 5a,d,g,j,m,p). LMM results showed that MAT was positively correlated with NPP_{stem} and NPP_{ag} (Table 3), which formed relatively high proportions of NPP_{tree} (Table 1). Although NPP_{bra} and NPP_{leaf} were also positively correlated with MAP (Table 3), NPP of these two components only formed relatively low proportions of NPP_{tree} (Table 1). In fact, tree growth in our study region is probably not often limited by precipitation. Moreover, the magnitude of variation in MAP was not large enough to lead to significant changes in forest productivity in this region, since *P. massoniana* is distributed in a large gradient zone of precipitation in south-east Yangtze River basin and the southeast rivers [43]. Liang et al. (2015) [44] found that air temperature was the dominant climatic factor that controlled the interannual variability in NPP throughout the country except for arid and semi-arid regions in the middle–north and northwest parts of China. Similarly, Wang et al. (2011) [45] found that forest NPP appeared to be primarily a function of temperature, not precipitation, in subtropical Louisiana, USA. Furthermore, Churkina and Running (1998) [46] reported that temperature appeared to be the primary control on evergreen needle-leaved forest NPP on a global scale. Overall, our findings were in line with previous studies, indicating that *P. massoniana* NPP was more sensitive to temperature than to precipitation.

4.1.2. Soil Effects

P. massoniana NPP_{stem} significantly increased with SOM and AP (Table 3), indicating that, although this species can be used as a pioneer species for afforestation in infertile soil, soil with sufficient nutrients can promote biomass accumulation of *P. massoniana*. AP was significantly positively correlated with NPP_{stem}, while AN and AK were not significantly correlated with *P. massoniana* NPP (Table 3). NPP of *P. massoniana* forests was more sensitive to soil AP than to soil AN and AK. This result may be attributed to the fact that southern China is dominated by acidic red soil, where phosphorus is mostly fixed in insoluble form, and the AP content directly absorbed by plants is extremely low [47]. Therefore, the response of *P. massoniana* to phosphorus deficiency has received extensive attention in the context of N deposition in recent years [48]. In our study, soil BD was significantly positively correlated with NPP_{stem}, NPP_{root}, and NPP_{tree} (Table 3). However, there were relatively few studies on the relationship between soil physical properties and tree growth of *P. massoniana*. Therefore, this result needs to be further confirmed.

4.1.3. Stand Characteristics Effects

During this study, *P. massoniana* NPP_{stem} significantly increased, whereas NPP_{leaf} significantly decreased with age (Table 3). This finding can be explained by the variation in biomass allocation patterns with tree growth. Stems, as support tissues, obtain more biomass investment with tree growth, at the expense of branch and leaf biomass [49]. Wood (dead cells) continuously accumulates in the stem as trees grow, whereas branch and leaf biomass decrease as the early death of lower branches is caused by mutual shading [49]. Additionally, following canopy closure, older and taller trees entail greater expenditures for their construction and maintenance and expend more energy than younger and shorter trees to supply leaves with the same amount of water [49]. *P. massoniana* NPP_{stem}, NPP_{root}, NPP_{ag}, and NPP_{tree} significantly increased with age (Table 3), probably because the forest stands in our database were mostly middle-aged plantations (Table S1), in which age-related decline in forest productivity has not yet occurred [50,51]. Stand density was significantly positively correlated with NPP_{stem}, NPP_{bra}, NPP_{root}, NPP_{ag}, and NPP_{tree} (Table 3), indicating that density had a marked impact on productivity estimation. This finding was similar to that of Bormann and Gordon (1984) [52], who found that aboveground mass and total mass per unit area were the greatest in the dense stands. In contrast to the trends in per-unit-area values, for average tree values, dry mass of all individual-tree components and totals were the highest in the open stands, and the lowest in the dense stands [52]. At lower stand densities, less growth per unit area is obtained, but this will be offset by greater growth per tree [53]. In a 42-year thinning experiment, Horner et al. (2010) [54] found that moderately thinned stands had the highest aboveground carbon storage rate and the highest aboveground carbon stocks rather than unthinned treatment (highest density). Therefore, although there was a significant positive correlation between NPP and density in this study, this relationship varied between stands due to the trade-off between maximizing individual tree size or stand yield [53]. Therefore, for long-term managed plantations, the initial density, thinning practice and the competition–density effect should be considered in productivity predictions [55,56].

4.2. Latitudinal Effects on NPP of P. massoniana Forests

In this study, *P. massoniana* NPP_{stem}, NPP_{bra}, NPP_{leaf}, NPP_{root}, NPP_{ag}, and NPP_{tree} exhibited inverse latitudinal trends, but no longitudinal and elevation trends (Table 2). Liu et al. (2016) [26] found that forest NPP show no clear relationship with longitude but negatively correlate with latitude in southwestern China. Ni (2003) [11] also concluded that forest NPP significantly decreased with increasing latitude based on forest inventory data in China between 1989 and 1993. In the study of Zhan et al. (2018) [27] on forests in eastern China, the functional relationship between NPP and latitude was: $y = 0.06x^2 - 3.91x + 73.71$ ($R^2 = 0.38$, $p < 0.001$), NPP decreased first and then increased with the increase in latitude, and monotonically decreased with increasing latitude when the latitude was below 35.0 °N. Thus, the results of our study are consistent with previous studies.

Latitude is not an environmental factor capable of having direct influence on *P. massoniana* growth [57,58], but rather an indirect variable that condenses a set of factors that vary from south to north, including MAT, LTCM, MAP, AP, PH, and SOM in this study (Tables 4 and 5). Climatically, the north-to-south and west-to-east gradients in China both reflect shifts from cold and dry to warm and moist conditions, although the thermal gradient is steeper in the former and the moisture gradient more pronounced in the latter [25,59]. The correlation analysis of latitude and climate indicators showed that MAT, LTCM, and MAP decrease with increasing latitude (Table 4), which is in line with the generally accepted idea that the southern part of China is warmer and moister than the northern part. It is worth noting that HTWM, which was significantly negatively correlated with *P. massoniana* NPP_{stem} and NPP_{tree} (Table 3), did not show a significant latitudinal trend (Table 4). This suggested that high temperatures were a significant climate stressor across the species geographical distribution and were not restricted to marginal populations at low latitudes. MAT and MAP, which were significantly positively correlated with NPP (Table 3), showed inverse latitudinal trends (Table 4). Therefore, both temperature and precipitation contributed to the formation of the inverse latitudinal trends of

P. massoniana NPP. Temperature was a key contributing factor to these trends, as NPP was highly sensitive to any change in temperature. AP and SOM, which were significantly positively correlated with *P. massoniana* productivity (Table 3), showed different latitudinal trends (Table 5). Latitude was significantly positively correlated with AP, but significantly negatively correlated with SOM (Table 5). In fact, low phosphorus availability is a limiting factor on *P. massoniana* growth not only at low latitudes, but also in the broad red soil areas in southern China [60]. In summary, the latitudinal trends of *P. massoniana* productivity are driven by the significant influences of climate and soil properties.

4.3. Uncertainty Analysis

In this study, some uncertainties in forest productivity assessment have been eliminated in the process of developing the biomass database. However, there was still a great deal of variability in influencing factors. Atmospheric N deposition, elevated CO_2 concentration, and climate warming have been confirmed to affect *P. massoniana* growth [61,62]. Moreover, they can affect other biological processes and thus indirectly affect forest productivity. For example, N deposition will further aggravate soil acidification and change soil nutrient conditions (e.g., N:P ratio) [63,64]. Furthermore, in the process of biomass data acquisition, it was found that the size classes for *P. massoniana* roots were not uniform across studies, which made it difficult to establish a complete database containing root sizes, although roots with diameter < 2 mm were usually defined as fine roots. Pan et al. (2018) [65] suggested that understory vegetation generated different effects on soil carbon and nitrogen processes in aerially seeded *P. massoniana* plantations. Therefore, the missing components in field measurements, including shrubs, herbs, and litterfall, not only lead to an underestimation of actual NPP, but their significant contribution to biological processes of tree growth should also be considered in future studies.

5. Conclusions

We established a new regional database of *P. massoniana* forest biomass. Tree components (stem, branch, leaf, root, aboveground organs and total tree) exhibited large NPP variation across sampling sites. Climate, soil, and stand characteristics have a marked impact on the NPP of *P. massoniana* forests, and the influence of these variables on the NPP of different tree components varied. Appropriate temperature and soil nutrients, especially soil AP, are beneficial to *P. massoniana* growth. NPP of all tree components of *P. massoniana* exhibited inverse latitudinal trends, which were driven by the significant influences of climate and soil properties. Temperature was a key environmental factor for the formation of these trends. Future studies should place a particular emphasis on the effects of various components of the forest ecosystem, particularly soil, litterfall, and understory vegetation on the biological processes of tree growth, and their interactions with climate change.

Supplementary Materials: The following are available online at http://www.mdpi.com/1999-4907/11/4/404/s1, Table S1: Primary data of site condition, plot size, stand age, density, component biomass (stem, B1; branch, B2; leaf, B3; root, B4; aboveground organs, B5; and total tree, B6) and measurement year for *Pinus massoniana* forests in southern China. Table S2: Parameters fitted to the final model, full model, and null model by means of linear mixed-effects models.

Author Contributions: Conceptualization, X.H. and M.T.; methodology, Z.Z. and P.W.; formal analysis, P.W.; writing—original draft preparation, X.H.; writing—review and editing, Z.Z.; visualization, X.H. and C.H.; funding acquisition, M.T. All authors have read and agree to the published version of the manuscript.

Funding: This research was funded by the National Key R&D Program of China (2016YFD0600201), the Major Scientific and Technical Innovation Project of Hubei Province, China (2018ABA074) and research platform support was provided by the Long-Term Track Research Program of the Forest Ecological Station in the Three Gorges Reservoir Region (Zigui) of the Yangtze River, China.

Acknowledgments: The authors gratefully acknowledge financial support from the National Key R&D Program of China (2016YFD0600201), the Major Scientific and Technical Innovation Project of Hubei Province, China (2018ABA074) and research platform support from the Long-Term Track Research Program of the Forest Ecological Station in the Three Gorges Reservoir Region (Zigui) of the Yangtze River, China.

Conflicts of Interest: The authors declare no conflict of interest.

References

1. Brockerhoff, E.G.; Barbaro, L.; Castagneyrol, B.; Forrester, D.I.; Gardiner, B.; González-Olabarria, J.R.; Lyver, P.O.B.; Meurisse, N.; Oxbrough, A.; Taki, H.; et al. Forest biodiversity, ecosystem functioning and the provision of ecosystem services. *Biodivers. Conserv.* **2017**, *26*, 3005–3035. [CrossRef]
2. Hamilton, S.E.; Friess, D.A. Global carbon stocks and potential emissions due to mangrove deforestation from 2000 to 2012. *Nat. Clim. Chang.* **2018**, *8*, 240–244. [CrossRef]
3. Keeling, H.C.; Phillips, O.L. The global relationship between forest productivity and biomass. *Glob. Ecol. Biogeogr.* **2007**, *16*, 618–631. [CrossRef]
4. Malhi, Y.; Doughty, C.; Galbraith, D. The allocation of ecosystem net primary productivity in tropical forests. *Philos. Trans. R. Soc. B Biol. Sci.* **2011**, *366*, 3225–3245. [CrossRef]
5. Cloern, J.E.; Abreu, P.C.; Carstensen, J.; Chauvaud, L.; Elmgren, R.; Grall, J.; Greening, H.; Johansson, J.O.R.; Kahru, M.; Sherwood, E.T.; et al. Human activities and climate variability drive fast-paced change across the world's estuarine-coastal ecosystems. *Glob. Chang. Biol.* **2016**, *22*, 513–529. [CrossRef]
6. Drake, J.E.; Gallet-Budynek, A.; Hofmockel, K.S.; Bernhardt, E.S.; Billings, S.A.; Jackson, R.B.; Johnsen, K.S.; Lichter, J.; Mccarthy, H.R.; Mccormack, M.L.; et al. Increases in the flux of carbon belowground stimulate nitrogen uptake and sustain the long-term enhancement of forest productivity under elevated CO_2. *Ecol. Lett.* **2011**, *14*, 349–357. [CrossRef]
7. Malhi, Y.; Girardin, C.A.J.; Goldsmith, G.R.; Doughty, C.E.; Salinas, N.; Metcalfe, D.B.; Huaraca Huasco, W.; Silva-Espejo, J.E.; del Aguilla-Pasquell, J.; Farfán Amézquita, F.; et al. The variation of productivity and its allocation along a tropical elevation gradient: A whole carbon budget perspective. *New Phytol.* **2017**, *214*, 1019–1032. [CrossRef]
8. Girardin, C.A.J.; Malhi, Y.; Aragão, L.E.O.C.; Mamani, M.; Huaraca Huasco, W.; Durand, L.; Feeley, K.J.; Rapp, J.; Silva-Espejo, J.E.; Silman, M.; et al. Net primary productivity allocation and cycling of carbon along a tropical forest elevational transect in the Peruvian Andes. *Glob. Chang. Biol.* **2010**, *16*, 3176–3192. [CrossRef]
9. Malhi, Y. The productivity, metabolism and carbon cycle of tropical forest vegetation. *J. Ecol.* **2012**, *100*, 65–75. [CrossRef]
10. Ni, J.; Zhang, X.S.; Scurlock, J.M.O. Synthesis and analysis of biomass and net primary productivity in Chinese forests. *Ann. For. Sci.* **2001**, *58*, 351–384. [CrossRef]
11. Ni, J. Net primary productivity in forests of China: Scaling-up of national inventory data and comparison with model predictions. *For. Ecol. Manag.* **2003**, *176*, 485–495. [CrossRef]
12. Yang, J.; Ji, X.; Deane, D.C.; Wu, L.; Chen, S. Spatiotemporal distribution and driving factors of forest biomass carbon storage in China: 1977–2013. *Forests* **2017**, *8*, 263. [CrossRef]
13. Xu, X.; Li, K. Biomass carbon sequestration by planted forests in China. *Chin. Geogr. Sci.* **2010**, *20*, 289–297. [CrossRef]
14. Silva, L.N.; Freer-Smith, P.; Madsen, P. Production, restoration, mitigation: A new generation of plantations. *New For.* **2019**, *50*, 153–168. [CrossRef]
15. Zhao, M.; Zhou, G.S. Estimation of biomass and net primary productivity of major planted forests in China based on forest inventory data. *For. Ecol. Manag.* **2005**, *207*, 295–313. [CrossRef]
16. Payn, T.; Carnus, J.M.; Freer-Smith, P.; Kimberley, M.; Kollert, W.; Liu, S.; Orazio, C.; Rodriguez, L.; Silva, L.N.; Wingfield, M.J. Changes in planted forests and future global implications. *For. Ecol. Manag.* **2015**, *352*, 57–67. [CrossRef]
17. Xi, W.; Wang, F.; Shi, P.; Dai, E.; Anoruo, A.O.; Bi, H.; Rahmlow, A.; He, B.; Li, W. Challenges to sustainable development in China: A review of six large-scale forest restoration and land conservation programs. *J. Sustain. For.* **2014**, *33*, 435–453. [CrossRef]
18. Zhou, Z.X. *Masson Pine in China*; China Forestry Publishing House: Beijing, China, 2001. (In Chinese)
19. Feng, Z.W.; Chen, C.Y.; Zhang, J.W.; Wang, K.P.; Zhao, J.L.; Gao, H. Determination of biomass of *Pinus massoniana* stand in Huitong county, Hunan province. *Sci. Silvae Sin.* **1982**, *18*, 127–134. (In Chinese)
20. Ali, A.; Ahmad, A.; Akhtar, K.; Teng, M.; Zeng, W.; Yan, Z.; Zhou, Z. Patterns of biomass, carbon, and soil properties in masson pine (*Pinus massoniana* Lamb) plantations with different stand ages and management practices. *Forests* **2019**, *10*, 645. [CrossRef]

21. Xiang, W.; Liu, S.; Deng, X.; Shen, A.; Lei, X.; Tian, D.; Zhao, M.; Peng, C. General allometric equations and biomass allocation of *Pinus massoniana* trees on a regional scale in southern China. *Ecol. Res.* **2011**, *26*, 697–711. [CrossRef]

22. You, Y.; Huang, X.; Zhu, H.; Liu, S.; Liang, H.; Wen, Y.; Wang, H.; Cai, D.; Ye, D. Positive interactions between *Pinus massoniana* and *Castanopsis hystrix* species in the uneven-aged mixed plantations can produce more ecosystem carbon in subtropical China. *For. Ecol. Manag.* **2018**, *410*, 193–200. [CrossRef]

23. Komiyama, A.; Ong, J.E.; Poungparn, S. Allometry, biomass, and productivity of mangrove forests: A review. *Aquat. Bot.* **2008**, *89*, 128–137. [CrossRef]

24. Liang, H.; Huang, J.G.; Ma, Q.; Li, J.; Wang, Z.; Guo, X.; Zhu, H.; Jiang, S.; Zhou, P.; Yu, B.; et al. Contributions of competition and climate on radial growth of *Pinus massoniana* in subtropics of China. *Agric. For. Meteorol.* **2019**, *274*, 7–17. [CrossRef]

25. Zhang, H.; Song, T.; Wang, K.; Wang, G.; Liao, J.; Xu, G.; Zeng, F. Biogeographical patterns of forest biomass allocation vary by climate, soil and forest characteristics in China. *Environ. Res. Lett.* **2015**, *10*, 44014. [CrossRef]

26. Liu, L.B.; Yang, H.M.; Xu, Y.; Guo, Y.M.; Ni, J. Forest biomass and net primary productivity in Southwestern China: A meta-analysis focusing on environmental driving factors. *Forests* **2016**, *7*, 173. [CrossRef]

27. Zhan, X.; Guo, M.; Zhang, T. Joint control of net primary productivity by climate and soil nitrogen in the forests of Eastern China. *Forests* **2018**, *9*, 173.

28. Forrester, D.I. The spatial and temporal dynamics of species interactions in mixed-species forests: From pattern to process. *For. Ecol. Manag.* **2014**, *312*, 282–292. [CrossRef]

29. Chen, G.; Hobbie, S.E.; Reich, P.B.; Yang, Y.; Robinson, D. Allometry of fine roots in forest ecosystems. *Ecol. Lett.* **2019**, *22*, 322–331. [CrossRef]

30. Zhang, W.J.; Wang, H.M.; Yang, F.T.; Yi, Y.H.; Wen, X.F.; Sun, X.M.; Yu, G.R.; Wang, Y.D.; Ning, J.C. Underestimated effects of low temperature during early growing season on carbon sequestration of a subtropical coniferous plantation. *Biogeosciences* **2011**, *8*, 1667–1678. [CrossRef]

31. Zheng, Y.; Yang, Q.; Xu, M.; Chi, Y.; Shen, R.; Li, P.; Dai, H. Responses of *Pinus massoniana* and *Pinus taeda* to freezing in temperate forests in central China. *Scand. J. For. Res.* **2012**, *27*, 520–531. [CrossRef]

32. Grueber, C.E.; Nakagawa, S.; Laws, R.J.; Jamieson, I.G. Multimodel inference in ecology and evolution: Challenges and solutions. *J. Evol. Biol.* **2011**, *24*, 699–711. [CrossRef] [PubMed]

33. Kuznetsova, A.; Brockhoff, P.B.; Christensen, R.H.B. lmerTest Package: Tests in Linear Mixed Effects Models. *J. Stat. Softw.* **2017**, *82*. [CrossRef]

34. R Core Team. *R: A Language and Environment for Statistical Computing*; R Foundation for Statistical Computing: Vienna, Austria, 2019.

35. Bartoń, K. MuMIn: Multi-Model Inference, R package version 1.43.15. 2019. Available online: https://CRAN.R-project.org/package=MuMIn (accessed on 20 December 2019).

36. Nakagawa, S.; Schielzeth, H. A general and simple method for obtaining R^2 from generalized linear mixed-effects models. *Methods Ecol. Evol.* **2013**, *4*, 133–142. [CrossRef]

37. Chen, F.; Yuan, Y.J.; Yu, S.L.; Zhang, T.W. Influence of climate warming and resin collection on the growth of Masson pine (*Pinus massoniana*) in a subtropical forest, southern China. *Trees-Struct. Funct.* **2015**, *29*, 1423–1430. [CrossRef]

38. Li, Y.; Liu, J.; Zhou, G.; Huang, W.; Duan, H. Warming effects on photosynthesis of subtropical tree species: A translocation experiment along an altitudinal gradient. *Sci. Rep.* **2016**, *6*, 1–14. [CrossRef]

39. Li, Y.; Zhou, G.; Liu, J. Different growth and physiological responses of six subtropical tree species to warming. *Front. Plant Sci.* **2017**, *8*, 1–11. [CrossRef]

40. Zhou, H.; Luo, Y.; Zhou, G.; Yu, J.; Shah, S.; Meng, S.; Liu, Q. Exploring the sensitivity of subtropical stand aboveground productivity to local and regional climate signals in South China. *Forests* **2019**, *10*, 71. [CrossRef]

41. Lindroth, A.; Grelle, A.; More, A. Long-term measurements of boreal forest carbon balance. *Glob. Chang. Biol.* **1998**, *4*, 443–450. [CrossRef]

42. Schuur, E.A.G. Productivity and global climate revisited: The sensitivity of tropical forest growth to precipitation. *Ecology* **2003**, *84*, 1165–1170. [CrossRef]

43. Zhang, Q.; Xu, C.Y.; Zhang, Z.; Chen, Y.D.; Liu, C.L. Spatial and temporal variability of precipitation over China, 1951–2005. *Theor. Appl. Climatol.* **2009**, *95*, 53–68. [CrossRef]

44. Liang, W.; Yang, Y.; Fan, D.; Guan, H.; Zhang, T.; Long, D.; Zhou, Y.; Bai, D. Analysis of spatial and temporal patterns of net primary production and their climate controls in China from 1982 to 2010. *Agric. For. Meteorol.* **2015**, *204*, 22–36. [CrossRef]

45. Wang, F.; Xu, Y.J.; Dean, T.J. Projecting climate change effects on forest net primary productivity in subtropical Louisiana, USA. *Ambio* **2011**, *40*, 506–520. [CrossRef] [PubMed]

46. Churkina, G.; Running, S.W. Contrasting climatic controls on the estimated productivity of global terrestrial biomes. *Ecosystems* **1998**, *1*, 206–215. [CrossRef]

47. Huang, W.; Liu, J.; Wang, Y.P.; Zhou, G.; Han, T.; Li, Y. Increasing phosphorus limitation along three successional forests in southern China. *Plant Soil* **2013**, *364*, 181–191. [CrossRef]

48. Hu, C.; Zhao, L.; Zhou, Z.; Dong, G.; Zhang, Y. Genetic variations and correlation analysis of N and P traits in *Pinus massoniana* under combined conditions of N deposition and P deficiency. *Trees-Struct. Funct.* **2016**, *30*, 1341–1350. [CrossRef]

49. Fang, Y.; Zou, X.; Lie, Z.; Xue, L. Variation in organ biomass with changing climate and forest characteristics across Chinese forests. *Forests* **2018**, *9*, 521. [CrossRef]

50. Ryan, M.G.; Binkley, D.; Fownes, J.H. Age-related decline in forest productivity: Pattern and process. *Adv. Ecol. Res.* **1997**, *27*, 213–262.

51. Pregitzer, K.S.; Euskirchen, E.S. Carbon cycling and storage in world forests: Biome patterns related to forest age. *Glob. Chang. Biol.* **2004**, *10*, 2052–2077. [CrossRef]

52. Bormann, B.T.; Gordon, J.C. Stand density effects in young red alder plantations: Productivity, photosynthate partitioning, and nitrogen fixation. *Ecology* **1984**, *65*, 394–402. [CrossRef]

53. Drew, T.J.; Flewelling, J.W. Stand density management: An alternative approach and its application to Douglas-fir plantations. *For. Sci.* **1979**, *25*, 518–532.

54. Horner, G.J.; Baker, P.J.; Mac Nally, R.; Cunningham, S.C.; Thomson, J.R.; Hamilton, F. Forest structure, habitat and carbon benefits from thinning floodplain forests: Managing early stand density makes a difference. *For. Ecol. Manag.* **2010**, *259*, 286–293. [CrossRef]

55. Xue, L.; Hagihara, A. Growth analysis on the C-D effect in self-thinning Masson pine (*Pinus massoniana*) stands. *For. Ecol. Manag.* **2002**, *165*, 249–256. [CrossRef]

56. Magruder, M.; Chhin, S.; Monks, A.; O'Brien, J. Effects of initial stand density and climate on red pine productivity within Huron National Forest, Michigan, USA. *Forests* **2012**, *3*, 1086–1103. [CrossRef]

57. López-Medellín, X.; Ezcurra, E. The productivity of mangroves in northwestern Mexico: A meta-analysis of current data. *J. Coast. Conserv.* **2012**, *16*, 399–403. [CrossRef]

58. Zhang, H.; Wang, K.; Xu, X.; Song, T.; Xu, Y.; Zeng, F. Biogeographical patterns of biomass allocation in leaves, stems, and roots in Chinas forests. *Sci. Rep.* **2015**, *5*, 1–12.

59. Han, W.X.; Fang, J.Y.; Reich, P.B.; Ian Woodward, F.; Wang, Z.H. Biogeography and variability of eleven mineral elements in plant leaves across gradients of climate, soil and plant functional type in China. *Ecol. Lett.* **2011**, *14*, 788–796. [CrossRef] [PubMed]

60. Xu, R.; Zhao, A.; Li, Q.; Kong, X.; Ji, G. Acidity regime of the red soils in a subtropical region of southern China under field conditions. *Geoderma* **2003**, *115*, 75–84. [CrossRef]

61. Sun, F.; Kuang, Y.; Wen, D.; Xu, Z.; Li, J.; Zuo, W.; Hou, E. Long-term tree growth rate, water use efficiency, and tree ring nitrogen isotope composition of *Pinus massoniana* L. in response to global climate change and local nitrogen deposition in Southern China. *J. Soils Sediments* **2010**, *10*, 1453–1465. [CrossRef]

62. Wu, T.; Lin, W.; Li, Y.; Lie, Z.; Huang, W.; Liu, J. Nitrogen addition method affects growth and nitrogen accumulation in seedlings of four subtropical tree species: *Schima superba* Gardner & Champ., *Pinus massoniana* Lamb., *Acacia mangium* Willd., and *Ormosia pinnata* Lour. *Ann. For. Sci.* **2019**, *76*, 1–11.

63. Zhang, Y.; Zhou, Z.; Yang, Q. Nitrogen (N) Deposition impacts seedling growth of *Pinus massoniana* via N:P ratio effects and the modulation of adaptive responses to low P (phosphorus). *PLoS ONE* **2013**, *8*, e79229. [CrossRef]

64. Huang, W.; Zhou, G.; Liu, J.; Duan, H.; Liu, X.; Fang, X.; Zhang, D. Shifts in soil phosphorus fractions under elevated CO_2 and N addition in model forest ecosystems in subtropical China. *Plant Ecol.* **2014**, *215*, 1373–1384. [CrossRef]
65. Pan, P.; Zhao, F.; Ning, J.; Zhang, L.; Ouyang, X.; Zang, H. Impact of understory vegetation on soil carbon and nitrogen dynamic in aerially seeded *Pinus massoniana* plantations. *PLoS ONE* **2018**, *13*, e0191952. [CrossRef] [PubMed]

Article

Small-Scale Forest Structure Influences Spatial Variability of Belowground Carbon Fluxes in a Mature Mediterranean Beech Forest

Ettore D'Andrea [1,*], **Gabriele Guidolotti** [2], **Andrea Scartazza** [3], **Paolo De Angelis** [4] and **Giorgio Matteucci** [1]

[1]   CNR-ISAFOM, Via Patacca, 85 I, 80056 Ercolano, Italy; giorgio.matteucci@isafom.cnr.it
[2]   CNR-IRET, Viale Guglielmo Marconi, 2, 05010 Porano, Italy; gabriele.guidolotti@cnr.it
[3]   CNR-IRET, Via Moruzzi 1, 56124 Pisa, Italy; andrea.scartazza@cnr.it
[4]   UNITUS-DIBAF, Via San Camillo de Lellis, 01100 Viterbo, Italy; pda@unitus.it
*   Correspondence: ettore.dandrea@isafom.cnr.it

Received: 23 January 2020; Accepted: 21 February 2020; Published: 26 February 2020

Abstract: The tree belowground compartment, especially fine roots, plays a relevant role in the forest ecosystem carbon (C) cycle, contributing largely to soil CO_2 efflux (SR) and to net primary production (NPP). Beyond the well-known role of environmental drivers on fine root production (FRP) and SR, other determinants such as forest structure are still poorly understood. We investigated spatial variability of FRP, SR, forest structural traits, and their reciprocal interactions in a mature beech forest in the Mediterranean mountains. In the year of study, FRP resulted in the main component of NPP and explained about 70% of spatial variability of SR. Moreover, FRP was strictly driven by leaf area index (LAI) and soil water content (SWC). These results suggest a framework of close interactions between structural and functional forest features at the local scale to optimize C source–sink relationships under climate variability in a Mediterranean mature beech forest.

Keywords: *Fagus sylvatica* L.; net primary production; fine roots; drought; soil CO_2 efflux

---

1. Introduction

Terrestrial ecosystems, especially forests, have an active role in the global carbon (C) cycle: forests cover about 4.2×10^3 Mha of the earth's land surface, accounting for about 45% of terrestrial carbon and contributing to about 50% of terrestrial net primary production (NPP) [1]. As we are following the climatic scenario characterized by the highest variations [2], forests play a crucial role to mitigate global climatic change by removing 2.4 ± 0.4 Pg C y^{-1} from the atmosphere through growth [3]. This amount corresponds up to 30% of anthropogenic CO_2 emissions from fossil fuel burning and deforestation [4]; hence, it is evident how changes in the productivity of the forest ecosystem affects the C-cycle.

Overall, the forest net effect on the carbon cycle is strictly related to net primary production (NPP), which is the small difference between the amount of C absorbed through photosynthesis and the C emitted by plant (autotrophic) respiration [5]. NPP is usually estimated as the new organic matter produced during a given period (generally one year), in both aboveground and belowground plant compartments, and it is affected by environmental drivers [6].

Within the belowground compartment, fine root production (FRP) plays a relevant role on NPP at both ecosystem and global levels accounting for up to 67% and 22% of NPP, respectively [7,8]. Moreover, at the ecosystem level, FRP affects both autotrophic and heterotrophic components of soil CO_2 efflux (SR) [9–11], contributing to 30%–80% of annual total ecosystem respiration [12].

Despite the importance of fine roots in the ecological processes, our understanding on their dynamics is still limited [13,14]. Studies were mainly focused on the role of environmental drivers on FRP [15,16], evidencing the impact of FRP on SR, both within and among ecosystems [11,17,18]. Therefore, a better identification of drivers regulating belowground processes through FRP is essential for a correct estimation of the ecosystem C budget [19].

The forest structure, here defined as the distribution of trees over an area, is determined by past management practices and represents one of the major drivers of the forest C cycle [20–22] and biodiversity [23]. Indeed, the forest structure interacts with tree physiological functionality [13,24] and climate [19], affecting C allocation and source–sink relationships [25,26].

In this context, the general objective of this study was to explore the intra-site relationships among forest structures (number of trees, basal area, maximum diameter, and leaf area index), soil characteristics, and spatial variability of SR and FRP in a mature beech stand in Mediterranean montane conditions, characterized by an almost total canopy closure. Hence, FRP, SR, soil proprieties, and forest structural parameters were measured in different randomized plots inside the stand.

Specific aims of the present study were to assess (i) the fraction of annual NPP partitioned to FRP; (ii) if, and which, intra-stand forest structural parameters and soil characteristics affect FRP; and (iii) the effect of FRP on the spatial variability of SR.

2. Materials and Methods

2.1. Site Characteristics

The experiment was carried out during 2007–2008 in a European beech (*Fagus sylvatica* L.) forest near Collelongo (Abruzzi Region, Central Italy, Figure 1A), where a permanent experimental facility (Selva Piana stand, 41°50′58″ N, 13°35′17″ E, 1560 m elevation) was established in 1991. The Selva Piana stand is located within a 3000 ha community forest that is part of a wider forest area, included in the external belt of the Abruzzi National Park. The environmental and structural conditions of the stand are representative of central Apennine beech forests. In 2007, the stand density was 825 trees ha^{-1}, and the basal area was 40.5 m^2 ha^{-1} with a mean diameter at breast height of 25 cm and a mean height of 21.5 m. Mean tree age in 2007 was estimated to be about 115 years.

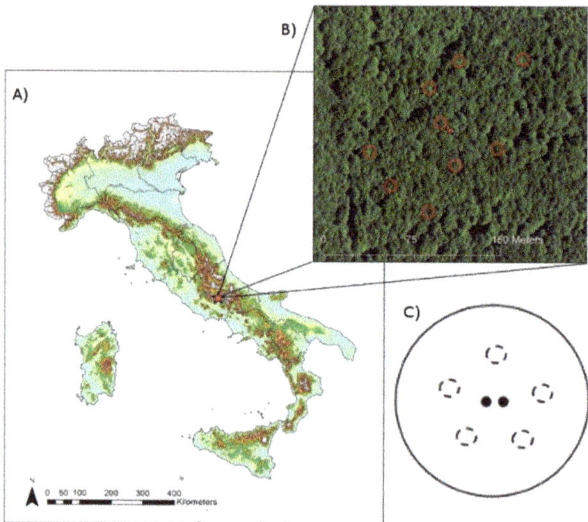

Figure 1. (**A**) Location of the Selva Piana experimental site. (**B**) Spatial distribution of the nine experimental plots (red circle) within the experimental site (the star identifies the location of the flux tower). (**C**) Schematic representation (not to scale) of the 5 m radius experimental plot (solid line), including the 5 soil CO$_2$ efflux (SR) collars (dashed circle) and the 2 ingrowth cores (black filled circle).

The forest structure is characterized by a sensible vertical stratification derived from a conversion of a beech coppice with standards to high stand [24,27] started after the middle of the 20th century.

The soil, developed on calcareous bedrock, has a variable depth (40–100 cm) and is classified as a humic Alisol [28]. Site topography is gently sloping. The climate is Mediterranean montane, with a mean annual temperature of 6.97 °C, and the mean temperatures of the coldest and warmest months are −1.04 and 16.3 °C, respectively (average of 1996–2014). Mean annual precipitation is 1116 mm, of which ~10% falls in summer. During the study, in 2007, the summer was extremely dry with only 3 mm precipitation in July and August (Figure 2).

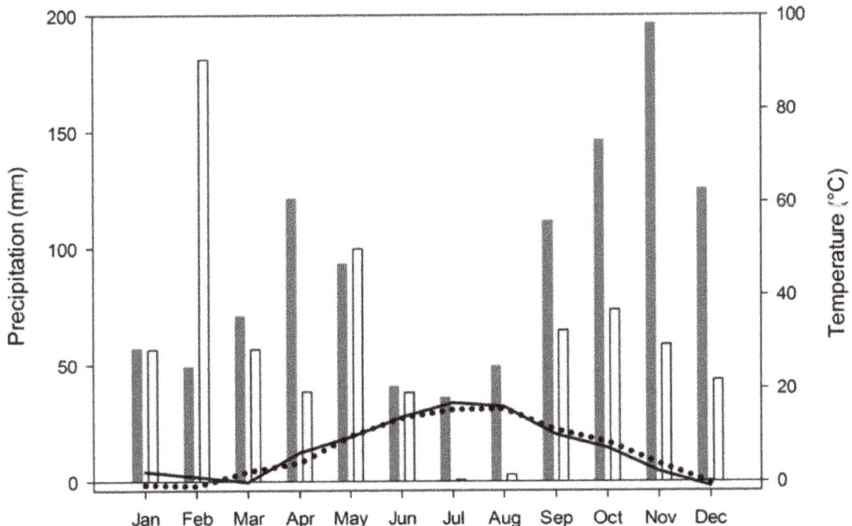

Figure 2. Climatic diagram of the Selva Piana stand. White bars represent the monthly sums of precipitation in 2007; grey bars represent mean monthly precipitation for 1996–2006; black and dotted lines represent the mean monthly temperatures in 2007 and for the period 1996–2006, respectively.

2.2. Experimental Design

To assess the role of local forest structure on the studied parameters, nine circular and relatively small experimental plots (5 m radius) were randomly established, maintaining a minimum distance of 15 m between plots centers (Figure 1B). Inside each plot, we measured forest structural parameters, soil CO_2 efflux, FRP, and soil characteristics (Figure 1C).

2.3. Fine Root Production (FRP)

The ingrowth core method (Ostonen et al. 2005) was used to estimate FRP. In each experimental plot, two ingrowth cores were installed at the beginning of April 2007. Cores were made of 0.4 cm mesh net of plastic material, to allow ingrowth of fine to medium roots (1 to 4 mm). The cores were cylindrical, with a base diameter of 5.5 cm and exploring a depth of 30 cm, as over 90% of fine roots are located at this soil depth [29–31]. Cores were filled using soil collected in the same stand near the experimental plots, air dried, and sieved at 0.4 mm to remove all the roots. One of the two cores was extracted 6 months later at the end of the growing season (October 2007), while the second core was collected one year after the installation, before the bud break (May 2008), because diffuse, porous ring species tend to produce a greater proportion of their roots after bud break [14]. After the extractions, cores were carried to the laboratory for fine roots collection (<2 mm). Hence, FRP for both annual (FRP_Y) and growing season (FRP_G) scales was estimated. Finally, FRP for the leafless period (FRP_{LP}, related to winter and early spring FRP) was calculated as the difference between FRP_Y and

FRP_G. In addition, other 11 plots were established, where only FRP was measured according to the above-described protocol. This additional dataset was used only to increase to 20 the sampling points used for NPP estimation.

2.4. Forest Structural Parameters

At the center of each sampling plot, leaf area index (LAI) was measured at the seasonal peak of 2007 (July) through two LAI 2000 Canopy Analyzers (Li-Cor) measuring above and below the canopy, respectively. LAI values were calculated using the software FV2200 (LICOR Biosciences, Lincoln, NE, USA) by considering only four of the five measuring rings to restrict the angle of view to better represent LAI of the sampling plots.

At the end of the experiment (May 2008), the diameter at breast height (DBH) of each tree inside the plots was measured, and the basal area (BA), representing the area (m^2) of the cross-section of the stem measured at 1.30 m height, mean, and maximum tree diameter of the plot (D_{max}) were derived.

2.5. Soil CO_2 Efflux, Microclimatic Condition, and Soil Characteristics

Inside the nine experimental plots, five PVC collars (10 cm diameter and 5 cm high, for a total of 45 points) were inserted in the soil with a circular distribution spaced at a minimum of 50 cm away from the neighboring trees (Figure 1C). A closed dynamic system (EGM 4, PP-System, Hitchin, UK), connected to a SRC-1 Soil Respiration Chamber (PP-System, Hitchin, UK), was used to measure SR. Measurements were performed from May 2007 until May 2008 for a total of 11 campaigns, 7 during growing season (from May to October 2007) and 4 in the leafless season (from November 2007 to April 2008) (see Guidolotti et al., 2013, for further information on SR measurements).

Soil temperature (T Soil) and soil water content (SWC) were measured at 0–10 cm by means of STP-1 (PP-System, Hitchin, UK) and time domain reflectometry techniques (Trime-FM, IMKO, gmbH, Ettlingen, Germany), respectively. All measurements were performed concurrently to the SR sampling.

In May 2008, litter and soil samples, down to 30 cm depth, were collected inside and below each PVC collar installed for SR measurements.

2.6. Carbon/Nitrogen Concentration

Litter, soil, and fine root C and N content were determined by an elemental analyzer (Model NA 1500, Carlo Erba, Milan, Italy). Soil samples were previously treated with HCl (10%) to remove carbonates.

2.7. Statistical Analysis

Analysis of the relationships between fine root production with both forest structural parameters and soil characteristics were performed only on annual values of FRP (FRP_Y) because soil sampling and DBH measurements to calculate the forest structural parameters were carried out at the end of the experimental period in May 2008.

Stepwise analysis was used to select the independent variables determining FRP_Y (Table 1 and Table S1). We tested data normality and constant variance using the Shapiro–Wilk test and the Spearman rank correlation between the absolute values of the residuals and the observed value of the dependent variable. We applied communality analysis (CA) to a multiple linear regression built with the variables identified by the stepwise analysis to disentangle the effects of each independent variable. Communality analysis shares the explained variance into pure and joint effects of predictors in order to assess the relative contribution of each predictor to the explained variance of the response variable [32].

Table 1. Descriptive statistics of forest structural and soil parameters used to assess the relation with the FRP in the 9 sampling plots. Maximum (Max), minimum (Min), mean (Mean), and coefficient of variation (CV) values are reported for each parameter. N tree plot-1 is the number of tree inside each sampling plot; basal area (m^2) is the sum of the stem cross-section areas of the n trees present in each plot; D_{max} (cm) is the maximum diameter measured in the sampling plot tree height; LAI (m^2 m^{-2}); T soil (°C) is the average soil temperature measured during the SR campaigns ($n = 11$); SWC (%) is the annual average soil water content measured during the SR campaigns, except February 2008 because of the snow cover ($n = 10$); SMN (%) is the organic nitrogen percentage in the mineral soil layer; SMC (%) is the organic carbon percentage in the mineral soil layer; SON (%) is the organic nitrogen percentage in the organic soil layer; SOC (%) is the organic carbon percentage in the organic soil layer; litter amount (g DW m^{-2}); litter N (%) and litter C (%) are the nitrogen and carbon percentages of the litter, respectively.

	Max	Min	Mean	CV
Forest Structure Parameter				
N tree plot^{-1}	24	5	14	0.42
Basal area (m^2)	0.37	0.21	0.30	0.20
D_{max} (cm)	46.90	20.90	31.99	0.25
LAI (m^2 m^{-2})	7.13	5.33	6.10	0.10
Soil Parameters				
T soil (°C)	9.39	8.59	8.92	0.03
SWC (%)	34.56	14.73	23.61	0.23
SMN (%)	1.32	0.44	0.82	0.35
SMC (%)	17.16	7.35	10.58	0.33
SOC (%)	28.04	16.14	21.18	0.18
SON (%)	1.91	1.14	1.46	0.17
Litter Amount (g m^{-2})	265.48	191.22	232.14	0.10
Litter N (%)	2.02	1.75	1.86	0.05
Litter C (%)	41.49	36.14	38.92	0.04

3. Results

3.1. Fine Root Production and Its Drivers

Fine root production during the growing season (FRP_G) was estimated at 7.63 ± 2.02 Mg ha^{-1}, ranging from 5.47 to 10.93 Mg ha^{-1}, with a coefficient of variation (CV) of 0.26. In May 2008, 12 months after in-growth cores installation, FRP_Y was 9.80 ± 1.97 Mg ha^{-1} y^{-1}, ranging from 7.32 to 13.50 Mg ha^{-1} y^{-1} and with a CV of 0.20. The FRP_{LP}, estimated as the difference between FRP_Y and FRP_G, was 2.17 ± 1.50 Mg ha^{-1}.

Considering that total carbon NPP for the Selva Piana experimental site in 2007–2008 was 11.01 MgC ha^{-1} y^{-1} [33], and that the amount of C allocated in fine roots was 3.92 MgC ha^{-1} y^{-1} (with C content of fine roots at 40.02 ± 1.92%), the contributions to total NPP by FRP_Y, stem and branches, leaves, and coarse roots were 36%, 33%, 22%, and 9%, respectively.

The step-wise analysis results indicated LAI, related to basal area (Figure S1), and SWC as the variables affecting FRP_Y ($FRP_Y = -7.504 + 0.145$ SWC $+ 2.342$ LAI, $R^2 = 0.928$, $p < 0.01$). The commonality analysis suggested that 44% of the whole variability was affected by the pure effect of LAI, 20% related to SWC, and 36% was affected by the joint effect of the two predictors.

3.2. Spatial Variability of Soil Respiration

In the study period, SR was 1.49 ± 0.22 µmol CO_2 m^{-2} s^{-1} and ranged from 1.04 to 1.83 µmol CO_2 m^{-2} s^{-1}. We observed relevant variability at both spatial (among the 9 experimental plots) and seasonal (among the 11 SR campaigns) scales with mean CVs of 0.15 and 0.46, respectively.

In the study site, significant relationships between SR and FRP were observed. We found a highly significant effect of FRP on SR among the different plots considering both annual mean value (FRP_Y,

$R^2 = 0.702$, $p < 0.01$, Figure 3) and the datasets including growing (FRP$_G$) and leafless periods (FRP$_{LP}$) ($R^2 = 0.842$; $p < 0.01$; Figure 4). Conversely, among plots, we did not find any significant relationships between annual average SR and soil parameters reported in Table 1 (data not shown).

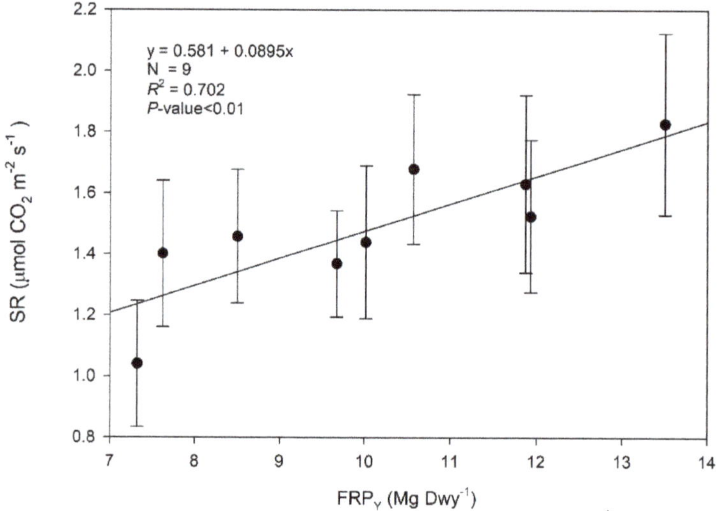

Figure 3. Relationship between annual fine root production (FRP$_Y$) and mean soil CO_2 efflux (SR). Each point is a different sampling plot, and error bars represent standard deviation.

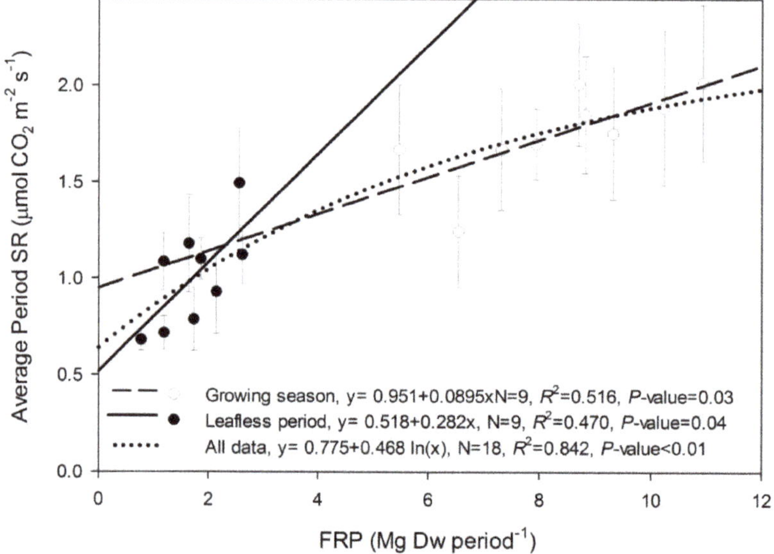

Figure 4. Relationship between fine root production (FRP) and mean soil CO_2 efflux (SR) in different periods of the study: from May 2007 to October 2007 (FRP$_G$, dashed line and empty circles); from November 2007 to April 2008 (FRP$_{LP}$, continuous line and black circles); dotted line shows overall relationship. Each point is a different sampling plot, and error bars represent standard deviation.

4. Discussion

4.1. Fine Root Production and Its Contribution to NPP

FRP values of this study were within the range of 2.9 and 9.6 Mg ha^{-1} yr^{-1} reported for several beech stands in Europe [31,34]. Indeed, in an independent experiment carried out in the same period and site using isotope-labelled soil in-growth cores, a net annual root-derived carbon input to soil was estimated at 4 Mg C ha^{-1} yr^{-1} [33].

Moreover, our study suggests that not accounting for FRP$_{LP}$ could lead to an underestimation of fine root contribution to C-cycle that could be relevant, confirming previous findings [35]. FRP$_{LP}$ could be supported by the mobilization and use of carbohydrate reserves demonstrated in several studies [36–38]. In our experimental site, this hypothesis is corroborated by the decrease of starch and soluble sugars during winter [39,40].

Our results indicate that FRP$_Y$ is mainly dependent on LAI, which represents a proxy of ecosystem productivity and ground coverage [41,42]. This suggests that above- and belowground compartments are strongly connected at both local [43–45] and regional scales [46]. The positive relationships between FRP$_Y$ with both LAI and SR demonstrate the crucial role of FRP in connecting the forest structure and soil C fluxes. The role played by fine roots on soil C fluxes could be dependent on the forest development stages. Indeed, a finding similar to that shown in the present study was reported for an old mature beech forest [25], while no relationship was found between SR and fine root biomass in a young beech forest [47].

A previous study carried out in 1996 in the Selva Piana stand [31] estimated an FRP$_Y$ of 3.8 Mg ha^{-1} yr^{-1}, less than half of the current study, although the ingrowth cores method estimated lower values of FRP [48]. Furthermore, in the cited study, FRP$_Y$ contributed less to annual NPP (28% vs. 36%). These results might be only partially explained by the 37% increment of aboveground biomass and could be affected by the strong differences in precipitation regimes in the two sampling years, especially during the July–August period. FRP, as well its contribution to NPP, can vary depending on environmental factors [49–51], as suggested by the optimal partitioning theory where C allocation to roots can increase when plant growth is limited by water and/or nutrients [52,53]. However, our results suggested a double effect of water availability on FRP at different temporal and spatial scales. Water limitations could stimulate allocation to fine roots (i.e., 1996 vs. 2007), but at the stand scale, in case of water shortage, FRP could be positively stimulated by SWC. If so, this result confirms the observed positive relationship found between beech fine root growth and water availability driven by different precipitation regimes during drought years [51].

4.2. Spatial Variability of Soil Respiration and FRP

A large intra-site variability in SR rates was observed in several ecosystem types ranging from savanna, tropical, boreal, to temperate forests [11,18,47,54–57]. In the present study, the spatial variability of SR was not related to the soil microclimatic environment, including T soil and SWC, confirming previous findings reported for three temperate European forests [58]. Hence, our results suggest that FRP plays a major role in determining the spatial variability of SR in a Mediterranean beech forest characterized by a closed canopy.

In addition, our data show a reduction of the FRP influence on SR (logarithmic regression) that could be related to the effect of soil water shortage on SR fluxes during the dry seasonal period, as previously demonstrated for the Selva Piana site [59].

5. Conclusions

This work described the spatial interactions among the forest structure and belowground C fluxes in a Mediterranean beech forest characterized by an almost complete canopy closure.

Fine roots played a relevant role in the ecosystem C-cycle, representing the main component of NPP (36%) and explaining about 70% of the annual soil CO_2 efflux variability inside the stand.

The results obtained in this study seem to indicate a functional mechanism to optimize source–sink C relationships in response to spatial variability of microclimatic drivers associated with changes of fine-scale forest structural traits. Forest structure and functionality are highly interactive; hence, an improved understanding of their relationships is fundamental to address forest adaptation and mitigation to climate change. Furthermore, as the structural features of the forest are derived from the past management, these results may inform adaptive forest management options.

Supplementary Materials: The following are available online at http://www.mdpi.com/1999-4907/11/3/255/s1, Figure S1: Relationship between the basal area, representing the sum of the area (m^2) of the cross-section of stems measured at 1.30 m height, and Leaf Area Index (LAI) measured at the centre of each sampling plot, Figure S2: Relationship between annual fine root production (FRP_Y) and leaf area index (LAI). Each point is a sampling plot. Table S1: Forest structural parameters, soil parameters, and FRP in the 9 sampling plots (SP). Basal area (m^2) is the sum of the stem cross section areas of the n trees present in each plot; D_{max} (cm) is the maximum diameter measured in the sampling plot tree height; LAI ($m^2 \ m^{-2}$); T soil ($°C$) is the average soil temperature measured during the SR campaigns ($n = 11$); SWC (%) is the annual average soil water content measured during the SR campaigns except February 2008 because of the snow cover ($n = 10$); SMN (%) is the organic nitrogen percentage in the mineral soil layer; SMC (%) is the organic carbon percentage in the mineral soil layer; SON (%) is the organic nitrogen percentage in the organic soil layer; SOC (%) is the organic carbon percentage in the organic soil layer; Litter amount (g DW m^{-2}); Litter N (%) and Litter C (%) are the nitrogen and carbon percentage of litter, respectively; FRP_Y, FRP_G, and FRP_{LP} (Mg Dw ha^{-1}) are fine root production estimated at annual scale, during the vegetative season, and during leafless period, respectively; SR_Y, SR_G, and SR_{LP} are the mean of the soil CO_2 effluxes measured during the whole study period, during the vegetative period, and during the leafless period, respectively.

Author Contributions: All the authors contributed to the manuscript: conceptualization, E.D., G.G., A.S., G.M., P.D.A.; methodology, E.D., G.G., A.S., G.M., P.D.A.; formal analysis, E.D., G.G.; investigation, E.D., G.G.; data curation, E.D., G.G.; writing—original draft preparation, E.D., G.G., A.S.; writing—review and editing, E.D., G.G., A.S., G.M., P.D.A.; supervision, G.M., P.D.A.; funding acquisition, G.M. All authors have read and agreed to the published version of the manuscript.

Funding: Researchers at the site in the year of this study were funded by FISR-CarboItaly; CarboEurope-IP, Contract n. GOCE-CT-2003-505572; Forest Focus Regulation 2152/2003 of the European Commission; CIRCE-IP, Contract no.036961.

Acknowledgments: The Collelongo-Selva Piana site was and is part of several international and national networks (CarboEurope, FluxNet, ICP-Forests, Conecofor) and since 2006 is one of the sites of the Italian Long Term Ecological Research network (LTER-Italy), part of the International LTER network (ILTER) and of eLTER Europe. Activity at the site is currently funded by the project "ForTer-Gestione Sostenibile e multifunzionale delle risorse forestali e territoriali" (DTA.AD002.486) and by resources available from Ministry of University and Research (FOE-2019). Contribution to the laboratory work by Ermenegildo Magnani is greatly acknowledged.

Conflicts of Interest: The authors declare no conflicts of interest.

References

1. Bonan, G.B. Forests and Climate Change: Forcings, Feedbacks, and the Climate Benefits of Forests. *Science* **2008**, *320*, 1444–1449. [CrossRef] [PubMed]

2. Bombi, P.; D'Andrea, E.; Rezaie, N.; Cammarano, M.; Matteucci, G. Which climate change path are we following? Bad news from Scots pine. *PLoS ONE* **2017**, *12*, e0189468. [CrossRef] [PubMed]

3. Pan, Y.; Birdsey, R.A.; Fang, J.; Houghton, R.; Kauppi, P.E.; Kurz, W.A.; Phillips, O.L.; Shvidenko, A.; Lewis, S.L.; Canadell, J.G.; et al. A Large and Persistent Carbon Sink in the World's Forests. *Science* **2011**, *333*, 988–993. [CrossRef] [PubMed]

4. Le Quéré, C.; Raupach, M.R.; Canadell, J.G.; Marland, G.; Bopp, L.; Ciais, P.; Conway, T.J.; Doney, S.C.; Feely, R.A.; Foster, P.; et al. Trends in the sources and sinks of carbon dioxide. *Nat. Geosci.* **2009**, *2*, 831–836. [CrossRef]

5. Collalti, A.; Prentice, I.C. Is NPP proportional to GPP? Waring's hypothesis 20 years on. *Tree Physiol.* **2019**, *39*, 1473–1483. [CrossRef]

6. Ciais, P.; Reichstein, M.; Viovy, N.; Granier, A.; Ogée, J.; Allard, V.; Aubinet, M.; Buchmann, N.; Bernhofer, C.; Carrara, A.; et al. Europe-wide reduction in primary productivity caused by the heat and drought in 2003. *Nature* **2005**, *437*, 529–533. [CrossRef]

7. Santantonio, D.; Grace, J.C. Estimating fine-root production and turnover from biomass and decomposition data: A compartment–flow model. *Can. J. For. Res.* **1987**, *17*, 900–908. [CrossRef]

8. McCormack, M.L.; Dickie, I.A.; Eissenstat, D.M.; Fahey, T.J.; Fernandez, C.W.; Guo, D.; Helmisaari, H.S.; Hobbie, E.A.; Iversen, C.M.; Jackson, R.B.; et al. Redefining fine roots improves understanding of below-ground contributions to terrestrial biosphere processes. *New Phytol.* **2015**, *207*, 505–518. [CrossRef]

9. Hanson, P.J.; Edwards, N.T.; Garten, C.T.; Andrews, J.A. Separating root and soil microbial contributions to soil respiration: A review of methods and observations. *Biogeochemistry* **2000**, *48*, 115–146. [CrossRef]

10. Gaudinski, J.; Trumbore, S.; Davidson, E.; Cook, A.; Markewitz, D.; Richter, D. The age of fine-root carbon in three forests of the eastern United States measured by radiocarbon. *Oecologia* **2001**, *129*, 420–429. [CrossRef]

11. Tang, J.; Baldocchi, D.D. Spatial-temporal variation in soil respiration in an oak-grass savanna ecosystem in California and its partitioning into autotrophic and heterotrophic components. *Biogeochemistry* **2005**, *73*, 183–207. [CrossRef]

12. Davidson, E.A.; Richardson, A.D.; Savage, K.E.; Hollinger, D.Y. A distinct seasonal pattern of the ratio of soil respiration to total ecosystem respiration in a spruce-dominated forest. *Glob. Chang. Biol.* **2006**, *12*, 230–239. [CrossRef]

13. Hopkins, F.; Gonzalez-Meler, M.A.; Flower, C.E.; Lynch, D.J.; Czimczik, C.; Tang, J.; Subke, J.A. Ecosystem-level controls on root-rhizosphere respiration. *New Phytol.* **2013**, *199*, 339–351. [CrossRef] [PubMed]

14. Mccormack, M.L.; Gaines, K.P.; Pastore, M.; Eissenstat, D.M. Early season root production in relation to leaf production among six diverse temperate tree species. *Plant Soil* **2015**, 121–129. [CrossRef]

15. Hertel, D.; Strecker, T.; Muller-Haubold, H.; Leuschner, C. Fine root biomass and dynamics in beech forests across a precipitation gradient—Is optimal resource partitioning theory applicable to water-limited mature trees? *J. Ecol.* **2013**, *101*, 1183–1200. [CrossRef]

16. Maeght, J.-L.; Gonkhamdee, S.; Clément, C.; Isarangkool Na Ayutthaya, S.; Stokes, A.; Pierret, A. Seasonal Patterns of Fine Root Production and Turnover in a Mature Rubber Tree (Hevea brasiliensis Müll. Arg.) Stand- Differentiation with Soil Depth and Implications for Soil Carbon Stocks. *Front. Plant Sci.* **2015**, *6*, 1–11. [CrossRef]

17. Saiz, G.; Green, C.; Butterbach-Bahl, K.; Kiese, R.; Avitabile, V.; Farrell, E.P. Seasonal and spatial variability of soil respiration in four Sitka spruce stands. *Plant Soil* **2006**, *287*, 161–176. [CrossRef]

18. Shibistova, O.; Lloyd, J.; Evgrafova, S.; Savushkina, N.; Zrazhevskaya, G.; Arneth, A.; Knohl, A.; Kolle, O.; Schulze, E.D. Seasonal and spatial variability in soil CO_2 efflux rates for a central Siberian Pinus sylvestris forest. *Tellus* **2002**, *54*, 552–567. [CrossRef]

19. Sevanto, S.; Dickman, L.T. Where does the carbon go?-Plant carbon allocation under climate change. *Tree Physiol.* **2015**, *35*, 581–584. [CrossRef]

20. Shugart, H.H.; Saatchi, S.; Hall, F.G. Importance of structure and its measurement in quantifying function of forest ecosystems. *J. Geophys. Res. Biogeosciences* **2010**, *115*, 1–16. [CrossRef]

21. D'Andrea, E.; Micali, M.; Sicuriello, F.; Cammarano, M.; Ferlan, M.; Skudnik, M.; Mali, B.; Čater, M.; Simončič, P.; De Cinti, B.; et al. Improving carbon sequestration and stocking as a function of forestry. *Ital. J. Agron.* **2016**, *11*, 56–60.

22. Collalti, A.; Trotta, C.; Keenan, T.F.; Ibrom, A.; Bond-lamberty, B.; Grote, R.; Vicca, S.; Reyer, C.P.O. Thinning Can Reduce Losses in Carbon Use Efficiency and Carbon Stocks in Managed Forests Under Warmer Climate. *J. Adv. Model. Earth Syst.* **2018**, 2427–2452. [CrossRef] [PubMed]

23. Bombi, P.; Gnetti, V.; D'Andrea, E.; De Cinti, B.; Vigna Taglianti, A.; Bologna, M.A.; Matteucci, G. Identifying priority sites for insect conservation in forest ecosystems at high resolution: The potential of LiDAR data. *J. Insect Conserv.* **2019**, *23*, 689–698. [CrossRef]

24. Rezaie, N.; D'Andrea, E.; Bräuning, A.; Matteucci, G.; Bombi, P.; Lauteri, M. Do atmospheric CO_2 concentration increase, climate and forest management affect iWUE of common beech? Evidences from carbon isotope analyses in tree rings. *Tree Physiol.* **2018**, *1975*, 1110–1126. [CrossRef]

25. Søe, A.R.B.; Buchmann, N. Spatial and temporal variations in soil respiration in relation to stand structure and soil parameters in an unmanaged beech forest. *Tree Physiol.* **2005**, *25*, 1427–1436. [CrossRef]

26. Collalti, A.; Tjoelker, M.G.; Hoch, G.; Mäkelä, A.; Guidolotti, G.; Heskel, M.; Petit, G.; Ryan, M.G.; Battipaglia, G.; Matteucci, G.; et al. Plant respiration: Controlled by photosynthesis or biomass? *Glob. Chang. Biol.* **2019**. [CrossRef]

27. Collalti, A.; Marconi, S.; Ibrom, A.; Trotta, C.; Anav, A.; D'Andrea, E.; Matteucci, G.; Montagnani, L.; Gielen, B.; Mammarella, I.; et al. Validation of 3D-CMCC Forest Ecosystem Model (v. 5. 1) against eddy covariance data for 10 European forest sites. *Geosci. Model Dev.* **2016**, *9*, 479–504. [CrossRef]

28. Chiti, T.; Papale, D.; Smith, P.; Dalmonech, D.; Matteucci, G.; Yeluripati, J.; Rodeghiero, M.; Valentini, R. Predicting changes in soil organic carbon in mediterranean and alpine forests during the Kyoto Protocol commitment periods using the CENTURY model. *Soil Use Manag.* **2010**, *26*, 475–484. [CrossRef]

29. Mainiero, R.; Kazda, M. Depth-related fine root dynamics of Fagus sylvatica during exceptional drought. *For. Ecol. Manag.* **2006**, *237*, 135–142. [CrossRef]

30. Ostonen, I.; Lohmus, K.; Pajuste, K. Fine root biomass, production and its proportion of NPP in a fertile middle-aged Norway spruce forest: Comparison of soil core and ingrowth core methods. *For. Ecol. Manag.* **2005**, *212*, 264–277. [CrossRef]

31. Scarascia-Mugnozza, G.; Bauer, G.A.; Persson, H.; Matteucci, G.; Masci, A. Tree Biomass, Growth and Nutrient Pools. In *Carbon and Nitrogen Cycling in European Forest Ecosystems*; Schulze, E.-D., Ed.; Springer: Berlin, Germany, 2000; pp. 49–62.

32. Huang, R.; Zhu, H.; Liu, X.; Liang, E.; Grießinger, J.; Wu, G.; Li, X.; Bräuning, A. Does increasing intrinsic water use efficiency (iWUE) stimulate tree growth at natural alpine timberline on the southeastern Tibetan Plateau? *Glob. Planet. Chang.* **2017**, *148*, 217–226. [CrossRef]

33. Alberti, G.; Vicca, S.; Inglima, I.; Belelli-Marchesini, L.; Genesio, L.; Miglietta, F.; Marjanovic, H.; Martinez, C.; Matteucci, G.; D'Andrea, E.; et al. Soil C:N stoichiometry controls carbon sink partitioning between above-ground tree biomass and soil organic matter in high fertility forests. *iForest-Biogeosciences For.* **2014**, *8*. [CrossRef]

34. Hendriks, C.; Bianchi, F. Root density and root biomass in pure and mixed forest stands of Douglas-fir and Beech. *Neth. J. Agric. Sci.* **1995**, *43*, 321–331.

35. Mao, Z.; Jourdan, C.; Bonis, M.; Pailler, F.; Rey, H.; Saint-andré, L.; Stokes, A. Modelling root demography in heterogeneous mountain forests and applications for slope stability analysis. *Plant Soil* **2013**, 357–382. [CrossRef]

36. Guo, D.L.; Mitchell, R.J.; Hendricks, J.J. Fine root branch orders respond differentially to carbon source-sink manipulations in a longleaf pine forest. *Oecologia* **2004**, *140*, 450–457. [CrossRef]

37. Schuur, E.A.G.; Trumbore, S.E. Partitioning sources of soil respiration in boreal black spruce forest using radiocarbon. *Glob. Chang. Biol.* **2006**, *12*, 165–176. [CrossRef]

38. Vargas, R.; Trumbore, S.E.; Allen, M.F. Evidence of old carbon used to grow new fine roots in a tropical forest. *New Phytol.* **2009**, *182*, 710–718. [CrossRef]

39. Scartazza, A.; Moscatello, S.; Matteucci, G.; Battistelli, A.; Brugnoli, E. Seasonal and inter-annual dynamics of growth, non-structural carbohydrates and C stable isotopes in a Mediterranean beech forest. *Tree Physiol.* **2013**, *33*, 730–742. [CrossRef]

40. D'Andrea, E.; Rezaie, N.; Battistelli, A.; Kuhlmann, I.; Matteucci, G.; Moscatello, S.; Proietti, S.; Scartazza, A.; Trumbore, S.; Muhr, J. Winter's bite: Beech trees survive complete defoliation due to spring late-frost damage by mobilizing old C reserves. *New Phytol.* **2019**, *224*, 625–631. [CrossRef]

41. Reich, P.B. Key canopy traits drive forest productivity. *Proc Biol. Sci.* **2012**, *279*, 2128–2134. [CrossRef]

42. Clark, D.B.; Paulo, C.; Oberbauer, S.F.; Ryan, G. First direct landscape-scale measurement of tropical rain forest Leaf Area Index, a key driver of global primary productivity. *Ecol Lett.* **2008**, *11*, 163–172. [CrossRef] [PubMed]

43. Chen, W.; Zhang, Q.; Cihlar, J.; Bauhus, J.; Price, D.T. Estimating fine-root biomass and production of boreal and cool temperate forests using aboveground measurements: A new approach. *Plant Soil* **2004**, *265*, 31–46. [CrossRef]

44. Finér, L.; Ohashi, M.; Noguchi, K.; Hirano, Y. Factors causing variation in fine root biomass in forest ecosystems. *For. Ecol. Manag.* **2011**, *261*, 265–277. [CrossRef]

45. Helmisaari, H.; Derome, J.; Nöjd, P.; Kukkola, M. Fine root biomass in relation to site and stand characteristics in Norway spruce and Scots pine stands. *Tree Physiol.* **2007**, *27*, 1493–1504. [CrossRef] [PubMed]

46. Reichstein, M.; Rey, A.; Freibauer, A.; Tenhunen, J.; Valentini, R.; Banza, J.; Casals, P.; Cheng, Y.; Gru, J.M.; Irvine, J.; et al. Modeling temporal and large-scale spatial variability of soil respiration from soil water availability, temperature and vegetation productivity indices. *Glob. Biogeochem. Cycles* **2003**, *17*, 320. [CrossRef]

47. Ngao, J.; Epron, D.; Delpierre, N.; Bréda, N.; Granier, A.; Longdoz, B. Spatial variability of soil CO_2 efflux linked to soil parameters and ecosystem characteristics in a temperate beech forest. *Agric. For. Meteorol.* **2012**, *154–155*, 136–146. [CrossRef]

48. Addo-Danso, S.D.; Prescott, C.E.; Smith, A.R. Methods for estimating root biomass and production in forest and woodland ecosystem carbon studies: A review. *For. Ecol. Manag.* **2016**, *359*, 332–351. [CrossRef]

49. Gill, R.A.; Jackson, R.B. Global patterns of root turnover for terrestrial ecosystems. *New Phytol.* **2000**, *147*, 13–31. [CrossRef]

50. Norby, R.J.; Jackson, R.B. Root dynamics and global change: Seeking an ecosystem perspective. *New Phytol.* **2000**, *147*, 3–12. [CrossRef]

51. Meier, I.C.; Leuschner, C. Belowground drought response of European beech: Fine root biomass and carbon partitioning in 14 mature stands across a precipitation gradient. *Glob. Chang. Biol.* **2008**, *14*, 2081–2095. [CrossRef]

52. Bloom, A.J.; Chapin, F.S.; Mooney, H.A. Resource Limitation in Plants-An Economic Analogy. *Annu. Rev. Ecol. Syst.* **1985**, *16*, 363–392. [CrossRef]

53. Merganičová, K.; Merganič, J.; Lehtonen, A.; Vacchiano, G.; Sever, M.Z.O.; Augustynczik, A.L.D.; Grote, R.; Kyselová, I.; Mäkelä, A.; Yousefpour, R.; et al. Forest carbon allocation modelling under climate change. *Tree Physiol.* **2019**, *39*, 1937–1960. [CrossRef] [PubMed]

54. Katayama, A.; Kume, T.; Komatsu, H.; Ohashi, M.; Nakagawa, M. Effect of forest structure on the spatial variation in soil respiration in a Bornean tropical rainforest. *Agric. For. Meteorol.* **2009**, *149*, 1666–1673. [CrossRef]

55. Kosugi, Y.; Mitani, T.; Itoh, M.; Noguchi, S.; Tani, M.; Matsuo, N.; Takanashi, S.; Ohkubo, S.; Rahim Nik, A. Spatial and temporal variation in soil respiration in a Southeast Asian tropical rainforest. *Agric. For. Meteorol.* **2007**, *147*, 35–47. [CrossRef]

56. Rodeghiero, M.; Cescatti, A. Spatial variability and optimal sampling strategy of soil respiration. *For. Ecol. Manage.* **2008**, *255*, 106–112. [CrossRef]

57. Sotta, E.D.; Meir, P.; Malhi, Y.; Nobre, A.D.; Hodnett, M.; Grace, J. Soil CO_2 efflux in a tropical forest in the Central Amazon. *Glob. Chang. Biol.* **2004**, *10*, 601–617. [CrossRef]

58. Matteucci, G.; Dore, S.; Stivanello, S.; Rebmann, C.; Buchmann, N. Soil Respiration in Beech and Spruce Forests in Europe: Trends, Controlling Factors, Annual Budgets and Implications for the Ecosystem Carbon Balance. *Ecol. Stud.* **2000**, *142*, 217–236.

59. Guidolotti, G.; Rey, A.; D'Andrea, E.; Matteucci, G.; De Angelis, P. Effect of environmental variables and stand structure on ecosystem respiration components in a Mediterranean beech forest. *Tree Physiol.* **2013**, *33*, 960–972. [CrossRef]

Article

Aboveground Biomass Response to Release Treatments in a Young Ponderosa Pine Plantation

Martin Ritchie *, Jianwei Zhang and Ethan Hammett

USDA Forest Service, Pacific Southwest Research Station, 3644 Avtech Parkway, Redding, CA 96002, USA;
jianwei.zhang2@usda.gov (J.Z.); ethan.hammett@usda.gov (E.H.)
* Correspondence: martin.ritchie@usda.gov; Tel.: +1-530-226-2551

Received: 8 August 2019; Accepted: 10 September 2019; Published: 12 September 2019

Abstract: Controlling competing vegetation is vital for early plantation establishment and growth. Aboveground biomass (AGB) response to manual grubbing release from shrub competition was compared with no release control in a twelve-year-old ponderosa pine (*Pinus ponderosa* Lawson & C. Lawson) plantation established after a wildfire in northeastern California. In addition, response to chemical release followed by precommercial thinning in an adjacent plantation was also examined as a growth potential from a more intensively managed regime, where shrub competition was virtually eliminated. We measured AGB in both planted trees and competing woody shrubs to partition the biomass pools in the plantation. The results showed a significant grubbing treatment effect on basal diameter (BD) at 10 cm aboveground ($p = 0.02$), but not on tree height ($p = 0.055$). Height and BD were 2.0 m and 7.4 cm in the manual release, respectively, compared to 1.7 m and 5.6 cm in the control. However, chemical release produced much greater rates of tree growth with a height of 3.6 m and BD of 14.7 cm, respectively. Tree AGB was 60% higher with the manual release of shrubs (1.2 Mg ha^{-1}) than with control (0.7 Mg ha^{-1}) ($p < 0.05$). The planted area without shrub competition yielded a much higher green tree biomass (16.0 Mg ha^{-1}). When woody shrub biomass was included, the total AGB (trees and woody shrubs) appeared slightly higher, but non-significant in the no release control (13.3 Mg ha^{-1}) than in the manual release (11.9 Mg ha^{-1}) ($p = 0.66$); the chemical release had 17.1 Mg ha^{-1}. Clearly, shrub biomass dominated this young plantation when understory shrubs were not completely controlled. Although the manual release did increase targeted tree growth to some degree, the cost may limit this practice to a smaller scale and the remaining shrub dominance may create long-term reductions in growth and a persistent fuels problem in these fire-prone ecosystems.

Keywords: chemical release; manual release; shrub biomass

---

1. Introduction

Release treatments are a common management practice for ponderosa pine (*Pinus ponderosa* Lawson & C. Lawson) plantations. Woody shrubs often compete with planted pines reducing growth and increasing mortality rates [1,2], and moisture is a primary limiting factor within the range of ponderosa pine [3]. Controlling competing vegetation reallocates more available soil water to planted trees [1], thus fostering successful establishment and more rapid canopy closure. Release treatments lead to increased rates of tree growth [4–6]. Wagner et al. [7] observed a growth release threshold of ~3 Mg ha^{-1} of shrub biomass in interior ponderosa pine. Through a study on a pine plantation at an advanced age (39 years) on a low-quality site, McDonald and Powers [8] found total aboveground biomass was greater without shrubs than with, although it was not clear if there was any statistical significance.

Although chemical release is a common and effective practice on industry lands, manual release is more often employed on federally-owned forests in the western United States. Although expensive, manual release can be effective at improving the growth of planted ponderosa pine [2,4].

Reducing competition from woody shrubs has consequences for estimating long-term productivity in ponderosa pine stands [9,10]. This is true both for volume estimation as well as biomass or carbon sequestration. In summarizing the results for 12 California long-term soil productivity installations, Zhang et al. [11] found that total aboveground biomass was significantly higher for non-vegetation-controlled plots than for plots with vegetation control at ages 5 and 10, but there was little difference by age 20. It appears that biomass or carbon in vegetation control plots will surpass that of non-vegetation control plots very early in stand development. This carbon-dynamics information is important when managing plantations for carbon storage and other ecosystem services. However, information is lacking on the biomass dynamics, especially with less intensive manual competing vegetation control efforts.

There are many factors that can vary among operational treatments. Timing of application(s), repetition of treatment, intensity of treatment or, in the case of herbicides, the chemical(s) chosen for a given application may all contribute to varying results. Thus, any study that took all these into consideration would be excessively complex, easily involving dozens of potential operational treatment combinations. A simpler approach, employed here, is to evaluate varying levels of treatment intensity that are representative of a range of management styles.

In this paper, we considered three different levels of management intensity in a young ponderosa pine plantation. The first was a control (no release) which represented a laissez-faire management approach to plantation establishment wherein planted trees fend for themselves with no assistance provided to establish any competitive advantage with non-crop vegetation. It is generally the least expensive in terms of operational costs, although there are indirect costs accrued from growth loss and higher mortality. The second approach, often favored on federal forest lands, was the use of manual grubbing around selected planted trees. It is the most expensive approach on a cost per unit area basis [2]. Recent figures from Region 5 of the Forest Service (personal communication, Joe Sherlock, a Pacific Southwest Region Silviculturist) show that an average of $5.6 million was spent annually on manual release treatments between 2002 and 2012. The advantage of this particular approach is primarily that it averts controversy associated with herbicide applications. Forest industries in California frequently use herbicides as a cost-effective release treatment which improves tree growth. While the cost of herbicide application is lower, this does not consider indirect costs related to other ecological values. Herbicide application is considered to be more effective at producing desired growth and survival for planted trees [2]. The third treatment employed herbicides to achieve a more complete control of competing vegetation. This was used as a reference point for growth potential at this site.

The objective of this study was to evaluate the tree growth and total aboveground productivity (including shrubs) of a twelve-year-old ponderosa pine plantation under three separate treatments representing a range of management intensities. We focused on aboveground biomass response in both trees alone and total aboveground woody vegetation. We did not include soil carbon due to the fact of a lack of observed differences between with and without vegetation controls in similar systems [12].

2. Materials and Methods

2.1. Study Site

The study was established at the Blacks Mountain Experimental Forest (BMEF) in the southern Cascade Range of northeastern California (40.72N latitude, 121.17W longitude). Elevations range from 1700 to 2100 m. The climate is montane Mediterranean characterized by warm, dry summers and cold, wet winters. Annual precipitation ranges from 231–743 mm and falls primarily as snow from November to May.

The BMEF is located in the Lassen National Forest in northeastern California (Figure 1); it is located in an endorheic basin with no year-round streams. The surrounding area is unpopulated; the nearest communities are Susanville 56 km to the southeast and Old Station 22 km to the west. The National Forest is managed for multiple uses and after severe wildfire, areas are typically replanted using native species well adapted to the area. Some adjoining parcels are privately owned and managed primarily for timber production. The forests in this area are dominated by ponderosa pine with a mix of white fir (*Abies concolor* (Gord. & Glendl.) Lindl. Ex Hildebr.) and incense-cedar (*Calocedrus decurrens* (Torr.) Florin) at higher elevations.

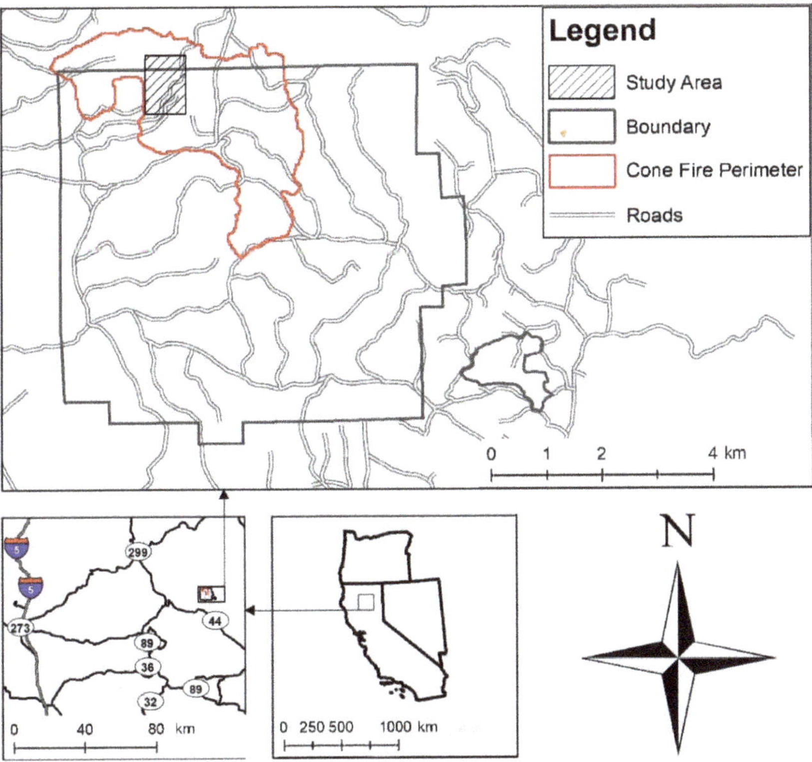

Figure 1. Geographic locations of the Blacks Mountain Experimental Forest, the Cone Fire perimeter, and study area in northeastern California, USA.

Approximately 650 ha of the BMEF and 200 ha of adjacent privately owned forest was burned by high-severity human-caused wildfire in 2002. Both ownerships were subsequently salvage-logged and planted with container-stock ponderosa pine at 3.7 m square spacing in the spring of 2004 (Figure 2). The experimental forest was planted with Styroblock 77/170 (164 ml volume per cavity; Beaver Plastics, Stuewe & Sons, Inc., Tangent, Oregon) while the adjacent privately owned land was planted with Styroblock 160/90 (90 ml) stock.

Figure 2. Post-fire regenerated plantations on public land at Blacks Mountain Experiment (left) and on private industry land (right), the latter formerly owned by Roseburg Forest Resources, now Sierra Pacific Industries. Both sides were planted with one-year-old ponderosa pine seedlings in 2005. The upper picture was taken at the photo point (*) prior to the precommercial thinning in 2009. The bottom 2014 image from Google Earth shows the post thinning plantation on industry land.

2.2. Blacks Mountain Release Treatments

Within the experimental forest plantation, a paired study of release treatments was established to compare the effects of manual grubbing on competing shrubs (primarily *Ceanothus velutinus* Douglas ex Hook and *Arctostaphylos patula* Greene) that germinated from seed after the wildfire. Manual grubbing is a common release treatment on national forest land in this region and herbicide release is used infrequently. Nine paired treatment units were established. At each of the nine locations, two 2.5 ha treatments were applied with random assignment: no release treatment and a release with all vegetation manually grubbed to a 1.5 m radius on half of the planted seedlings. Thus, 18 treatment units were regarded as our experimental units. Grubbing treatment was applied to 370 trees per ha (roughly half of the surviving planted trees in a unit) by assuming as future crop trees. Treatments were applied in the fall of 2007 and then again in the fall of 2009 after 4 and 6 growing seasons, respectively.

Circular measurement plots with a 3.6 m radius were established in a 16 × 13 m grid within each treatment unit. There was a total of 27 plots per treatment unit. The plots were established in 2008 and measured in 2015. All trees within the designated plot radius were measured for basal diameter (BD at 10 cm), total height (H), and crown width (CW). Crown width was measured on the long axis and perpendicular at the base of the live crown and then a geometric mean of the two was calculated. Basal diameter was recorded to the nearest mm with a caliper, height and crown widths to the nearest cm.

2.3. Industry Release Treatments

The adjacent plantation had a site preparation release treatment in 2003 of 22 kg ha^{-1} of Velpar (backpack application). This was followed with a directed release with two 4D targeting *Ceanothus prostratus* Benth. plants in 2005 (backpack application). A final release directed for *Ceanothus velutinus* plants with glyphosate was applied in 2007 (backpack application). Finally, a precommercial thin to 400 trees ha^{-1} was conducted in 2012 (at age nine).

A systematic array of 50 plots with the same measurements as the experimental forest (3.6 m radius circular plot) was established on the industry property in 2015. The same measurement standards were used with an augmented stump tally with measured basal diameters for all trees removed in the 2012 precommercial thin. Because we could not measure height directly on trees removed during the thinning, height was estimated using a regression relating H (m) and BD (cm) with other measured trees on both properties: H = 0.1 + a$_0$ BD^{a1} (Figure 3). The 0.1 m was added because BD was measured at 10 cm above ground.

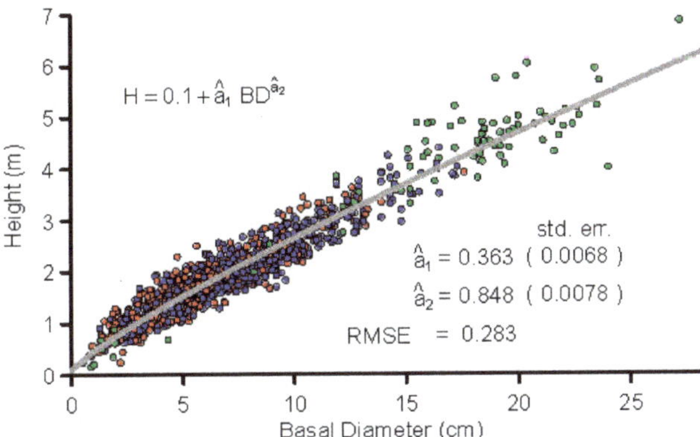

Figure 3. Height and basal diameter regression for combined non-linear fit of untreated (red), manual release (blue), and chemical release (green) trees.

A total of 81 standing trees were sampled, plus another 141 stumps of trees felled by precommercial thinning. These observations provide a reference for the site productivity potential of ponderosa pine on an identical site with trees in a free-to-grow condition.

2.4. Biomass Data

In both the experimental forest and on the adjacent property, nested rectangular shrub plots were established. Within each of the circular tree plots, a 1.5 m^2 shrub-biomass plot was established with the southwest corner at the center of the larger tree plot. This sub-plot was an efficient size to destructively sample in this vegetation structure. On the smaller plot, percent cover and height were obtained for each quadrant by species and averaged for the plot. Percent cover was an ocular estimate and vertical average height was measured to the nearest cm and averaged. There was a total of 486 plots on the BMEF experimental units and 50 on the neighboring privately owned parcel.

Because plots within the chemical release area had so little shrub cover, it was possible to do a complete inventory of shrub biomass by destructive sampling. Therefore, we recorded shrub cover and height and then removed and weighed shrub biomass on all plots where shrub biomass was found.

It was not feasible to destructively sample shrub biomass on all 486 plots in the experimental forest because of the high shrub cover, therefore subsampling of biomass was required for these plots. Within the experimental forest, observed cover values for each plot ranged from near zero to complete coverage of shrubs. This resulted in a total of 30 plots for biomass destructive sampling with a total of 51 separate species-biomass observations (Table 1).

Table 1. Summary statistics for cover (percent), height (cm), volume (m^3 m^{-2}), and shrub biomass (g m^{-2}) sampled by species (ARPA = *Arctosaphylos patula*, CEPR = *Ceanothus prostratus*, CEVE = *Ceanothus velutinus*, and OTSP = other species).

Species	n	Cover (%)	SD	Height (cm)	SD	Volume (m^3 m^{-2})	SD	Biomass (g m^{-2})	SD
ARPA	12	18.1	17.3	64.2	19.8	0.129	0.125	262	314
CEPR	11	29.0	35.4	6.8	3.0	0.020	0.026	87	109
CEVE	22	61.3	35.8	90.0	27.9	0.609	0.474	1366	1200
OTSP	6	9.1	10.9	41.0	28.8	0.028	0.021	25	24
ALL	51	38.0	36.3	60.2	39.5	0.301	0.416	673	1003

Tree biomass for all plots was estimated using equations presented by Powers et al. [13], Biomass = 278.1443 × $(BD)^2$ × Height + 0.4004, where biomass is in kg and basal diameter (BD) and height are in m.

Shrub biomass was calculated by oven-drying each sample until its weight stabilized at 80 °C. Leaves were separated to obtain the weight of woody biomass. From this sample, we fit woody (stem) biomass (g m^{-2}) as a function of observed crown volume (m^3 m^{-2}) using ordinary least squares regression and the natural log transformation of both variables (Figure 4). Using log-bias corrected parameter estimates [14], we then derived a woody shrub biomass estimate for each plot.

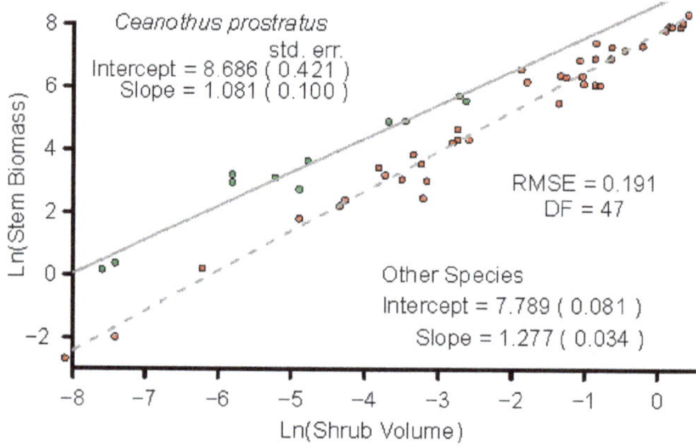

Figure 4. Model fit statistics and uncorrected parameter estimates for aboveground woody biomass (g m^{-2}) in shrubs as a function of shrub volume (m^3 m^{-2}), expressed for *Ceanothus prostratus*. (green, solid line) and other species (red, dashed line).

2.5. Analysis

Using plot mean height and BD, tree, woody shrub, and total biomass were estimated first. Then, we compared treatment effects using analysis of variance based on the experimental units in the experimental forest. Observations on industry land are presented as a reference for potential growth.

3. Results

Differences in height between the manual release treatment and the control on public land was 0.33 m, about 20% higher in the manual release treatment (Figure 5A), which is non-significant ($p = 0.055$) if the critical value for the comparison is based on $p = 0.05$. The average basal diameter was significantly higher in the manual release treatment (7.4 cm) than in the control (5.6 cm) (Figure 5B). As a reference point for the growth potential of the site, chemical release yielded much larger sized trees, 3.6 m in height and 14.7 cm in BD.

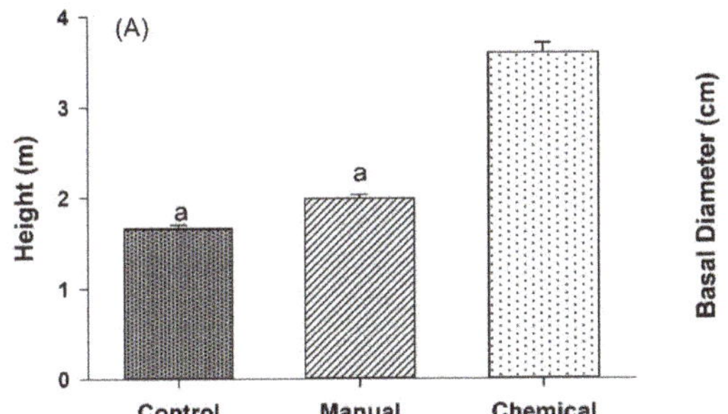
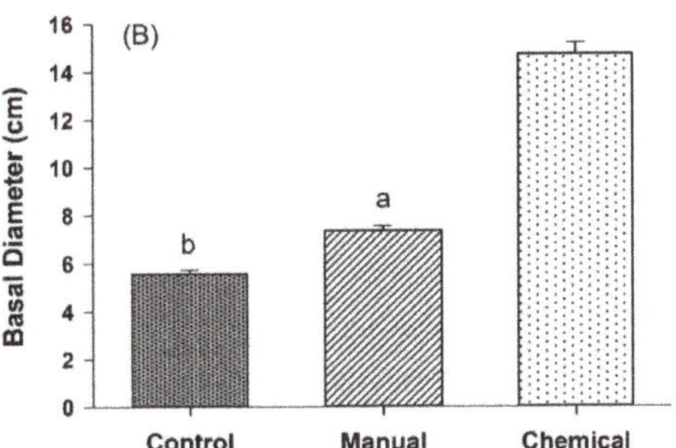

Figure 5. Means and standard errors of (**A**) height and (**B**) basal diameter for twelve-year-old ponderosa pine trees grown in no release control, manual release, and chemical release treatments. Different letters (a, b) refer to a significant difference between the two treatments ($p < 0.05$). Chemical release was not included in the analysis because it was not in the original design.

Similar trends were found in the area based on total tree biomass with 1.2 Mg ha^{-1} in the manual released treatment and 0.7 Mg ha^{-1} in the control; the difference was significant ($p = 0.045$). Chemical release treatment resulted 16.0 Mg ha^{-1} in green tree biomass (Figure 6A). However, when total vegetation AGB (trees + shrubs) was considered, the control treatment's production was slightly higher (but non-significant) in biomass than in the manual release treatment ($p = 0.656$) (Figure 6B). There were 17.1 Mg ha^{-1} on the adjacent private industry land with the chemical release.

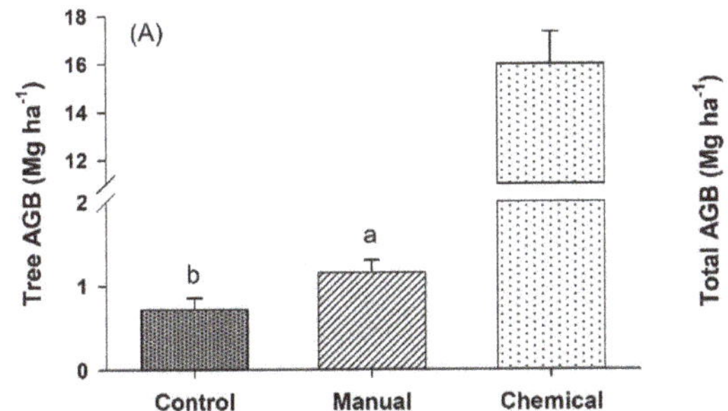
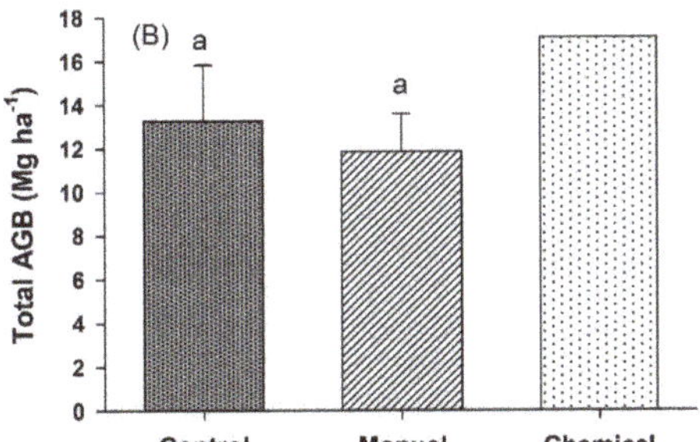

Figure 6. Means and standard errors of (**A**) tree aboveground biomass and (**B**) total aboveground biomass (i.e., trees and shrubs) on a twelve-year-old ponderosa pine plantation grown in no release control, manual release, and chemical release treatments. Different letters (a, b) refer to a significant difference between the two treatments ($p < 0.05$). Chemical release was not in the analysis because it was not in the original design.

4. Discussion

To the best of our knowledge, this is the first study to examine the long-term effects of partial grubbing release on shrub competition. Manual grubbing release has previously been studied mainly for seedling survival and individual tree growth with small plot size and few seedlings monitored during early plantation establishment [2,4]. Here, we took advantage of operational planting scale and designed the paired experimental units with a rather large plot size on both no release control and manual grubbing release. Fortunately, an adjacent private land was reforested using the same species and density in the same year as our experimental forest. Although we could not include it in our experimental design ten years ago, a sharp difference in plantation characteristics between management regimes brings into question whether statistical tests can tell us more than what we observed (Figures 2, 5 and 6).

While manual grubbing did increase tree growth, height and BD were not nearly as high as that in the chemical release treatment. Several reasons may explain this result. First, approximately only 50% of planted trees were released. In this dataset, there were 288 released trees and 257 non-released trees. The means the height, BD, and biomass presented in the results included all trees. Had we compared the released with non-released trees within the released treatment, we would find a larger mean size in the released trees (height 2.2 m and BD 8.5 cm) than in the non-released trees (1.8 m and 6.2 cm). Both were still greater than the no released control, but much smaller than trees in the adjacent land with chemical release. Second, 1.5 m radial grubbing around selected seedlings would have represented approximately 50% of the area, if all planted trees had survived and were subsequently released. Because we only released 288 trees, approximately only 25% of the area was grubbed. In contrast, chemical release in the industry land covered 100% of the area. Third, grubbing will not kill resprouting shrubs growing at this site. *Ceanothus velutinus*, *Ceanothus prostratus*, and *Arctostaphylos patula* are all resprouting species [15] and are difficult to kill with grubbing, as was evident in areas with and without manual grubbing where shrubs dominated the plantation (Figure 6). Because there were very few shrubs on the private industry land after treatments, tree biomass was the primary biomass pool, which was even bigger than the total AGB (trees and shrubs) in the other treatments, and it is worth noting that this does not include the additional 7.4 Mg ha^{-1} that was removed by precommercial thinning.

Planting trees and restoring forests following disturbances are still regarded as effective strategies for balancing CO_2 emission, as forests can store a large amount of carbon. However, within a forest, storing carbon in the tree bole is preferred to understory shrubs. Therefore, the objective for plantation management is to grow trees as quickly as possible, as larger trees are generally more resistant to future fire damage due to the fat of their thicker bark and taller stature [16,17]. Complete control of shrubs will provide the best results, as demonstrated in numerous studies [1,6]. Partial control of shrubs was difficult to implement, and the crown of remaining shrubs often quickly took over the space [5,18]; therefore, it was not as effective as the complete control at reducing competition.

Soil carbon, a larger carbon pool than aboveground vegetation, had a slow and complex sequestration process, which was not measured here. Based on studies at multiple sites with multiple years of understory control, Powers et al. [12] did not find any difference in soil carbon with and without understory control, including one installation inside the experimental forest. Because our plantations are of a similar age, we expect that the soil carbon does not differ among the two plantations.

Throughout the western United States, current reforestation efforts in the US Forest Service are conducted primarily in a post-burn environment as clearcutting practice has fallen off. Planning for regeneration is influenced by many considerations including fire severity, natural regeneration potential, soil erosion risk, water contamination, and wildlife habitat [19]. As a result, a relatively small proportion of burned areas are targeted for artificial regeneration. Furthermore, funding constraints limit the manual treatments for these plantations, where both trees and shrubs compete for the dominant position (Figure 2).

Cost was a critical limiting factor for manual grubbing; our cost per application was about US $700 ha^{-1}. The grubbing treatment was also ineffective, with two applications producing a large shrub biomass pool (Figure 6). It is difficult to know how many applications would be necessary to achieve complete control with grubbing, but given the propensity for these shrubs to re-sprout and abundant viable seeds in the soil seed bank, it would likely take at least six applications of manual grubbing to achieve a free-to-grow condition (Gary Fiddler, personal communication). If this is the case, then the cost for an effective manual grubbing regime would be on the order of US $4000 ha^{-1}.

Herbicide application is a highly regulated process by the State of California. Not only are applicators licensed by the California Environmental Protection Agency, but they also meet all laws and regulations related to the California Division of Occupational Safety and Health. More importantly, few herbicides have been registered for use in California forests, where applications are typically associated with plantation establishment on private industry lands.

The evaluation of total standing aboveground biomass, including both trees and shrubs, suggested a possible increase with chemical release (Figure 6). The difference between effective removal of shrubs and a free-to-grow environment through chemical release and manual release for green biomass was about 5.0 Mg ha^{-1}. But, if the tree AGB from precommercial thinning was considered, we would find a substantial increase in total aboveground biomass or carbon in the chemical release. This is likely due to the more efficient carbon sequestration by conifers versus shrubs after the overstory canopy closes [11]. In addition, ponderosa pine is a marketable product, so this carbon can be converted to wood products and maintained longer [20].

The observed productivity with respect to tree growth, using the OP-Yield projections [21], suggests that the intensively managed plantation is growing at a rate commensurate with a site index of 25 m at a base age 50 years, while the manually released stand is growing at a rate commensurate with a site index of 10 m at a base age 50 years; suggesting a significant reduction in tree productivity consistent with the findings of Newton and Hansen [10].

In this moisture-limited environment for tree growth, effective early control of competing vegetation is a key component of an effective management strategy for regenerating conifers [22–24]. One rule of thumb is that cover should be maintained at levels below 20 percent [2,18] and that early treatment is the key because growth losses can persist [6,25]. These results demonstrate the importance of liming the competition as it relates to aboveground biomass and carbon of trees in young plantations. Two applications of a manual release treatment were not able to produce a growth response approaching productive capacity. More frequent applications could conceivably be applied but these treatments are already cost-prohibitive on a unit area basis.

5. Conclusions

A significant effect of the manual grubbing release from shrub competition on tree growth was found when compared with the no release control. Yet, total aboveground biomass or carbon was only marginally influenced because shrub biomass dominated both sets of plots in this young plantation. On adjacent private industry land, the chemical release showed an order of magnitude more tree biomass or carbon than the public land treatments and even more total biomass including all shrubs. While manual grubbing commonly continues to be used to reduce the impacts of competing vegetation on tree growth and mortality on public land, this treatment does not allow trees to approach productive capacity of the site because these competitors are often not killed by the grubbing and grow back into the cleared space.

Here, in the western United States, especially in California, forest managers often face a reforestation challenge in forests burned by wildfire; last year, California wildfires consumed 0.75 million hectares. Only a small percentage of public lands are planted because of land-use restrictions. The success of artificial regeneration depends on how well competing shrubs are controlled. Although industry foresters often use herbicides to effectively control shrubs and achieve the potential productivity of plantations, as shown here, use of herbicides remains a contentious issue and tightly constrained by regulation. Clearly, the effects of herbicide use on the environment must be considered [26], which is also included in current environmental assessments required by the US National Environmental Policy Act. A broader tradeoff for controlling competing shrubs between using herbicides and grubbing or other means should be evaluated if biomass production or carbon sequestration is one of goals in the post-fire reforestation program.

Author Contributions: Conceptualization, M.R.; methodology, M.R., J.Z. and E.H.; formal analysis, M.R. and J.Z.; data curation, E.H.; writing—review and editing, M.R. and J.Z.

Funding: This research received no external funding.

Acknowledgments: The authors wish to acknowledge the cooperation of Sierra Pacific Industries in completing field work and providing information on treatment history. Use of trade names in this paper does not constitute endorsement by the USDA Forest Service.

Conflicts of Interest: The authors declare no conflict of interest.

References

1. Wagner, R.G. The role of vegetation management for enhancing productivity of the world's forests. *Forestry* **2006**, *79*, 57–79. [CrossRef]
2. McDonald, P.M.; Fiddler, G.O. Twenty-Five years of managing vegetation in conifer plantations in northern and central California: results, application, principles, and challenges. In *General Technical Report*; Department of Agriculture, Forest Service, Pacific Southwest Research Station: Albany, CA, USA, 2010; Volume PSW-GTR-231.
3. Powers, R.F.; Ferrell, G.T. Moisture, nutrient, and insect constraints on plantation growth: the "Garden of Eden" study. *N. Z. J. For. Sci.* **1996**, *26*, 126–144.
4. McDonald, P.M.; Fiddler, G.O. Feasibility of alternatives to herbicides in young conifer plantations in California. *Can. J. For. Res.* **1993**, *23*, 2015–2022. [CrossRef]
5. Zhang, J.W.; Oliver, W.W.; Busse, M.D. Growth and development of ponderosa pine on sites of contrasting productivities: relative importance of stand density and shrub competition effects. *Can. J. For. Res.* **2006**, *36*, 2426–2438. [CrossRef]
6. Zhang, J.W.; Powers, R.F.; Oliver, W.W.; Young, D.H. Response of ponderosa pine plantations to competing vegetation control in Northern California, USA: A meta- analysis. *Forestry* **2013**, *86*, 3–11. [CrossRef]
7. Wagner, R.G.; Petersen, T.D.; Ross, D.W.; Radosevich, S.R. Competition thresholds for the survival and growth of ponderosa pine seedlings associated with woody and herbaceous vegetation. *New For.* **1989**, *3*, 151–170. [CrossRef]
8. McDonald, P.M.; Powers, R.F. Vegetation trends and carbon balance in a ponderosa pine plantation: long-term effects of different shrub densities. In Proceedings of the 24th Annual Forest Vegetation Management Conference, Redding, CA, USA, 14–16 January 2003; University of California Cooperative Extension: Ventura, CA, USA, 2003.
9. Powers, R.F.; Reynolds, P.E. Ten-year responses of ponderosa pine plantations to repeated vegetation and nutrient control along an environmental gradient. *Can. J. For. Res.* **1999**, *29*, 1027–1038. [CrossRef]
10. Newton, M.; Hanson, T.J. Bias in site estimation from early competition. In Proceedings of the 19th Annual Forest Vegetation Management Conference, Redding, CA, USA, 20–22 January 1998; University of California Cooperative Extension: Ventura, CA, USA, 1999.
11. Zhang, J.W.; Busse, M.D.; Young, D.H.; Fiddler, G.O.; Sherlock, J.W.; Tenpas, J.D. Aboveground biomass responses to organic matter removal, soil compaction, and competing vegetation control on 20-year mixed conifer plantations in California. *For. Ecol. Manag.* **2017**, *401*, 341–353. [CrossRef]
12. Powers, R.F.; Busse, M.D.; McFarlane, K.J.; Zhang, J.; Young, D.H. Long-term effects of silviculture on soil carbon storage: does vegetation control make a difference? *Forestry* **2013**, *86*, 47–58. [CrossRef]
13. Powers, E.M.; Marshall, J.D.; Zhang, J.; Wei, L. Post-fire management regimes affect carbon sequestration and storage in a Sierra Nevada mixed conifer forest. *Forest Ecol. Manag.* **2013**, *291*, 268–277. [CrossRef]
14. Baskerville, G.L. Use of logarithmic regression in the estimation of plant biomass. *Can. J. For. Res.* **1972**, *2*, 49–53. [CrossRef]
15. McDonald, P.M. Adaptations of woody shrubs. In Proceedings of the Reforestation of Skeletal Soils in Southwest Oregon, Medford, OR, USA, 18–19 November 1981.
16. Ryan, K.C.; Reinhardt, E.D. Predicting postfire mortality of seven western conifers. *Can. J. For. Res.* **1988**, *18*, 1291–1297. [CrossRef]
17. Hood, S.M.; Varner, J.M.; van Mantgem, P.; Cansler, C.A. Fire and tree death: understanding and improving modeling of fire-induced tree mortality. *Environ. Res. Lett.* **2018**, *13*. [CrossRef]
18. Zhang, J.; Busse, M.D.; Fiddler, G.O.; Fredrickson, E. Thirteen-year growth response of ponderosa pine plantations to dominant shrubs (*Arctostaphylos* and *Ceanothus*). *J. For. Res.* **2019**. [CrossRef]
19. United States Forest Service. The principal laws relating to Forest Service activities. In *Agriculture Handbook*; U.S. Department of Agriculture, Forest Service: Washington, DC, USA, 1983; Volume 453, p. 591.
20. Bergman, R.; Puettmann, M.; Taylor, A.; Skog, K.E. The carbon impacts of wood products. *Forest Prod. J.* **2014**, *64*, 220–231. [CrossRef]
21. Ritchie, M.W.; Zhang, J. OP-Yield Version 1.0 User's Guide. In *General Technical Report*; USDA Forest Service, Pacific Southwest Research Station: Albany, CA, USA, 2018; Volume PSW-GTR-259, p. 26.

22. Oliver, W.W. *Early Response of Ponderosa Pine to Spacing and Brush: Observations on a 12-year-old Plantation*; USDA Forest Service, Pacific Southwest Forest and Range Experiment Station: Berkeley, CA, USA, 1979; Volume PSW-341, pp. 1–7.

23. Conard, S.G.; Radosevich, S.R. Growth response of white fir to decreased shading and root competition by montane chaparral shrubs. *For. Sci.* **1982**, *29*, 309–320.

24. Busse, M.D.; Cochran, P.H.; Barrett, J.W. Changes in ponderosa pine site productivity following removal of understory vegetation. *Soil Sci. Soc. Am. J.* **1996**, *60*, 1614–1621. [CrossRef]

25. McHenry, W.B.; Radoservich, S.R. Forest Vegetation Management. In *Principles of Weed Control in California*; California Weed Conference, Sponsors; Thompson Publications: Fresno, CA, USA, 1985; pp. 400–413.

26. Hively, W.D.; Hapeman, C.J.; McConnell, L.L.; Fisher, T.R.; Rice, C.P.; McCarty, G.W.; Sadeghi, A.M.; Whitall, D.R.; Downey, P.M.; Nino de Guzman, G.T.; et al. Relating nutrient and herbicide fate with landscape features and characteristics of 15 subwatersheds in the Choptank River watershed. *Sci. Total Environ.* **2011**, *409*, 3866–3878. [CrossRef] [PubMed]

Article

Growth and Tree Water Deficit of Mixed Norway Spruce and European Beech at Different Heights in a Tree and under Heavy Drought

Cynthia Schäfer [1,*], Thomas Rötzer [1], Eric Andreas Thurm [2], Peter Biber [1], Christian Kallenbach [3] and Hans Pretzsch [1]

[1]   Chair for Forest Growth and Yield Science, Department of Ecology and Ecosystem Management, Technical University of Munich, Hans-Carl-von-Carlowitz-Platz 2, 85354 Freising, Germany
[2]   Department Soil and Climate, Bavarian State Institute of Forestry (LWF), Hans-Carl-von-Carlowitz-Platz 1, 85354 Freising, Germany
[3]   Chair for Ecophysiology of Plants, Department of Ecology and Ecosystem Management, Technical University of Munich, Hans-Carl-von-Carlowitz-Platz 2, 85354 Freising, Germany
*   Correspondence: cynthia.schaefer@lrz.tum.de; Tel.: +49-8161-71-4711

Received: 11 May 2019; Accepted: 9 July 2019; Published: 11 July 2019

Abstract: Although several studies suggest that tree species in mixed stands resist drought events better than in pure stands, little is known about the impact on growth and the tree water deficit (TWD) in different tree heights at heavy drought. With dendrometer data at the upper and lower stem and coarse roots, we calculated the TWD and growth (ZGmax) (referring to the stem/root basal area) to show (1) the relationship of TWD in different tree heights (50% tree height (H50), breast height (BH), and roots) and the corresponding leaf water potential and (2) how mixture and drought influence the partitioning of growth and tree water. The analyses were made in a mature temperate forest of Norway spruce (*Picea abies* (L.) Karst.) and European beech (*Fagus sylvatica* (L.)). Half of the plots were placed under conditions of extreme drought through automatic closing roof systems within the stand. We found a tight relationship of leaf water potentials and TWD at all tree compartments. Through this proven correlation at all tree heights we were also able to study the differences of TWD in all tree compartments next to the growth allocation. Whereas at the beginning of the growing period, trees prioritized growth of the upper stem, during the course of the year the growth of lower stem became a greater priority. Growth allocation of mixed spruces showed a tendency of a higher growth of the roots compared to the BH. However, spruces in interspecific neighborhoods exhibited a lesser TWD in the roots as spruces in intraspecific neighborhood. Beeches in intraspecific neighborhoods showed a higher TWD in BH compared to H50 as beeches in interspecific neighborhoods. Mixture seems to enhance the water supply of spruce trees, which should increase the stability of this species in a time of climatic warming.

Keywords: tree water status; climate change; rainfall exclusion; *Picea abies* (L.) Karst.; *Fagus sylvatica* (L.); root–shoot allometry

1. Introduction

Climate models have predicted an increased number of drought events of longer duration and stronger intensity [1,2] that are likely to alter the growth and stability of forests [3–5]. Ciais et al. (2005) [6] gave evidence that precipitation deficits and extreme summer heat are capable of causing a Europe-wide reduction of ecosystem primary productivity. Increasing drought and accompanying changing resource availability lead to shifts in resource allocation within trees [7]. As predicted by functional equilibrium models [8,9] and proven by extensive studies [10], plants allocate additional biomass to those organs that acquire the most limiting resources. Consequently, plants allocate more biomass to the roots in such cases where belowground resources, such as water and nutrients, are limiting. When light or CO_2 are the limiting factors, plants allocate more biomass above ground. Tree species sensitive to drought can, therefore, respond to extreme drought with reduced stem growth and increased root growth [10,11].

Drought stress reactions can be determined, inter alia, via the leaf water potential (tree water status). Based on the difficult accessibility of branches, it is very laborious to measure leaf water status in tall trees. Continuous high-resolution measurements of stem radius variations meet this and provide an opportunity to gain deeper insights into the dynamics of tree water relations and growth patterns due to the opportunity to assess tree water status without a canopy crane or other circuitous methods for taking leaf water potential measurements in the tree crown. As such, they offer huge potential for ecological research under a changing climate. Stem radius variations are increasingly used in plant physiology to analyze stem growth and the tree water status [12–15] and have been analyzed for different tree species [14,16,17].

Usually, diurnal stem radius variations are measured by electronic, high-resolution point or band dendrometers [18,19]. The drought-induced changes can be recognized through modified characteristics of the bark tissue (decreasing cell turgor, which results in stem shrinking) and changes in radial growth [20,21]. When transpiration exceeds the water uptake from the soil, the tree relocates water storages—mainly located in the living cells within the cortex—to maintain the transpiration process. The coordination of stomatal and hydraulic regulations allows for an adjustment of the tree water use. Various environmental factors (e.g., temperature, soil water availability, vapor pressure deficit) control these mechanisms and thus the tree's water use. On a diurnal scale, shrinking and swelling of the stem is the result of these mechanisms and lead to alternating depletion and replenishment of the involved tissues. This process is driven by the transpiratory demand during daytime and overnight refilling of the living cells of the phloem tissue with water from the soil [22–24]. There are many ecophysiology models describing the dynamic radial and vertical water flow between the tree tissues [13,25–27]. Zweifel et al. (2000) [17] investigated stem radius changes and their relation to stored water in stems with truncated stem segments of living Norway spruces and were able to attribute the stem contraction to the living tissue outside of the cambium. Shrinking and swelling of the stem can, hence, be used as indicators for the whole tree water status [17,26,28] and can be measured for any species and any tree organ. How trees react during drought at the different tree compartments are very interesting, due to the species-specific strategies to cope with drought stress [29,30]. McCarthy and Enquist (2007) [31] showed that plants allocate biomass to the plant compartment which acquires the most limiting resource.

In addition to site and climatic conditions, the mixture of species also has a significant impact on the water supply and growth of a tree. Species mixture can improve forest ecosystem functions under changing climate through complementary interactions among a pair of species [32,33]. Complementary effects depend on the type of species and the changing resource availability [33–35]. The most widespread mixed forest stands in Central Europe consist of Norway spruce (*Picea abies* (L.) Karst.) and European beech (*Fagus sylvatica* (L.)). Mixtures of these tree species have been analyzed in many studies [36–39]. Evergreen spruce is considered to be particularly sensitive to drought stress [40,41], with a drought sensitive stomata closure [29] and correspondingly impeded photosynthesis. Deciduous beech is known to be more drought resistant as compared to spruce [41,42]. The mixture of these two tree species can have several advantages for both tree species. For example, Bolte and Villanueva (2006) [43] detected a deeper rooting system of beech in mixture with spruce compared to monocultures, and consequently, an enhanced water and nutrient availability for beech trees. The improved soil water storage due to the reduced interception of beech in mixture with spruce [44–46] can also have a positive effect on the water availability and change the entire stand's water balance [47].

The knowledge of species interactions in mixed forest stands has increased in recent years, with many investigations about beech and spruce trees [35,37,42,43]. However, most studies have focused on growth-related differences in mixture rather than on changes in tree water status and these were not observed at different tree positions. Further, the differences between stem and root show the shift of growth and tree water under drought conditions for the given species, where water drawn from internal stores during drought is not only from an ecophysiological perspective but also from a remote sensing perspective interesting. Remote sensing captures the water content of tree canopy (CWC), the product of leaf area index, leaf mass per area, and leaf water content [48]. A greater focus on water pools may improve our ability to understand and anticipate drought-induced changing traits or mortality in plants [49,50].

In the present study, we determined the basal area growth (ZGmax) and tree water status (tree water deficit, (TWD), as described by Zweifel et al. (2016) [51], also referred to the basal area) at three tree compartments: the upper stem (50% tree height—H50, at approximately 15 m tree height), the lower stem (breast height (1.3 m), BH), and at the coarse roots (roots) in the growing season. The TWD, as the measurement unit for the tree water status, was analyzed in relation to the leaf water. Subsequently, TWDs at H50, BH, and the roots were employed to analyze species-specific differences between beech and spruce in terms of drought-related changes in root–stem allometry.

The aim of the study was to investigate growth allocation as well as local tree water deficit in the tree compartments under heavy drought, in intra- and interspecific neighborhoods. Therefore, we used a rainfall exclusion experimental setup to provide drought stressed mature trees in the treatment plots and unstressed trees in the control plots. Naturally occurring drought was experimentally enhanced by means of stand scale rainfall exclusion, the Kranzberg ROOF Experiment (KROOF). We demonstrated how the allocation pattern of control and treatment trees or trees in intra- and interspecific neighborhoods during the growing season look.

We hypothesized that: (1) the stem basal area variations and the leaf water potential show a positive relationship at the different tree heights; (2) the relationship between growth response and the respective TWD is the same at the three different positions H50, BH, and root; and (3) interspecific neighborhood with beech trees facilitates spruce trees under drought stress.

2. Materials and Methods

2.1. Site Description

The study was located in southern Germany (longitude: 11°39′42″ E, latitude: 48°25′12″ N, altitude 490 m a.s.l), near Freising (Kranzberg forest) and approximately 35 km northeast of Munich. The soil of the Kranzberg forest is a luvisol developed from loess over tertiary sediments with high nutrient and water availability. The forest stand comprises European beech (*Fagus sylvatica* (L.)) and Norway spruce (*Picea abies* (L.) Karst.). The age of trees varies between 64 ± 2 years for spruce and 84 ± 4 years for beech (in 2015). In 2010, twelve plots were established with a total area of 1730 m^2 with 63 beech trees (mean height 26.1 m, mean diameter 28.9 cm at breast height) and 53 spruce trees (mean height 29 m, mean diameter 34.3 cm at breast height) (Table 1). On each plot, four trees were selected as monitoring trees (48 trees in total) (Table S1). Each of the 12 plots contained zones of spruce or beech trees in an intraspecific neighborhood and zones of spruce or beech trees in an interspecific neighborhood.

For the throughfall exclusion experiment (TE), roof structures were built in six plots below the crown of the trees at a height of about 3 m. The other six plots acted as control plots (CO). In spring 2010, the plots were trenched with a heavy-duty plastic trap to a depth of about 1 m to avoid external effects on and water intake in the experimental plots [52]. The roofs closed only during rainfall through a set of precipitation sensors, to avoid unintended micro-meteorological and physiological effects [38]. The drying cycles with closing roofs lasted from May to December 2014 (570 mm precipitation was excluded) and from March to November 2015 (480.2 mm precipitation was excluded). The annual precipitation average for the Kranzberg forest ranges between 750 and 800 mm for the entire year and between 460 to 500 mm year^{-1} in the growing season (mid-April to the end of October) (1971–2000) [53].

The annual average temperature is 7.8 °C and the average temperature for the growing season is 13.8 °C (detailed description provided by Pretzsch et al. (2012) [47]).

Table 1. Characteristics of the investigated stand where the treatment and control plots were located. (N: number of trees per ha; n: number of trees with dendrometers; BA: basal area per ha; V: total stem volume per ha; hq: mean height; dq: quadratic mean diameter at 1.3 m breast height).

	Area	N	n	BA	V	hq	dq
	(m^2)			(m^2)	(m^3)	(m)	(cm)
Drought Treatment							
Spruce		301	12	29.7	422	29.3	34.8
Beech		352	12	22.9	309	26.1	29.1
Total	145	653	24	52.6	730		
Control							
Spruce		310	12	28.8	400	28.7	33.8
Beech		356	12	22.6	305	26	28.7
Total	144	666	24	51.4	705		

2.2. Water Potential (Ψ Leaf)

Leaf water potentials at predawn (LWPpre) and midday (LWPmid) were measured on several sunny days during the growing season (April–October) in 2014 and 2015. Leaf water potential measurements were conducted in time windows from 2:00 h to 3:30 h CET for LWPpre and 13:00 to 15:00 h CET for LWPmid. The same experimental trees ($n = 31$) were used for the dendrometer measurements that could be conducted with the canopy crane. At a height of 25–30 m, south-exposed twigs of about 10–20 cm in length were taken from the sun crown (access through canopy crane) and were enclosed in humid plastic bags to prevent further water loss. The leaves were immediately measured with a pressure chamber (Model 3000 Pressure Extractor, Soil moisture Equipment Corp., Santa Barbara, CA, USA).

2.3. Stem Basal Area Variations (Growth and Tree Water Deficit)

On each of the 48 trees selected for measurement, three automatic dendrometers of two types (Ecomatik, Dachau, Germany) were installed at different tree heights. The DR-type dendrometer (DR, radius dendrometer) was installed at breast height (1.3 m, BH) and 50% tree height (H50). For measurements of the roots, circumference dendrometers of the DC2 type (DC, circumference dendrometer) were used and fixed on one main root per monitoring tree at most 20 cm depth and 30–50 cm distance to the tree. The thermal expansion coefficient of the sensor was <0.2/K µm. All dendrometers were fixed in a northeast direction to avoid environmental influences. From the spruce trees, the outermost tissues of the bark were removed to minimize hygroscopic effects of the outer bark. The frames of the dendrometers were fixed with stainless steel screws on the tree stem, with the linear transducer in direct contact with the stem/root surface. Measurements were recorded every 10 minutes. All measurement errors and proven outliers in the raw data were eliminated prior to further processing. Hourly means of the raw 10 minute measurements of stem radius variations were analyzed during the growing season (April to the end of September, days of the year (DOY) 91–273).

To describe how drought affects the tree organs, we used the tree water deficit (TWD), defined by Zweifel et al. (2016) [51]. First, the "pure" growth (further defined as zero growth, ZGmax) was extracted from the stem or root dendrometer measurements to determine the TWD (water signal). For separation, we used the zero growth concept of Zweifel et al. (2016) [51], which results in growth curves with a stepwise shape (Figure 1). When the current maximum of the stem basal area is exceeded, the increment increases. For our investigations, we used the maximum ZG value per day (ZGmax). The TWD was calculated as the difference between the growth-induced expansion of the stem and the daily shrinking and swelling. The negative values of the TWD revealed increasing shrinking of the stem basal area.

The radial measurements (ZG and TWD) were transformed into basal area fluctuations. This procedure is standard in examinations of tree ring data [54], but new for analysis of dendrometer data. We used this approach because the basal area fluctuation as a two-dimensional measurement better reflects the tree response than the radial fluctuation as a one-dimensional measurement which neglects the respective tree dimension.

Furthermore, we proved the relationship between the stem water signals and the leaf water potentials at midday and predawn. We found the best match for the relationship between LWPpre and TWD minimum (TWDmin, maximum daily shrinkage) during drought conditions (Supplementary Materials, Table S2) and used the LWPpre for further analysis of the relationship between TWDmin and LWPpre.

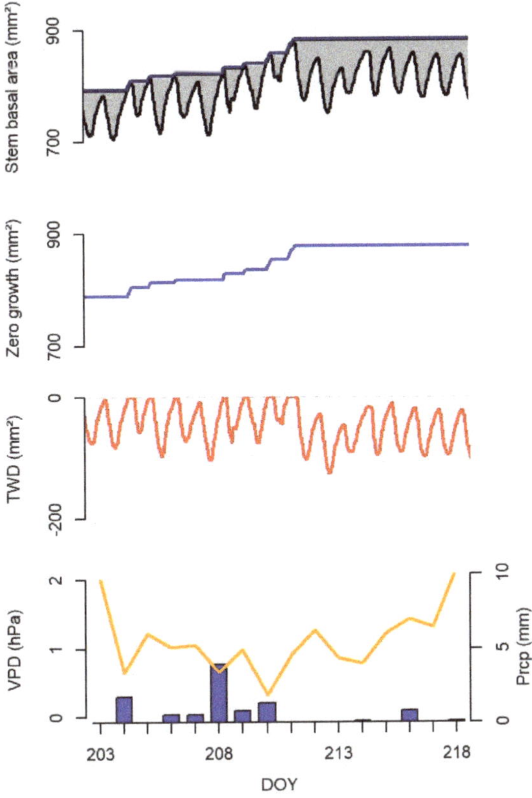

Figure 1. Exemplary illustration of the course of the stem basal area variation and the two applied indices, zero growth (ZG) and tree water deficit (TWD), for a period of 16 days in the growing season 2015 of an example spruce tree. The climatic graph of vapor pressure deficit (VPD) and daily precipitation sum (Prcp) illustrated how the deficit of water led to a stagnation of the zero growth and a decrease in the tree water deficit during the last five days of the example period.

2.4. Climatic Data

The weather data were collected from two sources. Temperature and relative humidity were measured at 10 min intervals in the forest stand and monitored with a temperature sensor (RFT-2, UMS) at a height of 27 m and stored in a datalogger (Logger Campbell CR100, Multiplexer AM16/32).

The sensor was protected against direct irradiation with a ventilated radiation shield. The vapor pressure deficit (VPD) was calculated with these data. Precipitation data were available from the nearby weather station, about 2 km from the study site in Kranzberg forest [55].

2.5. Statistical Analysis

Our experimental setup consisted of time series measurements of individual tree. The individual trees were grouped by different control and treatment plots. Consequently, the analysis was based on nested data. To consider this nesting, we applied linear mixed effect models (lmer) from the package lme4 [56]. The random effects b are the individual tree, with the index i, the plot with the index j and the year abbreviated with the index k. t represents the measurement at which we used the daily maximum value for zero growth (ZGmax) and the daily minimum value for the tree water deficit (TWDmin). ε always represents the residual error of the respective models.

To answer the question of whether the TWDmin at the various tree compartments is able to reflect drought stress, we examined the relationship between the TWDmin and LWPpre at the three different measurement positions. We pooled both years (2014 and 2015) into the same dataset for the analysis.

We applied linear mixed models in a logarithmic form (Equation (1)):

$$\ln\left(TWDmin\right)_{ijkt} = \beta_0 + \beta_1 \cdot \ln\left(LWPpre_{ijt}\right) + b_i + b_j + b_k + \varepsilon_{ijkt} \tag{1}$$

The applied logarithm led to a significantly better fitting of the data and considered the non-linear course of the analyzed relationship. The logarithm of the negative TWDmin values was enabled through a transformation by multiplying by -1. For the depiction, we adapted only the y-axis to negative values. All models where fitted species-specific for a straightforward interpretation.

To show how growth allocation or the TWDmin react between the three tree compartments under drought conditions, we examined the difference ($Diff$) of ZGmax and TWDmin at a measurement position above to the measurement position below (H50–BH and BH–Root). A value above zero would mean that the upper tree compartment profits, and a value below zero would indicate that the lower tree compartment had a higher growth. We chose to use the difference instead of the ratio because, when using the ratio, meaningful but very low TWDmin or ZGmax values in one compartment can lead to immoderate and meaningless outliers in the analysis.

The resulting difference value of the upper and lower measurements served as the independent variable. Because TWDmin and especially the ZGmax increase over the growing season, the difference between upper and lower compartments (*lowComp*) can be higher at the end of the growing period than at the beginning. Therefore, we always related the difference to the respective measurement of the lower compartment.

To show how mixture and drought treatment influence the growth allocation or TWDmin pattern of the upper and lower compartments, linear mixed models were applied. Mixture (*Mix*) and treatment (*Treat*) were included as fixed effects in a model (Equation (2)):

$$\begin{aligned} Diff_{ijkt} = \beta_0 &+ \beta_1 \cdot lowComp_{ijt} + \beta_2 \cdot Mix \cdot lowComp_{ijt} + \beta_3 \cdot lowComp_{ijt} \cdot Treat \\ &+ \beta_4 \cdot lowComp_{ijt} \cdot Mix \cdot Treat + b_i + b_j + b_k + \varepsilon_{ijkt} \end{aligned} \tag{2}$$

The significances of the fixed effects of the linear mixed models were tested by an F test with Satterthwaite's approximation ([57], R package *lmerTest*). To consider the large number of measurements points, we also calculated the conditional coefficient of determination (R^2) for the mixed-effect models with the command *r.squaredGLMM* from the *MuMIn* package. Additionally, the quality of the models was checked using the root mean square error (RMSE). All analyses were performed with R version 3.2.3 [58].

3. Results

3.1. Temperature and Precipitation in 2014 and 2015

There are clear differences in temperature and precipitation between the analyzed years 2014 and 2015 in the growing season (Figure 2a,b). The air temperature in 2015 was 1.1 °C above the average from 2001 to 2015 [59]. Compared to the year 2014, the summer months of the year 2015 had a higher number of days without rainfall or with low rainfall and simultaneously higher temperatures and higher vapor pressure deficits (VPDs).

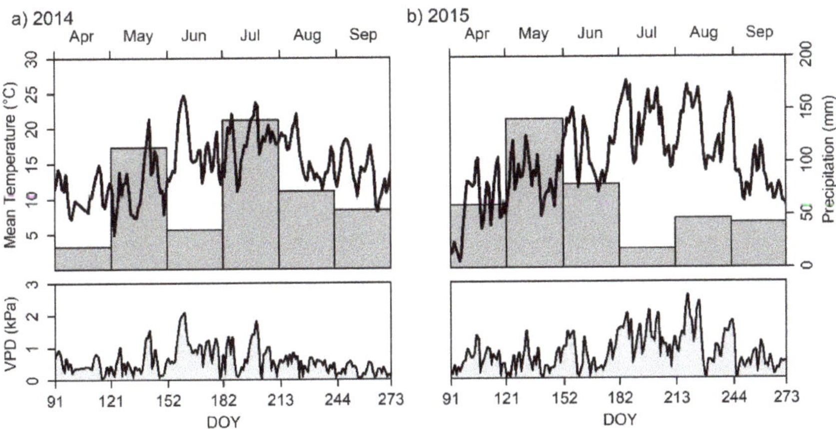

Figure 2. Monthly precipitation sums and daily mean temperature values (**a,b**, above) and daily vapor pressure deficit (VPD) (**a,b**, below) in the growing season (April–September) of the years 2014 and 2015.

3.2. Leaf Water Potentials

Comparing the midday water potentials (LWPmid) for the years 2014 and 2015, the LWPmid were significantly more negative for beech trees than for spruce trees (Supplementary Materials, Figure S1a–d, $p < 0.001$). Furthermore, we found significant differences between the trees of the drought treatment and control plots (Supplementary Materials, Table S3, $p < 0.01$), and the predawn water potentials (LWPpre) revealed clearer differences ($p < 0.001$) compared to LWPmid (Supplementary Materials, Figure S2a–d). The effect of the drought treatment was observable through more negative LWPpre compared to the control plots (Figure S3). No differences were observable between trees in intra- and interspecific neighborhoods.

3.3. Zero Growth and Tree Water Deficit

To determine growth and TWDmin, we separated the growth from the daily shrinking and swelling. Figure 3 provides an overview of the data (intra- and interspecific neighborhoods are shown in Figure S4, Supplementary Materials). Spruce trees had a higher diameter increment compared to beech trees in 2014 at all three measuring positions (H50, BH, and roots). The diameter increment of the spruce trees was smaller for the dry year 2015 compared to the year 2014. In beech, by contrast, relevant growth reduction of the trees could not be observed in 2015 compared to 2014. Drought-treated beech trees even showed higher growth in BH and the roots in 2015.

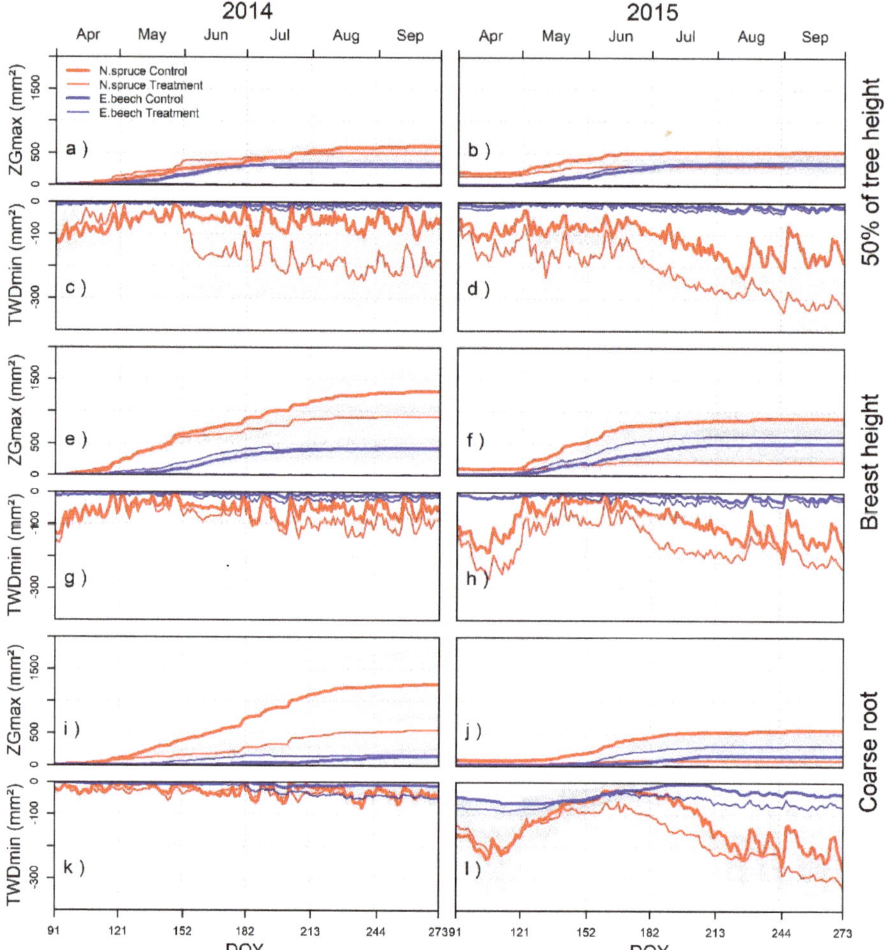

Figure 3. Mean TWDmin and zero growth (growth without the water signal) referring to the stem/root basal area (mm^2) for the years 2014 (left) and 2015 (right) for spruce (red) and beech (blue) at the control (thick line) and treatment (thin line) plots at 50% tree height (**a–d**), breast height (BH, **e–h**), and roots (**i–l**). Shaded regions are confidence intervals. Data are shown for the growing season.

Comparing the daily TWDmin of the growing season in 2014 and 2015, the effect of the drought year 2015 was observable in the intense shrinking in the summer months (DOY 152–243) (Figure 3d,h,l). Furthermore, there was a high shrinkage phase at H50 in 2014 for spruce trees (possibly through an adaptation reaction at the beginning of the drought treatment). The treatment plots indicated a more distinct stem shrinkage compared with the control plots for both years. Species-specific differences can also be seen in the magnitudes of the daily TWD. Spruce trees revealed more distinct stem water changes than beech trees (see daily values of TWD in Figure 3c,d,g,h,k,l). Overall, the stem shrinkage was highest at H50 and roots compared to BH.

3.4. Relationship between Tree Water Deficits and Water Potentials at Different Tree Heights

The TWDmin values were more negative when the LWPpre became more negative at the stem (BH, H50) and the roots (Figure 4a–c). Spruce trees revealed a higher TWDmin than beech trees at all positions. The *r*-squared (R^2) of the different models ranged between 0.38 and 0.73. The relationship was significant for both species and all positions ($p < 0.001$, Table 2). The TWDmin was highest in the crown, but the roots also showed high fluctuations when LWPpre reached −1.2 MPa.

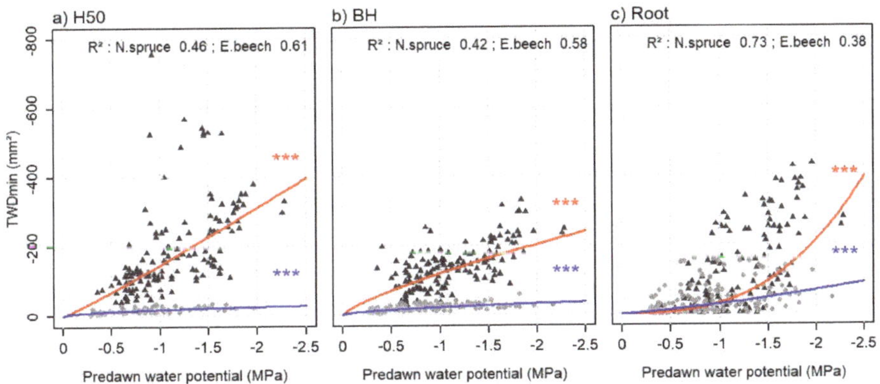

Figure 4. Relationship between TWDmin and predawn water potential (LWPpre) for spruce (red, black triangles) and beech (blue, light grey circles) trees for 50% tree height (**a**), breast height (**b**), and roots (**c**) (R^2 and significance levels based on models from Table 2; *** $p < 0.001$).

Table 2. Parameter estimates and statistics for the logarithmized relationship of the tree water deficit minimum (TWDmin) and the predawn water potential (LWPpre) at three different tree heights (50% tree height (H50), breast height (BH), and roots). All measurements of the TWDmin referred to the stem/root basal area (mm^2). The dependent variables are in the columns. Rows show the output of the model with the fixed variables (*N*: number of LWPpre and TWDmin measurements). Significance level: *** $p < 0.001$.

	log(TWDmin)					
	H50		**BH**		**Root**	
Species	**N. Spruce**	**E. Beech**	**N. Spruce**	**E. Beech**	**N. Spruce**	**E. Beech**
Intercept	4.995 ***	2.876 ***	4.786 ***	3.071 ***	3.412 ***	3.317 ***
log(LWPpre)	1.081 ***	0.501 ***	0.760 ***	0.563 ***	2.804 ***	1.292 ***
R^2	0.46	0.61	0.42	0.58	0.73	0.38
RMSE	126.93	6.17	50.40	7.93	112.27	44.47
N	138	79	144	134	138	140

3.5. Stem and Root Growth and TWDmin in Different Tree Compartments

How the allocation was oriented under control or treatment and inter- or intraspecific neighborhoods is illustrated in Figures 5 and 6. On the *y*-axis, the difference between BH and H50 or BH and the roots is given for the zero growth (panel above) or TWDmin (panel below). On the *x*-axis, the basal area growth or TWDmin of the respective lower tree compartment is given (BH, Figure 5 or roots, Figure 6). The *x*-axis of ZGmax is also a proxy for time within the growing period, while the *x*-axis of TWDmin is a proxy for increasing drought stress which was happening in the middle of the growing period. The significances of the respective linear mixed-effect models are summarized in Table 3.

The model quality can be check in the data provided in Supplementary Materials Figures S5–S12. All in all, the fitted linear regressions show good relationships between observed to predicted values (Figures S5–S12a,b). There was a strong overlapping within the data, especially close to the regression lines, which resulted in high model performances of the p-value and R^2. We revealed the overlapping close to the regression lines by hexagon density plots (Figures 5 and 6) and histograms of the residuals (see Supplementary Materials Figures S5–S12c). The model output was nearly normally distributed, which can be seen in the QQ-plots of the residuals (Figures S5–S12e). The deviation of the residuals from normal distribution was under 5%. Nevertheless, we checked these points very carefully to be sure that these outliers did come from measurement inconsistencies.

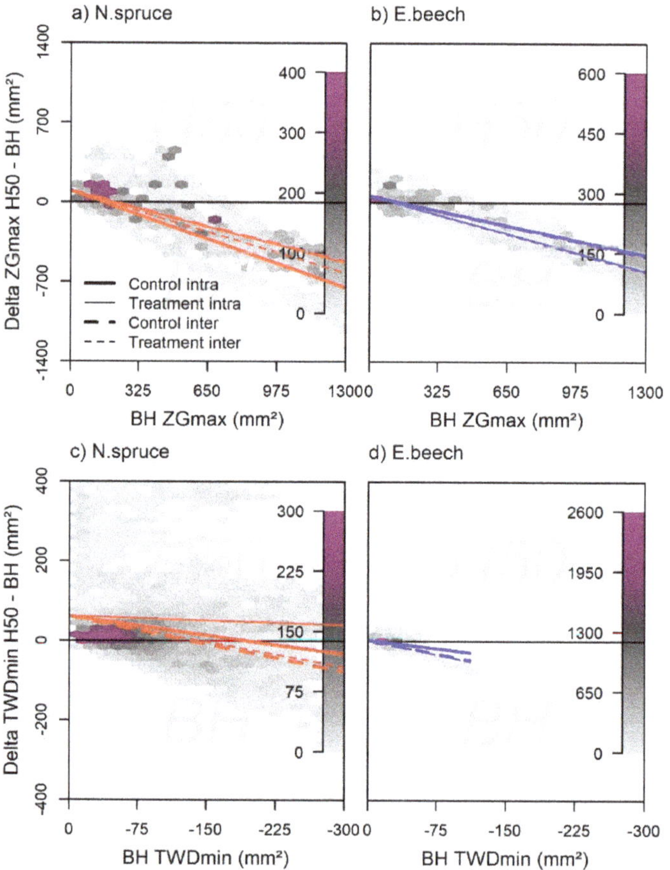

Figure 5. Relationship of growth (ZGmax) (**a,b**) and the tree water deficit (TWDmin) (**c,d**) represented by the difference of the measurements at 50% tree height (H50) and breast height (BH), dependent on the breast height measurement. Values below the zero line mean a growth allocation into the lower stem compartment at breast height. For the TWDmin, the negative values represent a lower tree water deficit at breast height. Linear mixed models depict how strongly the allocation was influenced by drought treatment (thick line—control, thin line—treatment) or mixture (intraspecific neighborhood—straight line, interspecific neighborhood—dashed line). The respective models are shown in Table 3.

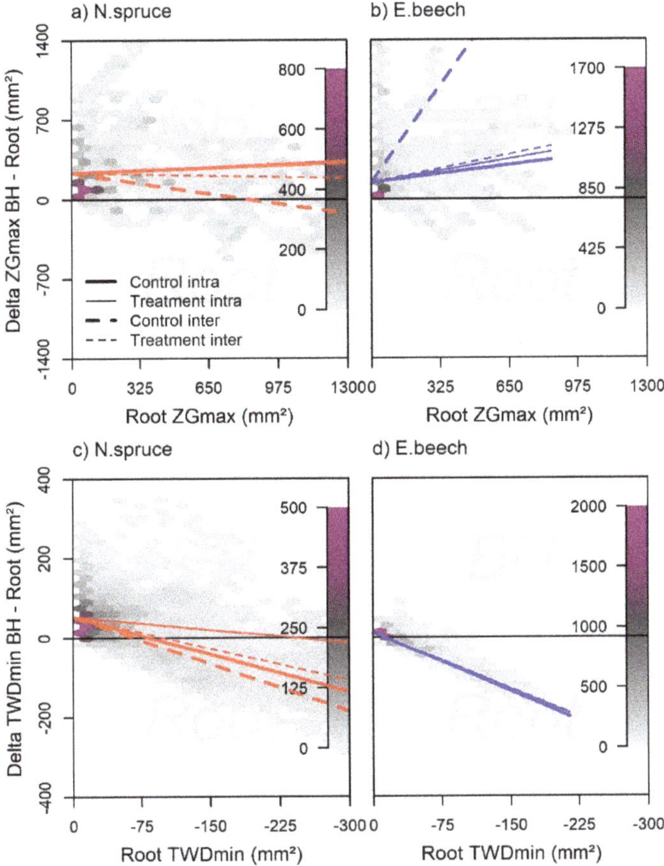

Figure 6. Relationship of growth (ZGmax) (**a,b**) and the tree water deficit (TWDmin) (**c,d**) represented by the difference of the measurements at breast height (BH) and at the roots, dependent on the root measurements. Values below the zero line mean a growth allocation into the roots. For the TWDmin, the negative values represent a lower tree water deficit in the roots. Linear mixed models depict how strongly the allocation is influenced by drought treatment (thick line—control, thin line—treatment) or mixture (intraspecific neighborhood—straight line, interspecific neighborhood—dashed line). The respective models are shown in Table 3.

In general, the interaction of treatment and mixture with the lower tree compartment (*x*-axis) was significant, except for beech trees in some cases. The interpretation of the influence of treatment and mixture always refers to their dependency on the lower tree compartment. We subsequently looked at this interaction or rather the orientation and position of the control to treatment curves (or intra- to interspecific curves).

The *x*-axis showed an increasing basal area increment over time within the growing season. We found that the growth in BH was always higher, except for at the beginning of the growing season, where growth in H50 was higher. The tree appears to invest in the upper trunk (H50) at the beginning of the growing season and then more in the BH (Figure 5a,b, Table 3(a),(b)). In the following, we first describe the ZG of spruce, followed by that of beech, and then the TWD of both species.

Treated intraspecific spruce trees showed a tendency to grow more in H50 as interspecific spruces of the control plots (Figure 5a, Table 3). Beech trees at the control plots showed an increasing increment of the upper stem, whereas beeches of the treatment plots revealed a higher growth in BH than in H50 (Figure 5b, Table 3(b)). A tendency of a higher TWD in H50 could be found for spruces in interspecific neighborhoods (control and treatment plots) compared to the intraspecific spruces (Figure 5c, Table 3(c)). The same pattern could be observed for beech trees (Figure 5d, Table 3(d)).

Overall, spruce trees showed a higher root growth than beech trees. Spruces in interspecific neighborhoods showed a higher tendency to root growth than in intraspecific neighborhoods (Figure 6a, Table 3(e)). In contrast, beech tress in intraspecific neighborhoods had a higher increment in the roots than in BH compared to beeches in interspecific neighborhood (Figure 6b, Table 3(f)). The TWD was higher in the roots for both species. Spruce trees of the treatment plots and in intraspecific neighborhoods had the highest TWD in the roots (Figure 6c, Table 3(g)). For beech trees in intra- and interspecific neighborhoods, no differences could be found (Figure 6d, Table 3(h)).

Table 3. Parameter estimates and statistics for the diameter growth (ZGmax) and tree water deficit (TWDmin) of the 50% tree height (H50) and stem at breast height (BH) from Figure 5 and stem at breast height (BH) and root from Figure 6, dependent on drought (treat) and species mixing (mixture). The dependent variables are in the columns. Rows show the output of the model with the fixed variables (N: number of measurements). Significance levels: *** $p < 0.001$; ** $p < 0.01$; * $p < 0.05$; (*) $p < 0.1$.

Position	H50–BH				BH–Root			
ZGmax/TWDmin	ZGmax		TWDmin		ZGmax		TWDmin	
	(a)	(b)	(c)	(d)	(e)	(f)	(g)	(h)
Art	N. Spruce	E. Beech	N. Spruce	E. Beech	N. Spruce	E. Beech	N. Spruce	E. Beech
Intercept	100.961 ***	59.753 *	62.534 **	2.908 (*)	235.5 **	149.09 **	53.718 ***	13.17 ***
BH	−0.481 ***	−0.408 ***	−0.468 ***	−0.463 ***				
Treat × BH	−0.075 ***	−0.109 ***	0.037	−0.028 (*)				
Mixture × BH	−0.173 ***	0.01	0.148 **	0.18 ***				
Mixture × Treat × BH	0.243 ***	−0.001	0.209 **	0.026				
Root					−0.279 ***	2.63 ***	−0.796 ***	−0.992 ***
Treat × Root					0.241 ***	−2.257 ***	0.258 ***	0.048 ***
Mixture × Root					0.347 ***	−2.398 ***	0.161 ***	0.001
Mixture × Treat × Root					−0.23 ***	2.338 ***	0.155 ***	−0.011
R^2	0.95	0.94	0.56	0.77	0.73	0.67	0.81	0.94
RMSE	69.6	49.9	59.9	4.2	182.1	194.8	42.9	9.6
N	7924	4866	7924	4866	8290	7940	8290	7940

4. Discussion

4.1. Relationship between Tree Water Deficit and Leaf Water Potential

We found a strong relationship between TWDmin and LWP for both species at the three tree positions (H50, BH, roots). Furthermore, the high-resolution TWD measurements could show the differences between intra- and interspecific neighborhoods clearer compared to the LWP. The LWP showed no significant differences between intra- and interspecific spruce and beech trees, but the effect of the drought treatment could be shown compared to the trees of the control plots. Spruce trees, as a more drought sensible tree species, showed the expected higher LWP compared to the more drought resistant beech trees.

We found the best match for TWDmin and LWPpre, but the relationship between TWDmin and LWPmid was also significant. The study of Remorini and Massai (2003) [60] confirmed that the LWPpre is a better tree water status indicator than LWPmid.

Early dendrometer studies, such as Cohen et al. (2001) [61], focused on the maximum daily shrinkage and compared the data with the water potential at midday and predawn. The Cohen study shows the link between maximum daily stem shrinkage (MDS) and predawn and midday LWP. The MDS was closely related to the predawn and midday water potential, similar to the present study. A more recent study by Dietrich et al. (2018) [15] showed the relationship between TWD and LWP of different tree species, which included Norway spruce and European beech. It was illustrated that the daily TWD displayed the tree water status better than the maximum daily shrinkage, in which the stem water signal was not separated from the growth signal. Particularly under dry conditions, the TWD was a consistent proxy for the tree water status of tall trees.

The temporal and spatial patterns of stem radius variations of Norway spruces were determined in Zweifel and Häsler (2001) [62]. As in the present study, they found similar but not identical dynamics of stem and root radius fluctuations at different heights (along the stem at 6, 10, 14, 18 m above ground and on roots). There were differences in the heights of the curves (amplitudes) at different tree heights and there was a time lag between the tree compartments. In the present study, the time offset was not investigated and could be an option for further investigations.

Similar to our study, Zweifel and Häsler (2001) [62] found the greatest fluctuations in the upper stem part, within the crown, due to the proximity to the tree crown and thus transpiration [63], next to the roots which also showed high radius fluctuations as in the present study. High root basal area variations could be caused by a high aridity, as in the drought experiment and the very dry year 2015. Hinckley et al. (1978) [64] stated that the water potential differences between the crown and the soil increases the water movement within the tree, and thus, increases the water movement from the internal stored water in the bark. The water storages near the crown can be reached faster than water storages from the soil due to the more negative leaf water potential than the root water potential. Thus, diurnal water potential fluctuations are larger in the branches than the roots, leading to larger daily shrinkage in the branches.

In the present study, the relationship between TWD and LWP could be illustrated at the stem (at H50, BH) and at the roots. This is in line with the first hypothesis that the stem basal area variations and the leaf water potential show a positive relationship at the different tree heights.

4.2. Root and Stem Growth and TWDmin in Different Tree Heights

The applied method—the difference between upper and lower tree compartments in relation to the lower tree compartment—led to a visible deviation of the measured TWDmin and ZGmax at the different tree compartments. We are aware that the results have to be interpreted with caution due to the included variation. Nevertheless, the applied models considering the offset of the individual trees showed significant relationships. Within our approach, we followed the principle of parsimony, with linear relationships and as few parameters as possible. Also, the R^2 and the RMSE provided evidence for a high model explanatory power. The QQ-plots showed us that there were only a few outliers and we checked them carefully to avoid any inconsistencies in our measurements.

We found that the trees invested more in the basal area growth of the upper stem (H50) at the beginning of the growing season. Later in the season, growth investment shifted increasingly to the lower stem (BH). An explanation for this might be the theory of the seasonal distribution of the growth hormone auxin. In spring, it is produced in the apical meristem and transported down to the stem in the phloem of the tree [65,66]. In this way, the strength of radial growth shifts from the top to stem base. During the growing period, the growth of spruce and beech trees were favored in BH than in H50. Beech trees at the control plots showed a tendency of a higher growth in H50 than in BH compared to trees of the drought treatment.

We also found that TWDmin was higher in the upper stem (H50) than in the lower stem (BH), but with an increasing stem shrinkage, the shrinkage in BH was higher. The dendrometer position at H50 was near the crown and more water reserves could be used for transpiration [63]. With increasing water stress, the water storage pools in H50 are possibly exhausted and more water reserves of the storage pool in BH are used. This could be related to the higher diameter growth at BH compared to H50 due to the lesser shrinkage and accordingly higher cell turgor which is important for cell expansion. Zweifel et al. (2016) [51] stated that periods of stem shrinkage allow for very little growth. In addition, Van der Maaten–Theunissen and Bouriaud (2012) [67] revealed a reduced growth of Norway spruce at all stem heights during summer drought in southwestern Germany with the greatest reduction in growth at breast height.

For beech trees, the diameter increments were higher at BH. Despite this, the TWDmin was also higher at BH than at H50. Overall, the TWD was less pronounced for beech trees than for spruce trees. This could be also observed at the LWP and this could be due to the higher resistance of beech trees to drought [14,42]. Beech trees have an anisohydric character and continue growing and transpiring under dry conditions in contrast to spruce trees. Furthermore, spruce and beech trees have different rooting systems [43]. Beech trees have a deeper rooting system than spruce trees and can reach water from deeper soil horizons, which may reduce the use of water from the storage tissue. Spruce trees as an isohydric tree species respond with a reduced stomatal conductance under drought stress, and thus reduced transpiration, which in turn reduces growth [29,68].

Furthermore, spruce and beech trees revealed a higher growth at BH than at the roots and a higher TWDmin in the roots. The roots are affected by drought conditions through the lack of soil water supply and more water from the storage tissue in the roots could be used to maintain the transpiration process. In addition, the growth differences shifted more to the roots for beech in intraspecific neighborhoods and for spruce trees in interspecific neighborhoods. More root growth could imply more water stress, which is explained in McCarthy and Enquist (2007) [31] and Ledo et al. (2018) [69]. The resource supply of plants is determined by the shape of the rooting system, the shape of the tree crown, site conditions, and proximity to other trees [39,70–72]. Depending on the prevailing conditions, biomass is allocated differently in the compartments of the tree (crown, stem, and root). The optimal partitioning theory of McCarthy and Enquist (2007) [31] indicates that a limited resource leads to a promotion of growth of the plant organ that receives this resource (roots). Our findings about the allometric relationships for growth—endorsed by the TWDmin results—support optimal partitioning theory and are consistent with the findings of other studies [73–75].

In contrast, Schall et al. (2012) [76] found a significant increase of the percentage of belowground compartments for beech seedlings but not for spruce seedlings. When considering the TWDmin of beech trees, no significant differences between control and drought treatment were obvious. Thus, the second hypothesis that the relationship between growth response and the respective TWD is the same at the three different positions H50, BH, and root could be confirmed for spruce trees and for beech trees at the BH–root consideration. Spruces showed a higher growth in BH with a shift to root growth with increasing increment and time at the control and treatment plots and the TWD was respectively higher in the roots.

4.3. Differences in Intra- and Interspecific Neighborhoods

Many studies have pointed out that species mixture can have positive effects on the biodiversity [77], productivity [78], and soil fertility [79] of the whole system and that these effects depend on which species are mixed. Spruce and beech trees have different physiological and morphological traits [43,80]. Pretzsch (2014) [80] stated that plasticity in crown and root architecture appears to be the key to understanding effects of mixed system productivity.

Our investigations showed differences in inter- and intraspecific neighborhoods under drought conditions when considering TWDmin and the root growth. For spruce trees, the stem basal area growth (H50–BH) at BH was highest in intraspecific neighborhood at the control plots. The TWDmin showed a similar pattern and was highest in intraspecific neighborhood at BH. The BH–root comparison indicated a higher growth in BH for spruces in intraspecific neighborhoods compared to the interspecific neighborhoods. The TWD of spruces was in interspecific neighborhoods lesser than in intraspecific neighborhoods at the control and treatment plots at the roots. An explanation for this pattern could be that the growth is influenced by several factors (e.g., soil characteristics, nutrient supply, light) in addition to water availability [33,81]. The TWDmin reflects the water status of the tree. Therefore, spruces are facilitated in an interspecific neighborhood in terms of the tree water status under increasing drought, but not for stem radius growth. Nevertheless, the stem radius growth could be influenced under extreme drought conditions due to the higher TWDmin. A lesser TWDmin in the roots supported growth and offers the opportunity to reach more water resources in deeper soil horizons. Spruce trees, with their mainly shallow rooting systems and only few sinker roots have limited access to deeper soil water resources. Therefore, spruces use more water reserves from the storage tissue within one day. In addition, Bolte and Villanueva (2006) [43] found that beech trees in a neighborhood with spruce trees rooted in deeper soil horizons than in intraspecific neighborhood. That can be a reason for the higher growth of beeches in interspecific neighborhoods in BH than in the roots compared to the intraspecific neighborhoods. Spruce trees may benefit from this favorable characteristic through the effect of hydraulic lift [82–84]. Hydraulic lift is the passive movement of water from moist to dry soil horizons by plant root systems. Usually at night when transpiration has ceased, water is released from the roots into the upper soil horizon [82]. Beech trees can redistribute water from deeper to shallower soil horizons with their rooting systems. The reallocated water in the dryer soil layers can be used by beech trees as well as by the surrounding spruce trees [83]. This could be an explanation for the lower TWD of spruce trees in the interspecific neighborhood compared to spruces in the intraspecific neighborhood.

In contrast, the effect of drought treatment and control on the TWD of beeches was not significant. In several studies, positive reactions of mixed beech trees have been detected [39,43,85], but we did not find this positive interaction.

Thus, the third hypothesis that an interspecific neighborhood with beech trees facilitates spruce trees under drought stress could be confirmed for the water status of spruce trees, but not for the radius growth.

5. Conclusions

The present study is one of the first to investigate changing growth at different tree compartments within a single year and with the help of TWDmin and ZGmax.

We found a relationship between the TWDmin and leaf water potentials at the three investigated tree compartments for spruce and beech. This confirms that dendrometer measurements are a good tool for drought stress analyses at different tree compartments. The dendrometer measurements were much easier to handle than the water potential measurements and they were able to provide water status information within a timely and high-resolution manner throughout the whole year. However, the relationship between TWDmin and leaf water potentials included a deviation. Additional measurements, like phloem thickness, might improve the understanding of the relationship among both parameters.

With our investigation of growth and TWDmin in 2014 and 2015, we showed the growth pattern and compared it with the TWDmin pattern. Nevertheless, several studies have shown that the biomass allocation of a plant changes across the life course. Therefore, it is important to extend the measurements and to also include a tree's juvenile and senescence stages.

Surprisingly, for spruce trees we found that an interspecific neighborhood resulted in a higher root growth and a lesser TWD in the roots than in BH. Beech trees were less affected by drought and showed a higher growth in BH in an interspecific than in an interspecific neighborhood compared to the roots. The TWDmin could showed the effect of neighborhood better than the LWP measurements. The LWP measurements showed no significant differences in intra- and interspecific neighborhoods.

To answer the question of whether spruce benefited from the mixture with beech under drought conditions, we considered the TWDmin and concluded that the mixture of beech could reduce the drought stress for spruce under future climatic warming.

Supplementary Materials: The following are available online at http://www.mdpi.com/1999-4907/10/7/577/s1, Figure S1: Mean water potential at midday for the years 2014 (left) and 2015 (right) for spruce trees in intra- and interspecific neighborhoods at the control and treatment plots (a,b) and for beech trees in intra- and interspecific neighborhoods at the control and treatment plots (c,d). Data is shown for the growing season, Figure S2: Mean predawn water potential for the years 2014 (left) and 2015 (right) for spruce trees in intra- and interspecific neighborhoods at the control and treatment plots (a,b) and for beech trees in intra- and interspecific neighborhoods at the control and treatment plots (c,d). Data is shown for the growing season, Figure S3: Mean predawn water potential for the years 2014 and 2015 for spruce trees in intra- and interspecific neighborhoods at the control and treatment plots (a) and for beech trees in intra- and interspecific neighborhoods at the control and treatment plots (b). Data is shown for the growing season, Figure S4: Mean TWDmin and zero growth (growth without the water signal) referring to the stem/root basal area (mm^2) for the years 2014 (left) and 2015 (right) for spruce (red) and beech (blue) in intraspecific (solid line) and interspecific (dashed line) neighborhoods at 50% tree height (a–d), breast height (BH, e–h) and the roots (i–l). Shaded regions are conficence intervals. Data are shown for the growing season, Figure S5: N. spruce ZG H50-BH. Model critism plots for the linear mixed effect models in Table 3 and Figures 5 and 6. The models critism plots are in the same order. Description of the single plots within each Figure: (a) plot of the outermost fitted values against the observed values of the response variable; (b) plot of the innermost fitted values against the innermost Pearson residuals; (c) histogram of the innermost residuals; (d) QQ-plot of the estimated random effects; (e) QQ-plot of the Pearson residual; (f) notched boxplot of the innermost Pearson residuals by the grouping variables plot:indivudal tree:year; (g) scatterplot of the variance of the Pearson residuals within the grouping variables, Figure S6: E. beech ZG H50-BH, Figure S7: N. spruce TWD H50-BH, Figure S8: E. beech TWD H50-BH, Figure S9: N. spruce ZG BH-Root, Figure S10: E. beech ZG BH-Root, Figure S11: N. spruce TWD BH-Root, Figure S12: E. beech TWD BH-Root. Table S1: Characteristics of the individual spruce (n = 24) and beech trees (n = 24), seperated by plot (n = 12, overall 48 trees), drought treatment (6 plots, 24 trees)/control (6 plots, 24 trees) and intra- and interspecific neighborhoods (six control plots and six treatment plots with respectively six intraspecific beech trees; six interspecific beech trees; six intraspecific spruce trees; six interspecific spruce trees) for the year 2014 (DBH: diameter at 1.3 m breast height), Table S2: Coefficient of determination (R^2) of TWDmin (tree water deficit, daily minimum), LWPpre (water potential at predawn) and of TWDmax (tree water deficit, daily maximum) and LWPmid (water potential at midday). The R^2 based on the relationship between leaf water potential (LWP) and tree water deficit (TWD) at the three different tree heights (H50, BH, Root). The respective models based on Equation (1). The last two rows contain the means of both species and of all tree heights, Table S3: Parameter estimates and statistics for the water potential at midday and predawn dependent on species and drought treatment. Standard deviations are in brackets. The dependent variables are in the columns. Rows show the output of the model with the fixed variables. Significance levels: ***, $p < 0.001$; **, 0.01; *, 0.05; (*), 0.1.

Author Contributions: Conceptualization, H.P., T.R. and C.S.; Methodology, C.S., E.A.T., T.R. and H.P.; Software, C.S. and E.A.T.; Validation, C.S., E.A.T., T.R., P.B., C.K., and H.P.; Formal Analysis, C.S., E.A.T.; Investigation, C.S. and C.K.; Resources, C.S. and C.K.; Data Curation, C.S.; Writing—Original Draft Preparation, C.S.; Writing—Review and Editing, C.S., E.A.T., T.R., P.B. and H.P.; Visualization, C.S.; Supervision, H.P. and T.R.; Project Administration, H.P. and T.R.; Funding Acquisition, H.P. and T.R.

Funding: This research was funded by the German Science Foundation (Deutsche Forschungsgemeinschaft), grant number PR 292/12-1, MA 1763/7-1, MU 831/23-1 "Tree and stand-level growth reactions on drought in mixed versus pure forests of Norway spruce and European beech", the Bavarian State Ministry for Nutrition, Agriculture and Forestry, and the Bavarian State Ministry for Environment and Consumer Protection.

Acknowledgments: We thank the German Science Foundation (Deutsche Forschungsgemeinschaft) for providing the funds for the projects PR 292/12-1, MA 1763/7-1, and MU 831/23-1 "Tree and stand-level growth reactions on drought in mixed versus pure forests of Norway spruce and European beech". Recognition is given to the Bavarian State Ministry for Nutrition, Agriculture and Forestry and to the Bavarian State Ministry for Environment and Consumer Protection for their generous support of the roof buildings. Special thanks go to the KROOF-Team for the great work in the field.

Conflicts of Interest: The authors declare no conflict of interest.

References

1. IPCC. *Climate Change 2013: The Physical Basis. Contribution of Working Group I to the Fifth Assessment Report of the IPCC*; Cambrige University Press: Cambridge, UK, 2013; ISBN 978-1-107-66182-0.

2. Meehl, G.A.; Tebaldi, C. More intense, more frequent, and longer lasting heat waves in the 21st century. *Science* **2004**, *305*, 994–997. [CrossRef] [PubMed]

3. Fuhrer, J.; Beniston, M.; Fischlin, A.; Frei, C.; Goyette, S.; Jasper, K.; Pfister, C. Climate risks and their impact on agriculture and forests in Switzerland. In *Climate Variability, Predictability and Climate Risks: A European Perspecive*; Wanner, H., Grosjean, M., Rastlisberg, R., Xoplaki, E., Eds.; Springer Netherlands: Dordrecht, The Netherlands, 2006; pp. 79–102. ISBN 978-1-4020-5713-7.

4. Allen, C.D.; Macalady, A.K.; Chenchouni, H.; Bachelet, D.; McDowell, N.; Vennetier, M.; Kitzberger, T.; Rigling, A.; Breshears, D.D.; Hogg, E.H.; et al. A global overview of drought and heat-induced tree mortality reveals emerging climate change risks for forests. *For. Ecol. Manag.* **2010**, *259*, 660–684. [CrossRef]

5. Rötzer, T.; Liao, Y.; Goergen, K.; Schüler, G.; Pretzsch, H. Modelling the impact of climate change on the productivity and water-use efficiency of a central European beech forest. *Clim. Res.* **2013**, *58*, 81–95. [CrossRef]

6. Ciais, P.; Reichstein, M.; Viovy, N.; Granier, A.; Ogée, J.; Allard, V.; Aubinet, M.; Buchmann, N.; Bernhofer, C.; Carrara, A.; et al. Europe-wide reduction in primary productivity caused by the heat and drought in 2003. *Nature* **2005**, *437*, 529–533. [CrossRef] [PubMed]

7. Dieler, J.; Pretzsch, H. Morphological plasticity of European beech (Fagus sylvatica L.) in pure and mixed-species stands. *For. Ecol. Manag.* **2013**, *295*, 97–108. [CrossRef]

8. Chapin, F.S. The Mineral Nutrition of Wild Plants. *Annu. Rev. Ecol. Syst.* **1980**, *11*, 233–260. [CrossRef]

9. Thornley, J.H.M. A balanced quantitative model for root: Shoot ratios in vegetative plants. *Ann. Bot.* **1972**, *36*, 431–441. [CrossRef]

10. Poorter, H.; Niklas, K.J.; Reich, P.B.; Oleksyn, J.; Poot, P.; Mommer, L. Biomass allocation to leaves, stems and roots: Meta-analyses of interspecific variation and environmental control. *New Phytol.* **2012**, *193*, 30–50. [CrossRef]

11. Rötzer, T.; Seifert, T.; Pretzsch, H. Modelling above and below ground carbon dynamics in a mixed beech and spruce stand influenced by climate. *Eur. J. Res.* **2009**, *128*, 171–182. [CrossRef]

12. Zweifel, R.; Item, H.; Häsler, R. Link between diurnal stem radius changes and tree water relations. *Tree Physiol.* **2001**, *2*, 869–877. [CrossRef]

13. Steppe, K.; De Pauw, D.J.W.; Lemeur, R.; Vanrolleghem, P.A. A mathematical model linking tree sap flow dynamics to daily stem diameter fluctuations and radial stem growth. *Tree Physiol.* **2006**, *26*, 257–273. [CrossRef] [PubMed]

14. Brinkmann, N.; Eugster, W.; Zweifel, R.; Buchmann, N.; Kahmen, A. Temperate tree species show identical response in tree water deficit but different sensitivities in sap flow to summer soil drying. *Tree Physiol.* **2016**, *36*, 1508–1519. [CrossRef] [PubMed]

15. Dietrich, L.; Zweifel, R.; Kahmen, A. Daily stem diameter variations can predict the canopy water status of mature temperate trees. *Tree Physiol.* **2018**. [CrossRef] [PubMed]

16. Ortuño, M.F.; García-Orellana, Y.; Conejero, W.; Ruiz-Sánchez, M.C.; Mounzer, O.; Alarcón, J.J.; Torrecillas, A. Relationships between Climatic Variables and Sap Flow, Stem Water Potential and Maximum Daily Trunk Shrinkage in Lemon Trees. *Plant. Soil* **2006**, *279*, 229–242. [CrossRef]

17. Zweifel, R.; Item, H.; Häsler, R. Stem radius changes and their relation to stored water in stems of young Norway spruce trees. *Trees* **2000**, *15*, 50–57. [CrossRef]

18. Zweifel, R.; Zimmermann, L.; Newbery, D.M. Modeling tree water deficit from microclimate: An approach to quantifying drought stress. *Tree Physiol.* **2005**, *25*, 147–156. [CrossRef] [PubMed]

19. Ortuño, M.F.; Conejero, W.; Moreno, F.; Moriana, A.; Intrigliolo, D.S.; Biel, C.; Mellisho, C.D.; Pérez-Pastor, A.; Domingo, R.; Ruiz-Sánchez, M.C.; et al. Could trunk diameter sensors be used in woody crops for irrigation scheduling? A review of current knowledge and future perspectives. *Agric. Water Manag.* **2010**, *97*, 1–11. [CrossRef]

20. Daudet, F.-A.; Ameglio, T.; Cochard, H.; Archilla, O.; Lacointe, A. Experimental analysis of the role of water and carbon in tree stem diameter variations. *J. Exp. Bot.* **2005**, *56*, 135–144. [CrossRef]

21. Vieira, J.; Rossi, S.; Campelo, F.; Freitas, H.; Nabais, C. Seasonal and daily cycles of stem radial variation of Pinus pinaster in a drought-prone environment. *Agric. For. Meteorol.* **2013**, *180*, 173–181. [CrossRef]

22. Klepper, B.; Browning, V.D.; Taylor, H.M. Stem Diameter in Relation to Plant Water Status. *Plant. Physiol.* **1971**, *48*, 683–685. [CrossRef]

23. Kozlowski, T.T. Water Deficits and Plant Growth. In *Shrinking and Swelling of Plant Tissues*; Kozlowski, T.T., Ed.; Academic Press: New York, NY, USA, 1976; pp. 1–64.

24. Zweifel, R.; Zimmermann, L.; Zeugin, F.; Newbery, D.M. Intra-annual radial growth and water relations of trees: Implications towards a growth mechanism. *J. Exp. Bot.* **2006**, *57*, 1445–1459. [CrossRef] [PubMed]

25. De Schepper, V.; Steppe, K. Development and verification of a water and sugar transport model using measured stem diameter variations. *J. Exp. Bot.* **2010**, *61*, 2083–2099. [CrossRef] [PubMed]

26. De Swaef, T.; De Schepper, V.; Vandegehuchte, M.W.; Steppe, K. Stem diameter variations as a versatile research tool in ecophysiology. *Tree Physiol.* **2015**, *35*, 1047–1061. [CrossRef] [PubMed]

27. Steppe, K.; Sterck, F.; Deslauriers, A. Diel growth dynamics in tree stems: Linking anatomy and ecophysiology. *Trends Plant. Sci.* **2015**, *20*, 335–343. [CrossRef] [PubMed]

28. Irvine, J.; Grace, J. Continuous measurements of water tensions in the xylem of trees based on the elastic properties of wood. *Planta* **1997**, *202*, 455–461. [CrossRef]

29. Klein, T. The variability of stomatal sensitivity to leaf water potential across tree species indicates a continuum between isohydric and anisohydric behaviours. *Funct. Ecol.* **2014**, *28*, 1313–1320. [CrossRef]

30. Rais, A.; van de Kuilen, J.-W.G.; Pretzsch, H. Growth reaction patterns of tree height, diameter, and volume of Douglas-fir (*Pseudotsuga menziesii* [Mirb.] Franco) under acute drought stress in Southern Germany. *Eur. J. Res.* **2014**, *133*, 1043–1056. [CrossRef]

31. McCarthy, M.C.; Enquist, B.J. Consistency between an allometric approach and optimal partitioning theory in global patterns of plant biomass allocation. *Funct. Ecol.* **2007**, *21*, 713–720. [CrossRef]

32. Pretzsch, H. Diversity and Productivity in Forests: Evidence from Long-Term Experimental Plots. In *Forest Diversity and Function: Temperate and Boreal Systems*; Scherer-Lorenzen, M., Körner, C., Schulze, E.-D., Eds.; Springer: Berlin, Germany; New York, NY, USA, 2005; pp. 41–64. ISBN 3-540-22191-3.

33. Forrester, D.I. The spatial and temporal dynamics of species interactions in mixed-species forests: From pattern to process. *For. Ecol. Manag.* **2014**, *312*, 282–292. [CrossRef]

34. Rötzer, T. Mixing patterns of tree species and their effects on resource allocation and growth in forest stands. *Nova Acta Leopold* **2013**, *114*, 239–254.

35. *Mixed-Species Forests. Ecology and Management*; Pretzsch, H.; Forrester, D.I.; Bauhus, J. (Eds.) Springer: Berlin, Germany, 2017; 653p.

36. Wiedemann, E. Der gleichaltrige Fichten-Buchen-Mischbestand. *Mitt. Aus Forstwirtsch. Forstwiss.* **1942**, *1*, 1–88.

37. Pretzsch, H.; Block, J.; Dieler, J.; Dong, P.H.; Kohnle, U.; Nagel, J.; Spellmann, H.; Zingg, A. Comparison between the productivity of pure and mixed stands of Norway spruce and European beech along an ecological gradient. *Ann. For. Sci.* **2010**, *67*, 712. [CrossRef]

38. Pretzsch, H.; Rötzer, T.; Matyssek, R.; Grams, T.E.E.; Häberle, K.-H.; Pritsch, K.; Kerner, R.; Munch, J.-C. Mixed Norway spruce (*Picea abies* [L.] Karst) and European beech (*Fagus sylvatica* [L.]) stands under drought: From reaction pattern to mechanism. *Trees* **2014**. [CrossRef]

39. Metz, J.; Annighöfer, P.; Schall, P.; Zimmermann, J.; Kahl, T.; Schulze, E.-D.; Ammer, C. Site-adapted admixed tree species reduce drought susceptibility of mature European beech. *Glob. Chang. Biol.* **2016**, *22*, 903–920. [CrossRef] [PubMed]

40. Zang, C.; Pretzsch, H.; Rothe, A. Size-dependent responses to summer drought in Scots pine, Norway spruce and common oak. *Trees* **2012**, *26*, 557–569. [CrossRef]

41. Pretzsch, H.; Schütze, G.; Uhl, E. Resistance of European tree species to drought stress in mixed versus pure forests: Evidence of stress release by inter-specific facilitation. *Plant. Biol.* **2013**, *15*, 483–495. [CrossRef] [PubMed]

42. Ammer, C.; Bickel, E.; Kolling, C. Converting Norway spruce stands with beech—A review of arguments and techniques. *Austrian J. For. Sci.* **2008**, *125*, 3–26.

43. Bolte, A.; Villanueva, I. Interspecific competition impacts on the morphology and distribution of fine roots in European beech (*Fagus sylvatica* L.) and Norway spruce (*Picea abies* (L.) Karst.). *Eur. J. For. Res.* **2006**, *125*, 15–26. [CrossRef]

44. Schume, H.; Jost, G.; Hager, H. Soil water depletion and recharge patterns in mixed and pure forest stands of European beech and Norway spruce. *J. Hydrol.* **2004**, *289*, 258–274. [CrossRef]

45. Cremer, M.; Prietzel, J. Die Mutter des Waldes und die Fremde: Douglasien-Buchen-Mischbestände: Aus bodenkundlicher Sicht eine attraktive Mischungsoption. *LWF Aktuell* **2017**, *2*, 24–25.

46. Rötzer, T.; Häberle, K.H.; Kallenbach, C.; Matyssek, R.; Schütze, G.; Pretzsch, H. Tree species and size drive water consumption of beech/spruce forests—A simulation study highlighting growth under water limitation. *Plant. Soil* **2017**, *183*, 327. [CrossRef]

47. Pretzsch, H.; Dieler, J.; Seifert, T.; Rötzer, T. Climate effects on productivity and resource-use efficiency of Norway spruce (*Picea abies* [L.] Karst.) and European beech (*Fagus sylvatica* [L.]) in stands with different spatial mixing patterns. *Trees* **2012**, *26*, 1343–1360. [CrossRef]

48. Brodrick, P.G.; Anderegg, L.D.L.; Asner, G.P. Forest Drought Resistance at Large Geographic Scales. *Geophys. Res. Lett.* **2019**, *46*, 2752–2760. [CrossRef]

49. Rosas, T.; Mencuccini, M.; Barba, J.; Cochard, H.; Saura-Mas, S.; Martínez-Vilalta, J. Adjustments and coordination of hydraulic, leaf and stem traits along a water availability gradient. *New Phytol.* **2019**, *223*, 632–646. [CrossRef] [PubMed]

50. Konings, A.G.; Gentine, P. Global variations in ecosystem-scale isohydricity. *Glob. Chang. Biol.* **2017**, *23*, 891–905. [CrossRef] [PubMed]

51. Zweifel, R.; Haeni, M.; Buchmann, N.; Eugster, W. Are trees able to grow in periods of stem shrinkage? *New Phytol.* **2016**, *211*, 839–849. [CrossRef] [PubMed]

52. Pretzsch, H.; Bauerle, T.; Häberle, K.H.; Matyssek, R.; Schütze, G.; Rötzer, T. Tree diameter growth after root trenching in a mature mixed stand of Norway spruce (*Picea abies* [L.] Karst) and European beech (*Fagus sylvatica* [L.]). *Trees* **2016**, *30*, 1761–1773. [CrossRef]

53. Hera, U.; Roetzer, T.; Zimmermann, L.; Schulz, C.; Maier, H.; Weber, H.; Kölling, C. Klima en detail—Neue hochaufgelöste Klimakarten zur klimatischen Regionalisierung Bayerns. *LWF Aktuell* **2011**, *86*, 34–37.

54. Biondi, F.; Qeadan, F. A Theory-Driven Approach to Tree-Ring Standardization: Defining the Biological Trend from Expected Basal Area Increment. *Tree-Ring Res.* **2008**, *64*, 81–96. [CrossRef]

55. Bavarian State Institute of Forestry (LWF). *Bavarian Forest Ecosystem Monitoring Plot Freising (WKS). Meteorological Data*; Bavarian State Institute of Forestry (LWF): Freising, Germany, 2016.

56. Bates, D.; Maechler, M.; Bolker, B.; Walker, S. Fitting Linear Mixed-Effects Models Using lme4. *J. Stat. Softw.* **2015**, *67*, 1–48. [CrossRef]

57. Kuznetsova, A.; Brockhoff, P.B.; Christensen, R.H.B. lmerTest: Tests in Linear Mixed Effects Models. R Package Version 2.0-29. *J. Stat. Softw.* **2017**, *82*, 1–26. [CrossRef]

58. R Core Team. *R: A Language and Environment for Statistical Computing*; R Foundation for Statistical Computing: Vienna, Austria, 2017.

59. Bavarian State Institute of Forestry (LWF). Umweltmonitoring. Available online: www.lwf.bayern.de/boden-klima/umweltmonitoring (accessed on 10 July 2016).

60. Remorini, D.; Massai, R. Comparison of water status indicators for young peach trees. *Irrig. Sci.* **2003**, 39–46. [CrossRef]

61. Cohen, M.; Goldhamer, D.A.; Fereres, E.; Girona, J.; Mata, M. Assessment of peach tree responses to irrigation water ficits by continuous monitoring of trunk diameter changes. *J. Hortic. Sci. Biotechnol.* **2001**, *76*, 55–60. [CrossRef]

62. Zweifel, R.; Häsler, R. Dynamics of water storage in mature subalpine Picea abies: Temporal and spatial patterns of change in stem radius. *Tree Physiol.* **2001**, *21*, 561–569. [CrossRef] [PubMed]

63. *Xylem Structure and the Ascent of Sap*, 2nd ed.; Tyree, M.T.; Zimmermann, M.H. (Eds.) Springer: Berlin, Germany; London, UK, 2011; ISBN 978-3-642-07768-5.

64. Hinckley, T.M.; Lassoie, J.P.; Running, S.W. Temporal and Spatial Variations in the Water Status of Forest Trees. *For. Sci Monogr.* **1978**, *20*, 1–72.

65. Kozlowski, T.T. *Tree Growth*; Ronald Press: New York, NY, USA, 1962.

66. Speer, J.H. *Fundamentals of Tree-Ring Research*; The University of Arizona Press: Tucson, AZ, USA, 2013; ISBN 0816526850.

67. Van der Maaten-Theunissen, M.; Bouriaud, O. Climate–growth relationships at different stem heights in silver fir and Norway spruce. *Can. J. Res.* **2012**, *42*, 958–969. [CrossRef]

68. McDowell, N.; Pockman, W.T.; Allen, C.D.; Breshears, D.D.; Cobb, N.; Kolb, T.; Plaut, J.; Sperry, J.; West, A.; Williams, D.G.; et al. Mechanisms of plant survival and mortality during drought: Why do some plants survive while others succumb to drought? *New Phytol.* **2008**, *178*, 719–739. [CrossRef]

69. Ledo, A.; Paul, K.I.; Burslem, D.F.R.P.; Ewel, J.J.; Barton, C.; Battaglia, M.; Brooksbank, K.; Carter, J.; Eid, T.H.; England, J.R.; et al. Tree size and climatic water deficit control root to shoot ratio in individual trees globally. *New Phytol.* **2018**, *217*, 8–11. [CrossRef]

70. Bayer, D.; Seifert, S.; Pretzsch, H. Structural crown properties of Norway spruce (*Picea abies* [L.] Karst.) and European beech (*Fagus sylvatica* [L.]) in mixed versus pure stands revealed by terrestrial laser scanning. *Trees* **2013**, *27*, 1035–1047. [CrossRef]

71. Nickel, U.T.; Weikl, F.; Kerner, R.; Schäfer, C.; Kallenbach, C.; Munch, J.C.; Pritsch, K. Quantitative losses vs. qualitative stability of ectomycorrhizal community responses to 3 years of experimental summer drought in a beech-spruce forest. *Glob. Chang. Biol.* **2017**. [CrossRef]

72. Kelty, M.J. The role of species mixtures in plantation forestry. *For. Ecol. Manag.* **2006**, *233*, 195–204. [CrossRef]

73. Thurm, E.A.; Biber, P.; Pretzsch, H. Stem growth is favored at expenses of root growth in mixed stands and humid conditions for Douglas-fir (*Pseudotsuga menziesii*) and European beech (*Fagus sylvatica*). *Trees* **2016**. [CrossRef]

74. McConnaughay, K.D.M.; Coleman, J.S. Biomass allocation in plants: Ontogeny or optimality? A test along three resource gradients. *Ecology* **1999**, *80*, 2581–2593. [CrossRef]

75. Nikolova, P.S.; Zang, C.; Pretzsch, H. Combining tree-ring analyses on stems and coarse roots to study the growth dynamics of forest trees: A case study on Norway spruce (*Picea abies* [L.] H. Karst). *Trees* **2011**, *25*, 859–872. [CrossRef]

76. Schall, P.; Lödige, C.; Beck, M.; Ammer, C. Biomass allocation to roots and shoots is more sensitive to shade and drought in European beech than in Norway spruce seedlings. *For. Ecol. Manag.* **2012**, *266*, 246–253. [CrossRef]

77. Paillet, Y.; Bergès, L.; Hjältén, J.; Odor, P.; Avon, C.; Bernhardt-Römermann, M.; Bijlsma, R.-J.; de Bruyn, L.; Fuhr, M.; Grandin, U.; et al. Biodiversity differences between managed and unmanaged forests: Meta-analysis of species richness in Europe. *Conserv. Biol.* **2010**, *24*, 101–112. [CrossRef] [PubMed]

78. Morin, X.; Fahse, L.; Scherer-Lorenzen, M.; Bugmann, H. Tree species richness promotes productivity in temperate forests through strong complementarity between species. *Ecol. Lett.* **2011**, *14*, 1211–1219. [CrossRef]

79. Rothe, A.; Binkley, D. Nutritional interactions in mixed species forests: A synthesis. *Can. J. Res.* **2001**, *31*, 1855–1870. [CrossRef]

80. Pretzsch, H. Canopy space filling and tree crown morphology in mixed-species stands compared with monocultures. *For. Ecol. Manag.* **2014**, *327*, 251–264. [CrossRef]

81. Pretzsch, H.; Forrester, D.I.; Rötzer, T. Representation of species mixing in forest growth models. A review and perspective. *Ecol. Model.* **2015**, *313*, 276–292. [CrossRef]

82. Caldwell, M.M.; Dawson, T.E.; Richards, J.H. Hydraulic lift: Consequences of water efflux from the roots of plants. *Oecologia* **1998**, *113*, 151–161. [CrossRef] [PubMed]

83. Siqueira, M.; Katul, G.; Porporato, A. Onset of water stress, hysteresis in plant conductance, and hydraulic lift: Scaling soil water dynamics from millimeters to meters. *Water Resour. Res.* **2008**, *44*. [CrossRef]

84. Dawson, T.E. Hydraulic lift and water use by plants: Implications for water balance, performance and plant-plant interactions. *Oecologia* **1993**, *95*, 565–574. [CrossRef] [PubMed]

85. Mölder, I.; Leuschner, C.; Leuschner, H.H. δ13C signature of tree rings and radial increment of Fagus sylvatica trees as dependent on tree neighborhood and climate. *Trees* **2011**, *25*, 215–229. [CrossRef]

MDPI
St. Alban-Anlage 66
4052 Basel
Switzerland
Tel. +41 61 683 77 34
Fax +41 61 302 89 18
www.mdpi.com

Forests Editorial Office
E-mail: forests@mdpi.com
www.mdpi.com/journal/forests

9 783036 504360